To order or receive additional information on these or any other McGraw-Hill titles, in the United States please call 1-800-822-8158. In other countries, contact your local McGraw-Hill representative.   BC14BCZ

ATM

Theory and Application

David E. McDysan
Darren L. Spohn

McGraw-Hill, Inc.

New York San Francisco Washington, D.C. Auckland Bogotá
Caracas Lisbon London Madrid Mexico City Milan
Montreal New Delhi San Juan Singapore
Sydney Tokyo Toronto

Library of Congress Cataloging-in-Publication Data

McDysan, David E.
    ATM : theory and application / David E. McDysan, Darren L. Spohn.
        p.   cm.—(McGraw-Hill series on computer communications)
    Includes bibliographies.
    ISBN 0-07-060362-6 (acid-free paper)
    1. Computer networks—Planning.   2. Asynchronous transfer mode.
    3. Telecommunication—Traffic.   I. Spohn, Darren L.   II. Title.
    III. Series.
    TK5105.5.M423   1994
    004.6′6—dc20   94-16750
                                                                        CIP

    3 4 5 6 7 8 9 0   DOC/DOC   9 0 9 8 7 6 5

ISBN 0-07-060362-6

*The sponsoring editor for this book was Jerry Papke, the editing super-
visor was Stephen M. Smith, and the production supervisor was
Donald F. Schmidt.*

Printed and bound by R. R. Donnelley & Sons Company.

This book is printed on acid-free paper.

This book is dedicated to my wife Debbie and my parents Lowell and Martha for their support and encouragement.

DEM

This book is dedicated to my best friend and love Becky Thomas for her continual support and encouragement.

DLS

Contents

Chapter 9. User, Control, and Management Planes   251

Part 4   ATM Hardware and Software   277

Chapter 10. ATM Hardware and Switching   279

Part 6   ATM-Based Protocol Interworking and Public Service Offerings   **459**

Preface

PURPOSE OF THIS BOOK

The primary objective of this book is to provide a working handbook for the theory and practical application of Asynchronous Transfer Mode (ATM) technology and ATM-based services. The book will serve as a reference guide that will assist the engineer or manager in understanding ATM technology and how other key services such as frame relay, SMDS, and IP can interoperate using ATM. The book is designed to introduce ATM as the newest technology in a suite of packet/cell-based protocols, demonstrating how ATM will provide a window of opportunity for over a decade of internetworking.

This book first presents the business drivers behind ATM and goes on to explore current computer and information networking directions. It then defines the precursor technologies, protocols, and services to ATM at a high level. Next, the book moves through a discussion of ATM and B-ISDN basics and standards. The reader is then stepped through ATM and B-ISDN protocols and concepts, followed by the hardware and software which make ATM possible. The next section of the book focuses on the ATM traffic contract, congestion control, traffic engineering, and design considerations in some detail. The coverage then moves to operation of other services such as frame relay, SMDS, and IP using ATM, along with a review of ATM-based services. The all-important aspects of operations and ATM network management are then covered in some detail. The book closes with a detailed discussion and comparison of other frame, packet, cell, and LAN technologies against ATM, and takes a look at the future of ATM.

The book also provides the source of each standard and information on where to find additional details within the standards or other published references. The discussions tend to reside on a high level, assuming that the reader has both a background in basic data communications and a working knowledge of transmission technologies, but quickly go into protocol details where pertinent.

Although this skill level is assumed, the writers go into some discussion on the basics.

The text focuses on the practical application of the technology, rather than a dry review of the standards or a parochial view of the technology from the protocol level. The book focuses on digital telephony starting in the 1950s and progresses quickly to present-day standards and a data services point of view.

INTENDED AUDIENCE

This book provides both a high-level summary and a detailed analysis of ATM protocols, operation, standards, technology, and services for use by the data communications manager, network design engineer, and student of data and computer communications. The book has been written to appeal to both groups of professionals, with the structure designed to first educate the reader on principles and then move on to the details. Specific details are provided for key points in understanding the technology, with technical references identified for the reader who wishes to pursue further study on a particular subject. A section has been added at the beginning of the text to define how the book could be used to teach a university course.

The reader should have some prior knowledge of telecommunications principles, although some of the basics of data transmission are briefly covered. Many principles set forth in the book *Data Network Design* by Darren Spohn (McGraw-Hill, 1993) are not reiterated, with references given to that text and other books for readers desiring to study or refresh their knowledge in certain background areas. Specific interest groups may include telecommunication and computer professionals, MIS staff, LAN/WAN designers, network administrators, capacity planners, programmers, and program managers. It is true that the book is structured for the technical professional, but many other groups will also benefit, including sales and marketing, the users of the technology, and even corporate executives who want an overview of ATM and the way they might use ATM technology and services in their businesses. Both LAN managers/administrators and data design engineers must be well versed in these principles. Any computing and engineering professionals must have a good understanding of current technology to make wise corporate decisions on their data communication needs, which will certainly involve ATM for many years to come. Thus, the book seeks to address the issues and needs of both the local area network, wide area network, and end application communities, highlighting commonalities as well as differences.

The book is written in a light, easy reading style in order to both retain the reader's interest during the dry topics and to make the learning process more enjoyable. Almost all formulas can be programmed into a spreadsheet, which is available from the authors at extra cost. Readers may find inconsistencies with descriptions of protocols and the ever-evolving ATM standards due to the changing state of affairs in many of the standards. The authors have made every effort to make the text up to the date as of the time of publication.

OVERVIEW OF THIS BOOK

This book is much more than a reproduction and summary of current standards. The standards are taken as the starting point, enhanced by illustrations, examples, and real-world applications to make the transition from theory to application. Particular areas that are covered in detail are a practical comparison of Time Division Multiplexing (TDM) and ATM, an easy-to-understand exposition of ATM, an in-depth comparison of currently available ATM systems, the mysteries of ATM traffic management revealed, the future of protocol interworking over ATM, inside ATM network management, a technology comparison, and finally, a vision of ATM's future.

The book in many ways presents an unconventional point of view. For example, it is not taken for granted that the paradigm of a voice-oriented telephony structure will be the future of data communications, just like ISDN has not transitioned from theory to application, while Ethernet and the Internet have taken the world by storm. The road from these current pillars of the data world has several obstacles; bridging the gap from application to ATM, routing and interconnection using ATM, and ubiquitous access are some of the key areas covered in detail.

As with other books in this series, this text covers the three critical aspects of ATM throughout the book: business drivers, technology, and application. The protocols are explained based on the industry standards, and recommendations are based on sound practical implementation. This is a very action-oriented text. The examples used are real-world examples. You will also see a slant toward designing networks, and references to many prerequisite and complementary texts that describe key concepts and technologies in greater detail, although this text easily stands on its own.

ATM: Theory and Application is divided into eight parts.

Part 1 provides a discussion of the business drivers for ATM and explores computer and information networking directions.

Chapter 1 summarizes the business drivers for ATM. These include defining the need for data communications and understanding the accelerating bandwidth principle, discussing ATM technology enablers, and then reviewing ATM application and business drivers. It concludes with a summary of ATM benefits and risks.

Chapter 2 sets ATM in the context of rapidly emerging computer and information networking directions. It begins with the history of voice and data networking, which has led to the data revolution. A view of the data network of the 2000s shows the move from private networks to virtual, intelligent data networks and ATM's role in the next battle of the war on technological obsolescence. This chapter compares and contrasts ATM with earlier data services and concludes with an analysis of outsourcing.

Part 2 provides a tutorial and overview of relevant transmission methods, networks, protocols, and services for data communications.

Chapter 3 presents protocol layering concepts, beginning with a definition of the OSI Reference Model. The discussion then moves to Digital Time Division Multiplexing (TDM), the Integrated Services Digital Network (ISDN) protocol, and circuit switching. The concepts of connection-oriented and connectionless network services are defined.

Chapter 4 covers basics of network topologies, circuit types and services, and asynchronous and synchronous transmission methods. The definitions of the terms asynchronous and synchronous are covered in detail. This chapter concludes with a comprehensive review of the principles of multiplexing and switching, with key points illustrated through detailed examples.

Chapter 5 introduces the key concept of packet switching, describing the reasons for its development, the basic principles, and history. Next four key packet switching protocols are summarized: X.25, the Internet Protocol (IP), frame relay, and Switched Multimegabit Data Service (SMDS). The chapter covers the origins, protocol structure, packet format, functions, traffic, and service aspects of each protocol, along with examples illustrating key points.

Part 3 introduces B-ISDN and ATM standards, basic and detailed protocol operation complete with examples, and protocol structure.

Chapter 6 covers the foundations of B-ISDN and ATM — the bodies and the manner in which they define the standards and specifications, complete with references to the major standards used in this book and how they can be acquired.

Chapter 7 provides a high-level introduction to ATM addressing and protocol operation, complete with descriptive analogies and examples.

Chapter 8 covers the operation of the three primary levels of ATM protocol reference model: the PHYsical layer, the ATM layer where cell structure occurs, and the ATM Adaptation layer that provides support for all higher layer services, such as circuit emulation, frame relay, IP, and SMDS.

Chapter 9 covers an overview of the user, control, and management planes and their relation to the ATM layers. The specifics of the control plane signaling protocols are covered.

Part 4 steps the reader through an overview of ATM hardware, software, and applications.

Chapter 10 presents a detailed analysis of ATM Central Office (CO) and premises switches, routers, hubs, and workstation interfaces. A detailed comparison of ATM switch vendors is provided, followed by an analysis of ATM in the PC, ATM-ready PBXs, and ATM chip set development. The chapter closes with an overview of IBM's ATM strategy.

Chapter 11 focuses on ATM networking and applications, with the central theme of IP over ATM. Two industry-leading examples are provided. ATM applications are analyzed in detail, open issues are identified, and industry standards work in progress is highlighted.

Part 5 provides the reader with an application-oriented view of the ATM traffic contract, congestion control, traffic engineering, and design considerations. This is the technical center of the book, providing practical guidelines and design formulas that afford considerable insight into the benefits and applicability of ATM for particular applications. Complex principles are presented in a manner intended to be more readable and understandable to a wider audience than other current publications.

Chapter 12 focuses on the ATM traffic contract — the agreement between the user and the network — that defines user throughput and the Quality of Service (QoS) that the user expects from the network. These components of the traffic contract are described in detail, highlighted by practical examples.

Chapter 13 focuses on the aspect of traffic control in ATM, which includes policing, shaping, and selective cell discard. Principles discussed include use of Usage Parameter Control (UPC), priority queueing, Generic Flow Control (GFC), Connection Admission Control (CAC), and fast resource management.

Chapter 14 covers the important topic of congestion control. The chapter presents a number of solutions and their range of applications, including the levels of congestion that can occur and the types of responses that exist; the standard method of selective cell discard using Cell Loss Priority (CLP); long-term control by use of resource allocation, network engineering, and management; and a set of possible flow control methods suitable to ATM.

Chapter 15 provides an in-depth overview of traffic engineering philosophy, basic queuing models, and approximate performance analysis for delay and loss in ATM switches and networks. The emphasis is not on mathematical rigor, but rather on usable approximations and illustration of key tradeoffs. References to more detailed publications are given for advanced readers.

Chapter 16 discusses additional practical design considerations such as the impact of loss and delay on application performance, the tradeoff between efficiency and features, the tradeoffs between private and public ATM networking, and choice of ATM device locations, network topology, and cost minimization.

Part 6 provides the reader with an in-depth study of the ATM interworking with higher layer protocols. It also covers the current and planned public service offerings based on ATM technology.

Chapter 17 covers the general principles of protocol interworking, and summarizes the current standards for frame relay, SMDS, and IP interworking using ATM. The ATM Data eXchange Interface (DXI) and the Broadband InterCarrier Interface (B-ICI) are also covered.

Chapter 18 provides a road map for public ATM service offerings. The decision factors for choosing an ATM service or network are presented after reviewing current ATM network architectures and services.

Part 7 provides the reader an overview of operations, network management, ATM layer management, and ATM performance measurement.

Chapter 19 covers the topics of Operations, Administration, Maintenance, and Provisioning (OAM&P). A comparison of the OSI and IETF network management approaches is provided. The ATM Forum Interim Local Management Interface (ILMI) and IETF ATM MIB (AToMMIB) are summarized.

Chapter 20 introduces ATM layer and fault management. ATM Operations and Maintenance (OAM) cell structures are discussed for use in fault detection and identification using the Alarm Indication Signal (AIS) and Remote Defect Indication (RDI) ATM OAM cells.

Chapter 21 defines reference configurations for specifying and measuring Network Performance (NP) and user Quality of Service (QoS). The chapter describes methods to ensure that these objectives are being met through the use of performance measurement techniques.

Part 8 presents a comparison with contemporary, competing technologies, such as frame relay, SMDS, IP, Fast Ethernet, and FDDI. It also contains the concluding chapter which speculates on future directions involving ATM.

Chapter 22 provides a critical comparison of ATM with other technologies including circuit switching, SMDS, FR, FDDI, IP, Ethernet, and Fast Ethernet. Technical, performance, and business aspects are covered.

Chapter 23 speculates on possible future directions of ATM, identifying harbingers and possible scenarios and summarizing the challenges ahead. The conclusion is that ATM will be successful, with questions remaining only as to the degree of success, what areas of ATM will be most widely used, and the time frame in which penetration by ATM will occur.

This book also contains two Appendixes. *Appendix A* lists the major acronyms and abbreviations used in the book. *Appendix B* provides a reference of national and international standards sources. A glossary of commonly used terms associated with the technologies, architectures, services, and protocols encountered throughout the book is provided, along with a detailed index at the end of the book.

INTRODUCTION

This book takes a new perspective on ATM — from both a theoretical and application point of view. After much media hype, Asynchronous Transfer Mode (ATM) is now making the transition from theory to application. The coincidence of exponentially increasing power of computers, coupled with increasing information transfer needs in the reengineering of the business and the revolution in fiber optic transmission and ATM switching, create a need for bandwidth that is accelerating. Technology, applications, and businesses now need the power of ATM.

With books, timing is everything. The same is true with technology. ATM technology, standards, and applications have finally come of age. Many standards for ATM signaling, addressing, traffic management, network management, the physical layer, ATM layer, and ATM Adaptation Layer (AAL) — where the true protocol intelligence resides in ATM — are now finalized. Therefore, now is a good time for a definitive book on ATM, covering not only the technology and standards, but the business drivers and applications as well.

This book provides all the information needed to determine if ATM will become part of your life. You will find that we do not present ATM from a myopic view, but instead offer you a look at competing and complementing technologies such as X.25, frame relay, SMDS, IP, and a variety of other higher layer protocols and services. In fact, ATM offers the capability to integrate these protocols and services into a single architecture that is scalable, flexible, and capable of handling bandwidths from several to hundreds of megabits per second, and even gigabits per second in the near future.

Many ATM books on the market today are either too technical — tied up in explaining every detail of the standard without relating to the reader what each aspect means to the user or provider — or too telephony-based — where the reader is dragged through the older telephony point-of-view without the benefit of viewing the technology through the eyes of the data and computer communications user. This book attempts to show the many aspects of ATM through the eyes of both the data and computer communications user and ATM-based service provider.

The demand for flexible, on-demand bandwidth for multimedia applications is growing. One technology for voice, data, and video integration, offering single access into a virtual data service, is a must. Data transfer bandwidths for text, video, voice, and imaging traffic are increasing exponentially, with data communications networks based on technologies such as ATM providing the intelligent network. We live in a distributed data world where everyone needs access to everyone else's data. Private lines are quickly becoming the exception as switched public and private data networks span the globe. Computers need to talk to one another the same way people pick up the phone and dial anyone else in the world. Because of this need, the market for high-speed data transport is exploding. The dawning of the age of gigabit-per-second data communication is upon us. Witness the explosive growth of the Internet. LANs, MANs, and WANs have already crossed the 100-Mbps range and are moving toward ATM as a technology and platform for gigabit-per-second, intelligent virtual data services.

LANs have become an integral element of almost every major corporation. The move toward visually oriented end user interfaces in computer software packages through the use of Graphical User Interfaces (GUIs) creates a tremendous need for flexible networking capabilities. As the number of LANs continues to grow, so does the requirement to interconnect them at native LAN speeds, and thus an emerging need for a technology like ATM. The low-speed private line bridge solutions are reaching their limits, further motivating the need for higher speed, cost-effective, flexible data networking. Frame relay and SMDS services have emerged to provide users with high-speed bandwidth-on-demand services which are now competing directly with private lines. IP has been widely adopted for conectionless networking information exchange and higher level protocol interworking. ATM is poised to integrate aspects of all of these services over a single intelligent architecture and technology.

Many business bandwidth requirements are exploding, such as medical institutions that transfer multimegabit imaging files, and film-making industries that transfer digitized video images which are stored and manipulated directly on a high-performance computer. An important aspect of these new-age networks is their ability to store and retrieve large image files. In the search for new technology to provide data communications on this scale, packet switching technology has seen a series of refinements that result in higher performance at lower costs. Frame relay supplants X.25 for higher speeds, efficient IP implementations support sophisticated networking, and cell-based multiplexing and switching are positioned to replace time division multiplexing and switching for data communications. ATM is appearing in trial networks across the world. As frame relay, SMDS, IP, and now ATM-based services gain increasing support from equipment vendors, local exchange carriers, and interexchange carriers, user acceptance will be a key factor as to the degree to which technology succeeds.

After reviewing the available technologies and services, many users ask the classical questions — "Which service do I use?" and "Do I need ATM or just want ATM?". This book shows that the answers to these questions are based on many factors and there may be multiple answers depending upon the specific user application questions. There is rarely a single solution, and the decision of technology and service generally comes down to what is best for the application and what is affordable — price versus performance, as well as migration and future expansion considerations. The decision to use ATM, or an ATM-based public network service, is also a complicated one. This book presents the business and technological cases for the use of ATM, explain its use, and offer methods of implementation.

HOW TO USE THIS BOOK FOR COURSES

This book can be used to teach a single-semester course focused on ATM or a two semester course on data communications with a focus in the second semester on the details of ATM. It is by no means an introductory text to data communications, but it can stand alone for some readers with some technical background. Readers are also referred to *Data Network Design* by Darren Spohn (McGraw-Hill, 1993) as an introductory text to techniques and principles needed as background to this book.

If the subject matter is to be taught over two semesters, it is recommended that the text be broken into two parts. Chapters to be taught in a first-semester course on a business introduction to data communications and basic architectures, protocols, technologies, and services could include Chapters 1 through 7, 22, and 23. Chapters of focus for a second-semester course on advanced ATM protocols and technologies would cover Chapters 8 through 21.

A single-semester course dedicated to data communications services (circuit switching, frame relay, SMDS, IP, and ATM) focusing on ATM should consider selections from Chapters 1, 2, 6, and 8 through 23. The student should have a minimum working knowledge of the material contained in Chapters 3 through 5 if this book is used in a single-semester course.

Labs should contain ATM design problems based on the cumulative knowledge gained from the class reading and outside reading assignments (recent technology updates). The exercises should involve multiple end system and intermediate system design problems. Due to the fluid nature of emerging ATM standards, students should be encouraged to use the text as a "working document," noting any changes as the standards are revised and updated in the book. This is your book — write in it! The authors plan to publish updated editions as appropriate available changes in technology and standards will likely warrant. Supplemental documentation and instructional tools can be obtained from the authors at extra charge.

AUTHORS' DISCLAIMER

Accurate and timely information was provided up to the date of publication. Many standards used were drafts at the time of writing and were assumed to become approved standards by the time of publication. At times, the authors present material which is practical on a large-scale design, but must be "scaled down" for a smaller business communications environment. Many data communications networks will operate, and continue to run, quite well on a dedicated private line network, but eventually the economics of switched technologies and services, even on the smallest scale, are worth investigating. Please excuse the blatant assumption that the user is ready for these advanced technologies — in many cases it will take some time before these technologies can be implemented. Also, please excuse any personal biases which may have crept into the text.

ACKNOWLEDGMENTS

Many people have helped prepare this book. They have provided comments on various drafts, information on products and services, and other value-added services. In particular, we would like to thank Mr. Lance Boxer, Mr. Scott Brigham, Mr. Curtis Brownmiller, Mr. Terry Caterisano, Mr. Ned Farinholt, Mr. John Fee, Mr. Herb Frizzell, Sr., Mr. Boris Gemarnik, Mr. Wedge Greene, Mr. Tom Hill, Mr. Thang Lu, Mr. Mori Manzoori, Mr. Paul Metzger, Mr. Patrick Milliken, Mr. John Navai, Mr. Rama Nune, Mr. Narayanasamy Rajagopal, Dr. Henry Sinnreich, Ms. Sylvia Thompson, Mr. Tony Toubassi, and Mr. Stephen vonRump of MCI Communications; Dr. Charles Baker of SMU; Mr. Arthur Henley of tti; Mr. Drew Perkins of Fore Systems; Dr. Mohan Kinra of GDC; Mr. Lou Wojnaroski of SRTI; Mr. Steve Walters of Bellcore; and other colleagues over the last several years who have shared their knowledge and expertise. They have helped us develop a greater understanding and appreciation for ATM and data communications.

This book does not reflect any policy, position, or posture of MCI Communications or Southwest Network Services (SNS). A caveat should be added that this work was not funded or supported financially by MCI Communications or SNS nor by MCI Communications or SNS resources. Ideas and concepts expressed are strictly those of the authors. Information pertaining to specific vendor or service provider products is based upon open literature or submissions freely provided and adapted as required. Our friends and associates at MCI Communications and Southwest Network Services supported the project in spirit, especially our management team at MCI of Mr. Lance Boxer, Mr. Paul Weichselbaum, Mr. Arthur Henley, Mr. Bill Halpin, and Mr. Scott Brigham and at SNS of Mr. Craig Tysdal, Mr. Ken Kieley, and Mr. John Merritt, and are hereby thanked.

Also, special thanks to Mr. Gary Kessler of MAN Technology Corporation for his detailed technical review of the entire book and many helpful suggestions, Mr. Herb Frizzell and Mr. Patrick Milliken for both the English style and syntax review, and Debbie McDysan for the review and editing of the graphics.

We would also like to thank the many ATM switch vendors who submitted information used in this book, including: Mr. Larry Lang of cisco Systems, Irfan Ali of Newbridge Networks Inc., Mohander Kukie Chawla and Tariq Hussein of Siemens Stromberg-Carlson, Lance McCallum of LightStream Corp., Eric Pratt of DSC, Brian Walck and Gene Wahlberg of Wellfleet Communications, Charan Khurana of IBM, Ed Braunston and Steve Brown of General DataCom, Bill Holbert of NEC America, Inc., and Bill Carpiel and Shawn Darcy of Synoptics.

Finally we would like to thank Debbie McDysan for her support, encouragement, and sacrifice of time spent with her husband while he was writing this book; and Becky Thomas for her never-ending support and love throughout this project.

The combined support and assistance of these people has made this book possible.

David E. McDysan
Darren L. Spohn

1

Precursors to ATM

Asynchronous Transfer Mode (ATM) and the Broadband Integrated Services Digital Network (B-ISDN) are now more than buzzwords — they are reality! This book shows that ATM is making the transition from theory to application. Before delving into the many facets of ATM and B-ISDN, Part 1 introduces the basics — the precursors to ATM. These include the business drivers and enablers, the move of many corporations away from centralized computing into distributed client-server processing, the drastic increase in Local, Campus, Metropolitan, and Wide area network bandwidth requirements, and the re-engineering of the corporation.

In order to understand the future, one must first understand the public and private data services of the 1990s, as well as transmission and protocol basics, before progressing into newer technologies, such as ATM. The computer and information networking directions that are shaping and driving these new communication technologies must also be understood. Once these basics are reviewed, a detailed discussion of computer and information networking directions can be undertaken.

1

Business Drivers for ATM

This chapter defines the business need for data communications, specifically Asynchronous Transfer Mode (ATM) and ATM-based services leading to B-ISDN. The charter is to understand the business reasons that make ATM a viable, attractive alternative. First, the need for data communications is defined, and in particular ATM and ATM-based services, through the perception of the evolving network of the next century, the move from centralized to distributed networks, and finally, the rapid acceleration or need for higher and higher bandwidth at continually decreasing costs. Next, the technologies that serve as enablers to ATM are examined: digital transmission facilities which provide low bit error rates and SONET/SDH transmission facilities which yield extremely low bit-error rates complemented by very high transmission speeds. Finally, there are the applications that drive the technology — not only one "killer application" but multiple applications spanning many communities of interest [5]. Lastly, an overview of the benefits and potential risks associated with ATM and B-ISDN is presented.

1.1   DEFINING DATA COMMUNICATIONS NEED

The first step in the process of corporate data identity is to define the need for data communications services, which should be prioritized with respect to other needs in the organization. The economic cost justification of data communications is covered in later sections. This section will concentrate only on the new technological factors driving this need and the various options available.

The most likely scenario is as follows: an increasingly flatter, leaner organization is emerging, driving the need for each individual to have access to more data faster and to communicate with more entities. This is driving businesses to increase their reliance on less expensive, user controllable, multiple platform communications access in a multi-vendor environment. This is borne out by the move from hierarchical SNA and DECNET host-based networks to LANs at the local level and peer-to-peer internetworking of the Internet at the wide area level. As computers proliferate, so does the need for data communications. The business becomes increasingly reliant on data communications, and the "network" becomes a force to be reckoned with. The corporate or government communications budget becomes larger each year, while the per unit cost of bandwidth for available data services decreases daily. Possible scenarios are growth in LAN interconnectivity between departments, movement of mainframe applications to a distributed client-server architecture, or the emergence of a new class of image-intensive interactive applications. There are a large number of possible scenarios, each leading to the same conclusion: the explosion of data traffic that requires communication throughout the business to meet ever increasing demands for high-performance computing to every desktop.

How can these increased data requirements, phenomenal growth of user applications, processing power, and demand for connectivity and bandwidth be defined and understood? How can this parallel shift from hierarchical to flat, decentralized networks and organizations be quantified? This chapter provides answers to these questions and more by first reviewing some perspectives on recent data communication growth that create business drivers for ATM.

1.1.1 Transmission Infrastructure Evolution

International and national transmission infrastructures in modern countries have largely replaced digital microwave facilities with fiber optics even more rapidly than analog facilities were previously displaced by digital transmission systems. Satellite communication has evolved as a high-quality digital transmission medium for connectivity to remote areas or as backup to terrestrial facilities. Along heavy traffic routes fiber has replaced microwave, and in thinner routes digital microwave has replaced analog microwave. The deployment of fiber in major metropolitan areas is well under way, while significant deployment of fiber to residences is predicted to begin in the late twentieth century. Modern digital transmission communications will usher in a new baseline for the performance of digital data communications just as as digital transmission has made long distance calling sound quality as good as local calls just 10 years ago. The impact and benefits of high-performance digital transmission are recurring themes throughout this book.

1.1.2 From Centralized to Distributed Networks

Computer communications networks have evolved from centralized mainframe computing, through the minicomputer era, into the era of the distributed personal computer and workstation processing in the last 30 years. The first data computer communications networks resembled a hierarchical, or star, topology. Access from remote sites homed back to a central location where the mainframe computer resided (usually an IBM host or DEC VAX). Figure 1.1 shows this centralized, hierarchical computer network topology. Note that the star and hierarchy are actually different ways of drawing the same centralized network topology.

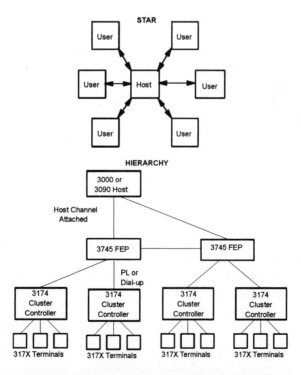

Figure 1.1 Centralized Star and Hierarchical Topology

There is a great parallel happening here between networks and organizations. Networks and entire corporate organizations are making the transition from hierarchical to distributed structures, both requiring greater interconnection and more productivity from each element or individual. This move from a hierarchical structure to a distributed structure is called *flattening*. In networks, flattening means there are fewer network elements with greater logical interconnection. In organizations, flattening is often referred to as the elimination of middle management, which requires greater horizontal communication and interaction within the organization. This trend continues as the corporation is re-engineered, also requiring re-engineering of the supporting computing and communication network infrastructure. Both scenarios create the need for increased connectivity and communications.

a. Five-Level Hierarchy

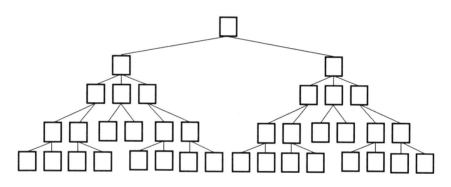

b. Three-Level Hierarchy

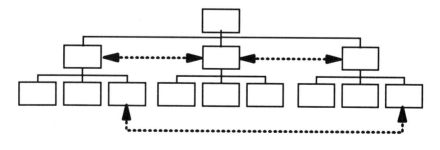

Figure 1.2 Flattening of Network and Organizational Structure

Figure 1.2 illustrates this parallel between organizational and network flattening, where an organization or network goes from a five-tier hierarchical design (a) to a more distributed three-tier (b) network which provides for nonhierarchical interactions, as shown by the horizontal arrows.

Today, more and more computing is accomplished through distributed processing and client-server relationships. Distributed processing is defined as the distribution of network intelligence and processing to many network sites, where each site communicates on a peer-to-peer level, rather than through a centralized hierarchy. The client-server architecture is a major trend in distributed processing, where client workstations communicate with distributed servers for their core information. Servers provide the means for multiple clients to share applications, with a license fee required only for the actively used applications. Client workstations then operate only a shell of the original application retrieved from the server. Servers also are used to share expensive resources, such as printers, CD-ROM juke boxes, mass storage, and databases. Most sites now have the intelligence and capability to communicate with many other sites directly on a peer-to-peer level, rather than through a centralized computer. The host can be appraised of what took place and maintain the database from which upper level management reports can be generated so that operations and maintenance issues are addressed, using systems such as IBM's NetView.

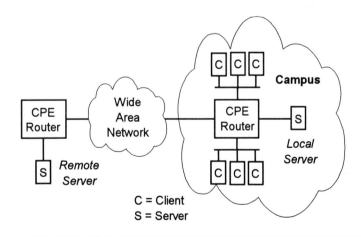

Figure 1.3 Distributed Client-Server Processing Network

Client-server computing distributes the actual storage and processing of information among many sites as opposed to storing and processing all information at a single, centralized location. This model reflects the increasingly flatter organizational structure and the need for increased connectivity and communications shown above. Figure 1.3 illustrates a computer communications network supporting the distributed client-server architecture. A server may be locally attached to a router, or remotely accessed across the WAN.

1.1.3   The Need for LAN/MAN/WAN Connectivity

There are two general scenarios found in corporations and growing businesses that drive the need for greater LAN connectivity across the wide area. The first is increased need for interconnection between distributed computing devices on LANs. The second is the logical extension of the LAN across wider geographic areas. The geographically dispersed LANs now have a range of connectivity choices — ranging from dedicated circuits to switched wide area and metropolitan area networks. The choice of technology and services is based upon many factors other than cost. In this section the paramount goal is to define the business need to connect these disparate networks. Computer networking has been defined in many ways. The following five definitions are used throughout the book:

Local Area Network (LAN): distance on the order of 0.1 km (350 ft); providing local connectivity, typically within a building.

Campus Area Network (CAN): distance on the order of 1 km (1.6 mi); providing connectivity between buildings in the same general area.

Metropolitan Area Network (MAN): distance on the order of 10 km (30 mi); providing regional connectivity typically between campuses over the geographic area associated with a major population center.

Wide Area Network (WAN): distance on the order of 100 to 1000 km (60 to 600 mi); providing national connectivity.

Global Area Network (GAN): distance on the order of 1000 to 10,000 km or more (600 to 6000 mi); providing connectivity between nations.

Typical LAN interconnect speeds, which once were limited to thousands of bits per second (kbps), are now moving quickly toward millions of bits per second (Mbps) and even billions of bits per second (Gbps). Once the LAN is established and operational, many factors drive the LAN to expand in physical and logical size. Data traffic is now growing at rates close to 30 percent per year, with bridges, routers, gateways, and switches making LAN expansion easy and cost-effective. Worldwide LAN interconnection will eventually be achieved. The business drivers for expanding local area networking usually fall into one or more of the following categories:

- Increased inter-LAN traffic
- Higher available transmission rates
- Increased application function and performance
- Cross-domain routing or cross-mainframe access capabilities
- Need to expand boundaries of the LAN
- Low installation and capital costs of new technology
- Expansion of the business through growth or acquisition

The technology drivers of internetworking LANs lead to the hybrid use of private data networks in conjunction with public WAN services in large corporations. This is occurring with a strong emphasis on delivering bandwidth-on-demand. One interpretation of bandwidth-on-demand arose in the LAN environment, where many users shared a single high bandwidth medium. At any instant only one user was likely to be active and hence had the entire shared medium bandwidth available for their use. Hence, bandwidth was not dedicated and was available in a sense to users "on demand." The realization of this concept within networks other than those on a shared medium can be seen in ATM technology and architecture. Another interpretation is analogous to the switched telephone network, where a call (demand) is placed for bandwidth. The call attempt usually succeeds, failing only with a small blocking probability, and hence is also interpreted as bandwidth-on-demand.

Users are spending much more on service and support budgets than hardware and software budgets. This statistic shows the decreasing cost of the equipment, as opposed to the increasing cost of support systems required to run the LAN. As further evidence of this trend the use of high-bandwidth circuits and services to support these networks doubles each year.

Another opposite trend is the actual decrease in the need for bandwidth due to more efficient coding and compression schemes. The most common evidence of this trend is the availability of modems for communication over voice-grade lines that now exceed the capacity of ISDN using compression. Also note the decrease in video conferencing, North American Television Standard Coding (NTSC), and High Definition TeleVision (HDTV) coding rates over time. Acceptable videoconferencing for business can be achieved at DS0 (64 kbps) rates today. NTSC coding was achieved at DS3 (45 Mbps) rates in the late 1980s, and is now approaching the DS1 (1.5 Mbps) rate for noninteractive programming. The need for 150 Mbps for HDTV transmission has also evaporated due to similar increases in coding efficiencies. Of course, the improvements in coding efficiencies are limited by the actual information content of the signal. In general, these schemes for efficient coding and compression arise when bandwidth is inordinately expensive, or a competitive niche occurs that justifies the expense of such coding or compression.

1.1.4 The Accelerating Bandwidth Principle

The number of Millions of Instructions Per Second (MIPS) sitting on the average desktop would have filled a medium office building 20 years ago. Not only are the MIPS of yesteryear's centralized computing power on the desktop today, but the need to interconnect them is growing as well. The growth in the need to communicate data is driven by new distributed processing applications, like groupware, shared databases, desktop videoconferencing, shared workspaces, multimedia, and electronic mail. These two trends, an increase in computing power and the need to communicate, combine to result in the accelerating bandwidth principle.

Current network bandwidths cannot handle the accelerating need for bandwidth for very long. Witness Ethernet, where initially 10 Mbps, even utilized at 40 percent efficiency, offered a tremendous amount of bandwidth in the local area. As workstation power and application demands increased, however, Ethernet LANs had to be segmented and re-segmented until in some cases there is only one workstation per Ethernet segment. FDDI was invented to provide 10 times more bandwidth (100 Mbps) than Ethernet, but on the leading edge of workstation technology even FDDI has already become a bandwidth constraint. Furthermore, FDDI was initially designed for support of only data applications. As a further illustration of the explosive growth that can be caused by open interconnections, observe the tremendous growth rate of traffic on the Internet — 20 to 25 percent per month! Furthermore, with the advent of audio and video multicast the demand outstrips Internet capacity on occasion.

Required connectivity is driven by the method in which business is performed and the nature of the organizational structure. In a flatter, empowered organization each individual may send, or provide access to, information that would only have been sent upwards in the preceding, hierarchical organization. There is a limit to how much interconnection can grow; however, there is a large amount of room for growth in an organization with a deep hierarchy and even a modest number of direct reports.

The following shows how the combined exponential growths in computing power and the nonlinear growth in intercommunications create an overall demand for data communications growth that is greater than exponential. We call this the *accelerating bandwidth principle*. Before exploring this concept, let's first review Amdahl's law. Amdahl's law states that the average application requires processing cycles, storage, and data communications in roughly equal proportion, namely 1 MIPS, 1 MByte and 1 Mbps. In the accelerating bandwidth principle we claim that this rule no longer will hold true.

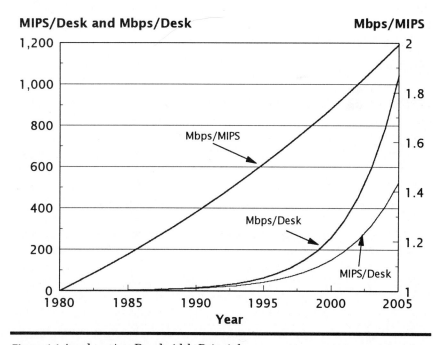

Figure 1.4 Accelerating Bandwidth Principle

Figure 1.4 illustrates the accelerating bandwidth principle. The curve labeled MIPS/Desk represents the exponential growth in computing power at the desktop at a rate of approximately 175 percent growth every two years. The curve labeled Mbps/MIPS represents the nonlinear growth of the required data communication of approximately 3 percent per year, resulting in a doubling of Amdahl's law of the proportion of bandwidth to processing power over 25 years. The curve labeled Mbps/Desk, which is the product of MIPS/Desk and Mbps/MIPS giving units of Mbps/Desk, represents the data communications bandwidth predicted by the accelerating bandwidth principle.

The accelerating bandwidth principle points to the inadequacy of Ethernet, FDDI, and illustrates the need for true Gigabit-per-second (Gbps) networking in the not too distant future. Another way to offset the increasing need for communications bandwidth is improved compression, which reduces the amount of data requiring transmission. However, compression only yields improvement up to the information theoretic limits.

Applications and business needs expand to fill the opportunities that technology can provide cost-effectively. The usage of computing and communications by applications is largely driven by cost. Basic computing and communications costs have been declining exponentially for the last 20 years.

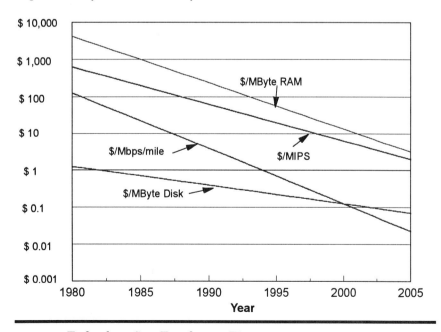

Figure 1.5 Technology Cost Trends over Time

Figure 1.5 plots the costs of processing, storage and communications over time. The decreasing $/MIPS, $/MByte of RAM, $/MByte of Disk and $/Mbps/mile over time shown in this figure are all exponentially decreasing since they are nearly straight lines on the semi-log scale. The $/MIPS and $/MByte are derived for microprocessor technology and semiconductor memory [2], [1]. The exponential increase in computing speed and accessible memory is a well-established trend, driven by integrated circuit technology, that will extend into the next century. Recent trends in processor and memory technology may even exceed exponential growth! The $/Mbps/mile curve is for fiber optic transmission [3]. Note that the cost of transmission is decreasing slightly faster than that of computer processing. The cost of mechanical disk storage is decreasing at the slowest rate.

The fact that transmission costs are decreasing more rapidly than mass memory and processing costs provides economic reinforcement to the accelerating bandwidth principle. It is becoming less expensive to have information transmitted over a data network than it is to store it locally, or perform calculations to generate it. When the cost value of copyrighted or licensed information content such as in distance learning, information on demand, and other interactive services is taken into account, this becomes an even greater justification for the accelerating bandwidth principle.

1.2 ATM TECHNOLOGY ENABLERS

The business factors driving a need for a high-performance data communications network, in particular a high-bandwidth wide area network, have been reviewed. This section now explores the technology enablers. The principal enablers for Asynchronous Transfer Mode (ATM) are:

- Enhancements to protocols that guarantee the effective transmission of data
- High performance digital transmission media
- Cost-effective high-performance electronics
- Worldwide support for standardization from all sectors of the industry through the ATM Forum
- Distributed computing
- The advent of the personal computer

Before going any further, ATM is defined from a 30,000-ft view, or macro level. ATM is a technology which allows for the transmission of data, voice, and video traffic simultaneously over high bandwidth circuits, typically on the order of hundreds of Megabits per second (Mbps) to Gigabits per second (Gbps) in the foreseeable future. ATM hardware and software platforms form a communications architecture based on the switching and relaying of small units of data called *cells*. The primary differentiator between ATM-based services and other existing data communications services, such as frame relay, SMDS, and FDDI, is that it is the first technology and protocol structure to effectively integrate voice, data, and video over the same communications channel at speeds in excess of DS3 (45 Mbps) and E4 (140 Mbps).

1.2.1 Protocol Enhancements

One enabler is the continuing decentralization of computing power from the centralized host to the desktop, and the associated requirement for more peer-to-peer networking. Not only do these desktop machines now have the processing power to run applications that once only the centralized host had, they also have more control of how information is passed, thus employing a wider range of controlling network and transport protocols such as TCP/IP and DECNET. These enhancements have pushed network intelligence outward to the end user, allowing the network to employ less intelligent, more cost-effective protocols which require less overhead. All of these protocols can be carried transparently across an ATM architecture.

Specifically, increased storage in the end stations allows larger retransmission windows to be maintained. The window size of TCP has been increased in RFC 1323 [4] from 64 kBytes to over 1 GByte for this very reason. Also, increased processing power allows more sophisticated flow control and windowing mechanisms to be implemented in the end systems. One example of protocol complexity is demonstrated in the sophisticated TCP flow control algorithms in the end station versus the relatively simple Internet Protocol (IP) used in routers. Chapters 11 and 14 cover this subject in more detail.

1.2.2   High-Performance Digital Transmission

Older network protocols implemented complex procedures just to ensure that a packet could be reliably sent from node to node, sometimes requiring multiple retransmissions over noisy analog links. The simplification of network switching protocols is primarily a result of essentially error-free physical layer communications over digital facilities versus the older error-prone analog facilities of the past. The infrequent occurrence of errors and associated retransmission is then achieved cost-effectively in end systems. Simpler network protocols, such as frame relay, SMDS, and ATM, rely on the performance of fiber optic transmission, which provides very low error rates, typically on the order of 10^{-12}.

The cost-effective availability of plesiochronous digital transmission rates such as DS1, DS3, and SONET rates of STS-Nc is a key enabler for high-speed ATM. Chapter 3 explains plesiochronous, SONET, and SDH rates and transmission structures in detail.

1.2.3   Worldwide Industry Support

Most sectors of industry support ATM. This began with the telecommunications industry defining initial B-ISDN and ATM standards. The late 1980s saw the development of the early prototype Central Office (CO) type of ATM switches. The traditional customer premises multiplexer and switch vendors then adopted ATM in the early 1990s. Next, router and hub manufacturers began building ATM interfaces for routers and hubs. Now computer vendors have started building ATM interface cards for workstations and personal computers. The final piece of the puzzle is falling in place with the initiation of operating system and application software development for ATM-based systems. These efforts point toward a strong commitment to the success of ATM because it addresses the user's need on an end-to-end basis. Many of these efforts are being stimulated and facilitated by the ATM Forum founded in 1992, which Chapter 6 reviews.

1.2.4 Power to the Desktop

Obviously, one of the biggest trends influencing the success of ATM is the proliferation of computing power to the desktop and its subsequent use in distributed computing. Personal Computer (PC) and workstation processing power (MIPS), memory size (Mbytes), and display size (Mpixels) are increasing at an exponential rate, with distributed networks taking these workstation attributes closer to every user on the network — in fact to the desktop and even to the application. As discussed earlier, organizational re-engineering creates a demand for increased bandwidth to interconnect these PCs and workstations. This increase is occurring at a rate faster than the computing performance metrics due to the accelerating bandwidth principle.

Personal Computers (PCs), such as the Apple computer, were born in "garage shops" and were initially known as toys for games and other amusements. That has changed with the PC and workstation now taking the premier position in the computing industry. However, as software applications were developed, and speeds and memory increased as costs dropped, the larger computer manufacturers began to see PCs on the desks of users who previously only had a "dumb" terminal connected to the host. The personal computer was legitimized by the IBM announcement of their Personal Computer (PC) in 1983. Since 1983, the PC has been the industry standard for corporate and government microcomputing. Figure 1.6 shows the growth in the number of professionals using PCs. This trend is likely to continue, with a PC on every desktop (and possibly every wrist!) by the year 2000. In an ironic twist of fate IBM attempted to set a proprietary, de facto industry standard by introducing the microchannel architecture, not making it open as the initial PC Industry Standard Architecture (ISA) was. Since then several other manufacturers have surpassed IBM in PC market share, including the upstart "garage shop" Apple!

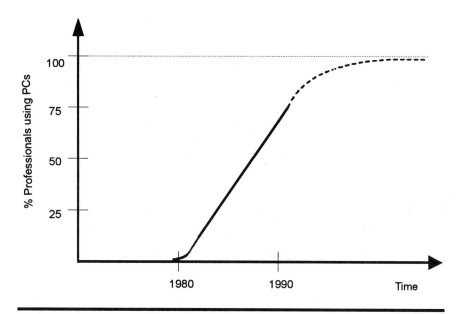

Figure 1.6 Increasing Use of Personal Computers

The PC has provided the user with the device for desktop access to the world of data. Mass storage of information has shrunk to a fraction of its original size and cost. Now the critical element is no longer hardware, but software. Indeed, the cost of software on a modern PC can easily exceed that of the hardware investment.

1.2.5   Standards and Interoperability

Fast realization and industrywide agreement on a common set of standards have also been prime enablers of ATM. These standards have led to a technology interoperability at all levels of the network and across many hardware platforms. The ATM Forum in particular has set a precedent by fostering cooperation between a wide range of industry segments to develop interoperable specifications from the desktop to the world. Chapter 6 provides a review of the ATM standards players, the processes, results to date, and future plans.

1.3 ATM APPLICATION DRIVERS

ATM is a multiplexing and switching technology that is designed for flexibility at the expense of efficiency. For any *single* application it is usually possible to find a better data communications technique, but ATM excels where it is desirable for applications with different performance, quality, and business requirements to be performed on the same computer, multiplexer, router, switch, and/or network. For these applications the flexibility of ATM can result in a solution that is more cost-effective than several separate, individually optimized technologies. Furthermore, the flexibility of ATM can *futureproof* investments since unenvisioned, future applications may also be supported by ATM.

A word of caution is also in order. It is certain that some other technology will follow ATM, such as Wavelength Division Multiplexing (WDM) described in Chapter 4, garnering the hype and support of computer, communication, and industry visionaries. But now ATM is at center stage.

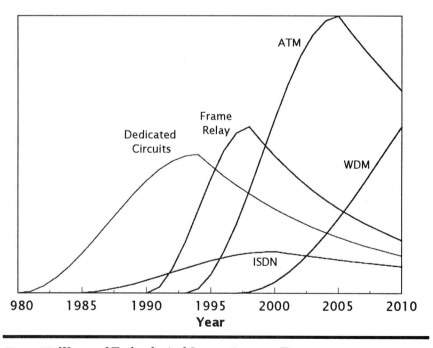

Figure 1.7 Waves of Technological Innovation over Time

Indeed, the measure of any technology's success might be in how well it is able to catch the wave of technological innovation illustrated in Figure 1.7. As time evolves, a technology can only meet the needs of a portion of the overall market. In this example, the need is bandwidth-related. An example of a possible next technology wave after ATM is the use of optical and/or wavelength switching at very high speeds. At lower speeds, ATM may have technology competition from improved compression, modems, or variable length packet switching. The relative values and timing of each technology are not intended to be precise, but merely to illustrate a possible future scenario. No technology will last forever, and legacy technology generally takes a long time to disappear from the scene.

1.3.1 Consumer Applications

Examples of consumer service applications requiring the flexibility and performance of ATM are:

- Entertainment imaging
- Work at home — telecommuting
- Home shopping services employing multimedia voice, data, image, and video using on-line databases and catalogues
- Video on demand for popular movies
- E-mail and multimessaging systems
- Interactive multimedia applications and games

Examples of broadcast public service applications requiring the flexibility of ATM are:

- Distance learning
- On-line video libraries for home study
- Video desktop training courses
- Videoconferencing

A key potential business driver for ATM is the support for consumer services, most notably that of *video-on-demand*. Several cable companies and telecommunications service providers have announced trials of video-on-demand using ATM switches at the head-end of cable distribution networks with sophisticated video servers. ATM is well suited to the ever-changing marketplace of video coding in providing a flexible, multiple-rate, integrated switching vehicle that can handle today's fixed video coding rates, as well as future variable video coding rates and differing speeds for video coding. Video programming can be further processed at these server sites to yield exciting consumer applications. Much as in the business arena, the rallying cry is for more bandwidth-on-demand; the hope is that consumers will exclaim "I want my ATM TV!"

1.3.2 Commercial Applications

Examples of commercial public service applications requiring the flexibility and performance of ATM are:

- LAN/MAN/WAN seamless interconnectivity and internetworking
- Medical imaging
- Distributed data access
- Large file transfer and real-time access to files
- Electronic publishing
- Finance industry reports
- Graphic-intensive industrial engineering applications (e.g., CAD/CAM/CAE/CIM) on-line
- Collaborative computing such as groupware
- Cooperative computing such as concurrent CAD/CAM engineering
- Integrated voice, video, data multiplexing and switching
- Terminal-to-mainframe communications
- Videoconferencing
- CD ROM Servers
- Inventory control systems
- Multimedia applications to the desktop (e.g., E-Mail)
- Desktop publishing
- Remote database access
- Electronic Funds Transfer (EFT)
- Financial modeling
- Leading edge ATM work groups
- Collapsed backbone campus networks
- Seamless interworking with legacy systems using ATM

Chapter 11 elaborates on the example of how ATM interconnects work groups of leading edge users empowered with high performance workstations (which can then be interconnected via an ATM campus backbone). Chapters 11 and 17 cover the concept of seamless interworking using ATM to enable interoperation with Ethernet, Token Ring, and other data communication protocols, such as X.25, frame relay and SMDS.

1.3.3 Application Demand for Bandwidth

The above applications range from providing cost consolidation efficiencies to actually re-engineering the business. Many applications show how people increasingly rely on visual or image information rather than audio or text information, as predicted by the old, but accurate adage that a picture is worth a thousand words. The increase in telecommuting and conducting business from the home also illustrates this trend. The recent partnering and buyout of cable firms by information transport and telecommunications providers also portend major changes in the infrastructure for providing multimedia, interactive networking to the home.

There is no single "killer application" for ATM [5]. Instead, there are many applications that together make ATM a viable technology for the desktop, local, and campus areas, as well as making ATM-based services a viable metropolitan and wide area network solution. Looking at the many types of information, apparently images and high-quality video require drastically increased bandwidth.

There are two generic types of information that applications require: an object of fixed size that must be transferred, and a stream of information that can be characterized by a certain data rate or bandwidth. Multimedia involves combinations of these basic information transfers. We illustrate the tradeoffs in response time, throughput, and the number of simultaneous applications that can be supported by ATM through several simple examples.

Figure 1.8 shows the time to transfer an object of a certain size at a particular transfer rate. Along the horizontal axis a number of common objects are listed as an illustration of the object size in millions of bytes (MBytes). The general trend is that the time required to transfer the information that represents the object decreases as the transfer rate is increased. A real-time service would require transfer in tens of milliseconds. The utility of the service in an interactive, near-real-time mode is usually viewed as requiring a transfer time of no more than a few seconds. A non real-time or batch application may require many seconds up to minutes, or even hours, for transfer of an object.

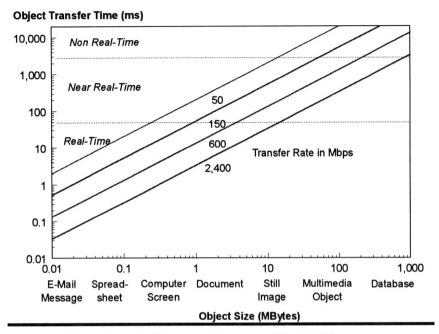

Figure 1.8 Object Transfer Time as a Function of Bandwidth

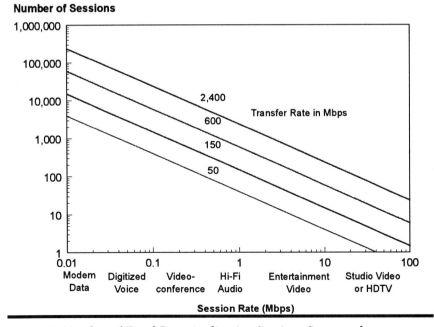

Figure 1.9 Number of Fixed Rate Application Sessions Supported

We now look at applications such as audio and video that use a certain amount of bandwidth. The bandwidth may be a fixed, continuous amount, or an average amount. This example does not consider this detail.   Figure 1.9 plots the number of these applications requiring a certain fixed bandwidth that can be supported by an allocated ATM bandwidth.   In general, as the bandwidth required by each application increases, the number of simultaneous applications supported decreases. Of course, allocating more overall ATM bandwidth increases the number of fixed-rate applications that can be supported.   ATM is flexible enough to support many applications of different rates in the same network; however, these are not illustrated in this figure.

This all points to a move toward visual reality — a dynamic, interactive visual representation of information rather than just textual or simple, static graphical representation. *Multimedia* is a term often used to represent the combination and presentation of multiple forms of data to the user.

Time to Transfer 1 MByte File (ms)

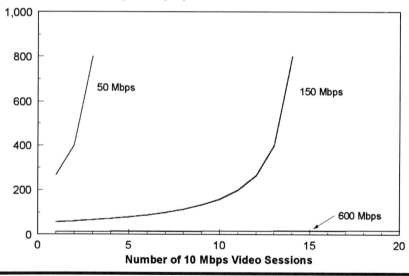

Figure 1.10 Bandwidth Rate Required in kbps for Multimedia

Figure 1.10 illustrates the bandwidth requirements and response time requirements of multimedia by showing how much time is required to transfer a high-resolution graphical image in the presence of a number of high-quality video sessions. Here is an example of the flexibility in ATM allowing an application to tradeoff the number of fixed rate 10 Mbps high-quality video windows that are active, versus the transfer time for a 1 MByte file for various ATM network bandwidths. Observe how the transfer time of the large file increases as the number of active video sessions increases. The fixed rate application must have continuous bandwidth in order for the video to appear continuous, while the object transfer application can simply take more time to transfer the object as the available bandwidth is decreased.

1.4 ATM BUSINESS DRIVERS

Each organization will eventually ask "why do I need ATM?" ATM business drivers take on many forms, from the need for higher speeds, increased flexibility, improved efficiency, and support for multiple traffic types to support entirely new applications. The following examples illustrate some of the key business drivers for ATM that subsequent sections discuss:

> ☆ The Death of Shared Media?
> ☆ Virtual Networking
> ☆ Seamless Interworking
> ☆ More Bandwidth for Less Bucks
> ☆ Futureproofing Investment
> ☆ Enabling Brave New Applications

1.4.1 The Death of Shared Media?

Normal LAN and MAN technologies — Ethernet (IEEE 802.3), Token Ring (IEEE 802.5), FDDI (ANSI X3.139 or ISO 9314), and the Distributed Queue Dual Bus (DQDB) (IEEE 802.6) — all connect to network devices through a shared medium, as shown in Figure 1.11a. Bandwidth is shared between all users on the shared medium with each user potentially having access to the entire bandwidth. A problem occurs when more than a few users are active on a 10 Mbps Ethernet, resulting in a usable throughput of less than 4 Mbps. Token ring, FDDI, and DQDB achieve better utilization through a more sophisticated bandwidth sharing protocol, but when the users' desktop rate begins to approach the shared medium's speed, there is no choice but to move to the next higher-speed shared medium LAN solution. The consequence is that the number of users per LAN segment has continually decreased, creating the market for hubs and bridges, as shown in Figure 1.11b. As workstation power increases, this example reduces to a single user per LAN segment connected to a LAN switch, as shown on the right-hand side of Figure 1.11c. This is where ATM comes in as a common logical interface technology that can scale in speed without requiring changes in software to support a new shared-medium solution, as shown in Figure 1.11d.

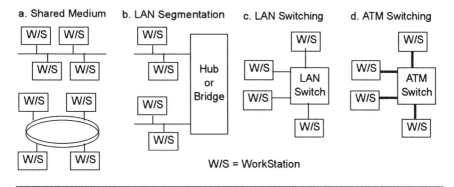

Figure 1.11 Evolution from Shared Medium to High-Performance Switching

1.4.2 Virtual Networking

Users expect the same level of support, ease of use, and connectivity that are provided in the LANs and internetworks of today. The standards needed to enable ATM to provide similar capabilities and interwork with the legacy shared-medium LAN networks will emerge in the mid 1990s. The vision of high-performance ATM workgroups interconnected with embedded legacy networks is illustrated in Figure 1.12. Existing LAN users are connected to hubs, routers, or switches which interwork with the ATM workgroup through a set of ATM-based interworking functions. Functions that must be supported in virtual networking are LAN emulation, routing, and Application Program Interfaces (APIs) that Chapter 11 covers in detail.

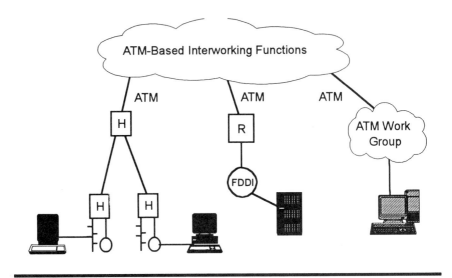

Figure 1.12 ATM-Based Virtual Networking

1.4.3 Seamless Interworking

Of course, most enterprises have more than one location and also have more applications than just data. The original premise of Broadband ISDN (B-ISDN) was to provide the capability to serve voice, video, and data using the same technology. This concept has been dubbed *seamless interworking* across the LAN and WAN. Figure 1.13 illustrates this vision of seamless interworking. Voice, video, and data are converted to ATM at the user site, where they are interconnected to the WAN via ATM. Access to voice and other legacy services is shown in the WAN. It is likely that ATM will not replace the older T1 multiplexers initially, but will provide a migration path through the support of circuit emulation in the interim, as illustrated in Figure 1.13. In the longer term, B-ISDN and ATM may even provide direct transport of voice across the WAN, most likely on transoceanic cables first. ATM may potentially integrate what currently are completely separate data and voice networks, using different carriers and services to form a single unified network. There are other advantages to this approach such as a single network management infrastructure and reduction of access and transmission costs.

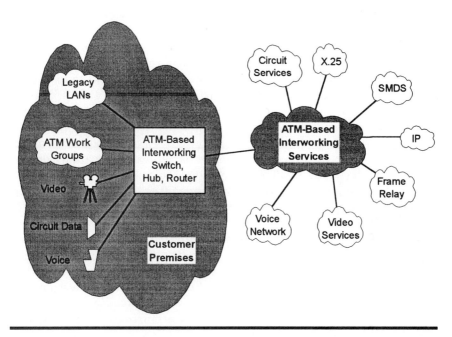

Figure 1.13 Seamless Interworking

Currently only 10 to 20 percent of the LAN traffic goes over the WAN. Current LANs can only achieve a sustained 30 to 70 percent load on the actual LAN speed in order to keep queuing delays reasonable. ATM can extend the LAN across the WAN with degradation in access time limited essentially by only the speed of light. The tradeoff between bandwidth and latency limits that applications encounter is described in Chapter 16.

This argument of seamless interworking also extends to mobile computing and access via standard telephone lines to these services. Mobile communications and phone links that extend the LAN to the automobile, train, meeting room, or hotel room will become commonplace as business travelers take with them mobile-communications-equipped PCs and computing devices. Lower-speed access will perform sophisticated data compression for interconnection into the seamless internetwork.

1.4.4 More Bandwidth for Less Bucks

The widespread use of T1 multiplexers in the 1980s was predicted to be a precursor to a wave of "T3" (which is the common name for DS3, as explained in Chapter 3) multiplexer deployment, which has not occurred. Understanding the reasons for the T1 multiplexer success and the lack of adoption of T3 multiplexers is central to placing the potential benefits of ATM in perspective. T1 multiplexers allowed high-performance, relatively low cost DS1 (colloquially called "T1") facilities to be shared between a variety of applications on a quasi-static basis using Time Division Multiplexing (TDM). As shown throughout this book, TDM bandwidth allocation is not well suited to high-performance, bursty data communications. The growth in demand for data communications has increased dramatically, but the demand for TDM-based service has not kept pace with the overall demand for bandwidth. DS3 speeds are over 28 times that of the DS1, but cost only 5 to 10 times more. The economics and restrictions of TDM of the T3 multiplexer were simply never justified for most users because better choices for public services were available within the planning horizon, such as frame relay, SMDS, and ATM.

ATM offers the capability to extend the LAN or MAN across the WAN at speeds comparable to the LAN or MAN (currently 10 to 100 Mbps) for less cost, because the bandwidth and switches are economically shared across many users, as shown in Figure 1.14. Instead of having to funnel the bandwidth of interconnected LANs down to the lower bandwidth provided by the static allocation of TDM connecting sites via DS1s in the DS3 access line, as shown in Figure 1.14a, ATM provides each LAN with the capability to burst at the full LAN access speed across the WAN on the DS3 access line, as shown in Figure 1.14b. This figure shows how TDM LAN interconnection takes much longer to transfer data, as shown by the time plots of actual usage on the access lines. Since all users do not burst simultaneously, and indeed are controlled so that they cannot, access to peak bandwidth on demand can be accommodated almost all of the time. Chapters 15 and 16 cover the concepts of statistical multiplexing and the achievable economies of integrating multiple applications on the same network.

The need for public network ATM service is growing as users quickly find that the capital cost of building a private ATM WAN is huge, compared with using a shared, partitioned, public network service. This will be even more evident as public network providers add multiple protocol feature support to their services. A finanical interest can also play a part here, too, since expense dollars are paid for public services in lieu of capital dollars for equipment. Within many corporations the expense-oriented approach can be more readily justified than the capital-oriented approach.

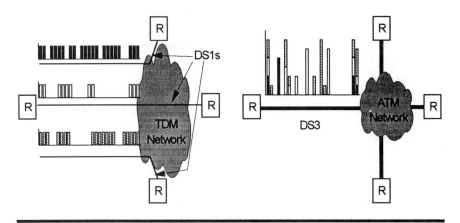

a. LAN Interconnectivity via TDM Network   b. LAN Interconnectivity via ATM Network

Figure 1.14 More Useable Bandwidth for Less Cost

1.4.5 Futureproofing Investments

When public service providers make ATM a core part of their backbone and service offerings, they achieve the benefit of flexibility and scalability. Flexibility and scalability take on many forms. Being able to increase the number and size of switches, users, and circuits is a key benefit. Support for a distributed architecture for high reliability is another benefit. The capability to upgrade network elements to faster processors, and upgrade routers, switches, hubs and workstations to the same standards-based architecture, is also a potential benefit. No technology will remain efficient and cost-effective forever, but the flexibility and scalability of ATM hardware and software potentially provide a longer life cycle than other MAN and WAN technologies, which "futureproofs" the ATM investment.

1.4.6   Enabling Brave New Applications

One of the primary attributes of ATM covered extensively in this book is its capability to support true voice, data, and video traffic integration with guaranteed Quality of Service (QoS) for each traffic type. This is done through ATM *virtual connections,* which allow bandwidth and connectivity to be flexibly and dynamically allocated. This book covers ATM virtual connections in great detail, beginning in Chapter 7. Existing services such as frame relay and SMDS were not designed to handle time-sensitive or delay-sensitive traffic; thus, unacceptable delays can be introduced into voice and video traffic with these data communications technologies.

Not only does ATM offer the capability to support time-sensitive traffic, typically voice or video, but it also offers enhanced delivery options such as point-to-multipoint and eventually broadcast. High-speed ATM multicast applications may include a broadcast of a videoconferencing application that includes text and video.

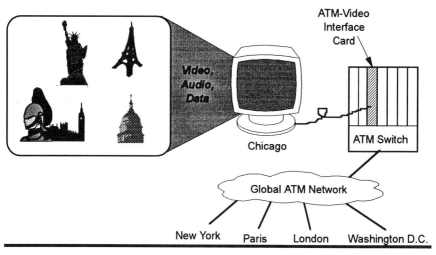

Figure 1.15 Multipoint, Multimedia ATM Application of the Future

Figure 1.15 shows a multimedia desktop workstation in Chicago. The person is holding a videoconference with four other individuals in New York, London, Paris, and Washington, D.C. In this futuristic example, an automatic translation server could be connected for language conversion between the parties speaking different languages (not shown in the figure). An ATM interface card in each workstation combines the video of the built-in monitor camera, the telephone, and text data into a single 100 Mbps ATM transmission stream to a local ATM switch. The ATM switch then broadcasts the voice, data, and video to switches at all four locations through intervening ATM WANs in the global ATM network. Of course, each of the sites could do the same in return. In order for this application to be effective, all attendees have four individual pop-up screens on their workstations so that they can see all the other participants, as illustrated in Figure 1.15. Multicast videoconferencing connections can be controlled by a conference option, where the choice of recipients is dynamically controlled by the originator. A future possibility could be a "meet me" type conference. Although this may seem futuristic, it will likely be commonplace before the end of the twentieth century.

1.5   BENEFITS AND RISKS OF ATM SUMMARIZED

This introductory chapter concludes with a summary of the potential benefits and risks associated with ATM. There are many potential benefits, but realistically there are also risks and competing technologies.

1.5.1   ATM Benefits

In summary, ATM handles a mix of delay-insensitive, loss-insensitive, delay-sensitive, and/or loss-sensitive traffic over the same ATM interface and network infrastructure. ATM combines the high speeds of circuit switching with the flexibility of packet switching over a single simplified network infrastructure. ATM is also highly flexible and scalable — allowing support ranging from small private networks to very large public networks. ATM offers the following benefits:

☺   Integration of multiple traffic types
☺   Efficient bandwidth use by statistical multiplexing
☺   Guaranteed bandwidth and resource allocation
☺   Dynamic bandwidth management
☺   High service availability
☺   Multiple Quality of Service (QoS) class support
☺   Suitability for both delay or loss sensitive and delay or loss insensitive traffic
☺   Seamless private and public network technology

☺ Automatic configuration and failure recovery
☺ Cost-effective fixed length cell processing
☺ Improved transmission utilization
☺ Futureproof investment (longer life cycle)

ATM will allow companies and users to build networks based on the future vision of uniting voice, data, and video communications on ATM-technology-based equipment. The potential of ATM to improve performance and lower overall network, equipment, and operating costs in the long term, and in many cases the short term, is the theme and reason for this book.

1.5.2 ATM Barriers and Risks to Success

On the other hand, ATM still has some barriers to overcome. The lack of applications development has discouraged some early adopters of ATM. Cost is no longer a major issue, as the cost of equipment and communications lines is always decreasing. Operational costs remain high, however, and together with the cost of managing the network may exceed the hardware and software costs.

ATM also has some competitive risks with technologies such as 100 Mbps FDDI, 100-Mbps Ethernet, Switched Ethernet, and fast Ethernet, although these technologies only handle data, while ATM allows for multimedia — voice, data, and video. Only time will tell which will succeed — probably combinations of all of them will be used for many years to come.

1.6   REVIEW

This chapter first discussed the four driving forces causing a need for data communications: evolution of the transmission infrastructure to Gigabit speed fiber optics, evolution from centralized to distributed networks, the need for extending LAN and MAN speeds into the WAN, and the accelerating bandwidth principle. A view of technology enablers such as protocol and transmission enhancements, standards and industry support, and distributed computing and the personal computer was presented. Next, a look at the consumer and commercial applications driving ATM introduced a discussion of the business drivers of ATM. Our approach in reviewing the business drivers started with the need for higher performance networking moving through virtual networking with legacy LAN technology. Next, the discussion moved to how buildings or campuses can be interconnected across wide areas in a vision of seamless interworking which promises to provide more bandwidth at less cost. Next, what may be the greatest benefit of ATM — enabling brave new applications including ones not yet conceived — was presented. The chapter closed with a review of the benefits and risks of ATM.

1.7 REFERENCES

[1]   P. Gelsinger, P. Gargini, G. Parker, A. Yu, "Microprocessors circa 2000," *IEEE Spectrum*, October 1989.

[2]   G. Myers, A. Yu, D. House, "Microprocessor Technology Trends," *Proceedings of the IEEE*, December 1986.

[3]   I. Chlamtac, W. Franta, "Rationale, Directions and Issues Surrounding High Speed Networks," *Proceedings of the IEEE*, January 1990.

[4]   V. Jacobson, R. Braden, D. Borman, "IETF RFC 1323, TCP Extensions for High Performance," May 1992.

[5]   J. McQuillan, "Where are the ATM Applications?," *Business Communications Review*, November 1993.

2

Computer and Information
Networking Directions

This chapter begins with a historical discussion of computer and networking directions through the end of the 1990s and into the twenty-first century. A short review of recent history shows an accelerated move of most business budgets and resources from primarily voice networks to networks that are primarily data. This was precipitated by the data revolution and an increased dependence on data communications, which in turn has led to the data network of the 2000s. This new network paradigm is based on: virtual public networking, increased network intelligence, and decentralized and ubiquitous networks and services. It is based on the premise that the corporation is continually at war with technological obsolescence. This chapter concludes with a discussion of outsourcing and downsizing (or "right-sizing") trends.

2.1   FROM VOICE NETWORKS TO DATA NETWORKS

A dramatic shift from voice communications networks to data communications networks has occurred over the past 20 years. Data communications now affects many aspects of our lives: controlling stock market transactions, medical research networks, electronic mail at work, and even Automatic Teller Machines (the other "ATM"!) that provide us with money on practically every street corner. What chain of events made data communications become so integral to our daily life in the last generation? How did the move occur so quickly from voice networks using analog transmission facilities to data networks riding digital transmission facilities? With these questions in mind, this chapter now begins with a review of the evolution of communications. As background, the history of voice communications is presented as the precursor to data communications.

2.1.1 A Brief History of Communications

Figure 2.1 depicts a view of the history of communication along the dimensions of analog versus digital encoding and synchronous versus asynchronous timing, or scheduling. The beginnings of spoken analog human communication are estimated to be over 50,000 years old. Graphic images have been found in caves over 20,000 years old. Written records over 5,000 years old have been found. Digital long distance optical communications began when the ancient Greeks used digital, optical communications to relay information using placement of torches on towers at relay stations before the birth of Christ. The Greeks and Romans also popularized *scheduled* public announcements and speeches as early examples of broadcast communications, as well as, individual, *unscheduled* communication in forums and debates.

In the seventeenth and eighteenth centuries, optical telegraphy was extensively used in Europe. Later, electrical telegraphy was invented by Samuel F. B. Morse in 1846, marking the beginning of modern digital electromagnetic communications. Radio telegraphy was invented shortly afterwards by G. Marconi. Broadcast analog radio communications of audio signals followed in the late nineteenth and early twentieth centuries. This technology was also applied to analog voice communication in the same time frame. Television signal broadcasting became commercially viable in the late 1940s. Then, in the 1950s, the conversion of analog voice to digital signals in the Plesiochronous Digital Hierarchy (PDH) began in large metropolitan areas to make better use of installed cabling. This was followed by the invention of packet switching in the 1960s as an offshoot from research into secure military communication networks. Fiber optic transmission and the concept of synchronous digital transmission were introduced in the early 1980s. Analog transmission of voice had a brief renaissance using Single SideBand (SSB) in the 1980s. ATM moves the wheel of technology around the circle back into the domain of digital asynchronous communication. The next major leap in technology will likely be Wavelength Division Multiplexing (WDM), which is analog and asynchronous. The speed of digital communication has increased geometrically over time through each of these evolving phases of technology.

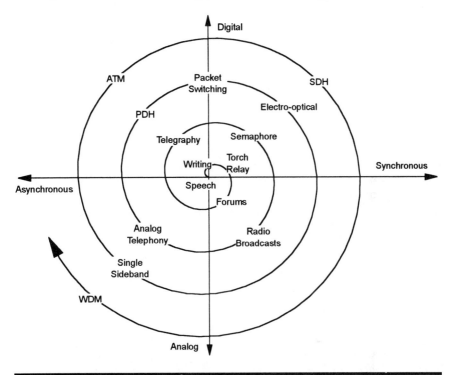

Figure 2.1 Data Communications "Wheel of Reincarnation"

This is not a perfect analogy — sometimes the wheel spins faster than at other times, for example, in the current day we move from Asynchronous Transfer Mode (ATM), an asynchronous digital technology, all the way around to Wave Division Multiplexing (WDM), an asynchronous, analog technology.

The military has been a key user of data communications networks throughout history. The telegraph was significant to the Union forces role in the Civil War. Many data processing and early computer systems were developed during World War II, when systems integration was necessary due to complexity. After the war the command and control centers, weapons and sensor systems, voice networks, and the computers which ran these systems needed to be centrally controlled within one interconnected communications network. This was the beginning of the Department of Defense (DoD) telecommunications architecture. Today, the United States DoD architecture comprises one of many data communications platforms in use. The latest government standard for protocols at DoD is the Government OSI Protocols (GOSIP) that are now a requirement in many Requests for Proposals (RFPs).

The next major advance by the DoD was the establishment of the Advanced Research Projects Agency NETwork (ARPANET). ARPANET was established in 1971 as the first packet-switched network. This data network connected military and civilian locations, as well as universities. In 1983 a majority of ARPANET users, including European and Pacific Rim contingents, were split off to form the Defense Data Network (DDN) — also referred to as MILNET. Some locations in the United States and Europe remained with ARPANET, and are now merged with the DARPA Internet, which provides connectivity to many universities and national telecommunications networks. The original ARPANET was decommissioned in 1990. Many of the advances in computer communications, including routing and networking, have been developed through experience on the Internet. The explosive growth of the Internet has to some extent been fueled by traffic generated regarding how to build larger, faster, and more feature-rich networks in the future. The premier example of the largest Internet in the world is the National Science Foundation NETwork (NSFNET) shown in Figure 2.2.

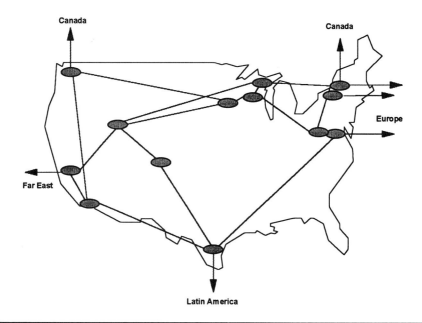

Figure 2.2 NSFNET Backbone for the Internet

Host-based networks accessed by local and remote terminals evolved through the use of private networks and packet-switched services. The primary example is the IBM Systems Network Architecture (SNA). This architecture provides the platform for many dumb terminals to communicate with an intelligent host or mainframe in a hierarchical, or star, fashion. This hierarchy developed because collecting expensive intelligence at the host and allowing the terminals to have little resident intelligence was the most cost-effective solution.

Local area networks (LANs) were the next major development in the computer communications networking environment, with the advent of Ethernet by Xerox in 1974. The advent of client-server architectures and distributed processing is the beginning of modern data communications. This ends the short course on the History of Data 101 — now we move on to recent data history.

2.1.2 Voice as Low-Speed Data

A standard voice-grade channel can be accurately represented by a 64 kbps (or 56 kbps) data stream. Nyquist's sampling theorem states that the number of samples taken from an analog signal must be taken at a rate no less than *twice* the bandwidth of that signal for it to be accurately reproduced. Therefore, the minimum sampling rate for a 4000-Hz bandwidth voice channel is 8000 samples per second. Employing 8 (or 7) bits per sample yields a 64-kbps (or 56-kbps) data stream. The coding of each voice sample is performed using one of two different nonlinear companding (COMpression/exPANDING) schemes, called μ-Law in North America and A-Law elsewhere. In fact, while voice is typically transmitted at 56-kbps, many digital encoding techniques now enable a voice channel to be transmitted at speeds as low as 8 kbps.

The 64-kbps representation of voice was first used for engineering economic reasons in large metropolitan areas to multiplex more voice conversations onto bundles of twisted pairs in crowded conduits, mainly in urban areas of the United States in the mid 1950s. Twenty-four voice channels were multiplexed onto a single twisted pair in what was known as a T1 repeater system, using a DS1 signal format. Chapter 3 reviews the DS1 signal format in detail. The scarce resource of twisted pairs was now utilized at 2400 percent of its previous capacity, a tremendous enhancement! The fact that these multiplexing methods could be used for the purpose of data communications came later.

Voice is very sensitive to delay, and somewhat loss-sensitive. Users do not tolerate appreciable delay during a full duplex or half duplex conversation, because it can inhibit interaction or result in annoying echo, nor will they accept sentence-flow garbling by the loss of entire syllables. Variations in delay can cause the speaker's voice to become unrecognizable, or render the speech unintelligible. The loss of small portions of syllables or words in voice communications is usually acceptable, however. Satellite delay, which most people have experienced, is a good example of the effects of large delay (in terms of echo) and the impact on interactive conversation. Certain new technologies such as ATM can handle delay-sensitive voice traffic, while other technologies such as frame relay cannot.

2.1.3   Voice and Data Network Architectures

The public telephone network naturally evolved from a narrowband to a broadband hierarchy based upon engineering economic considerations. Figure 2.3 depicts the classical five-level public telephone network hierarchy developed to minimize cost and achieve traffic engineering economies of scale, which resulted in a corresponding increase in bandwidth at higher levels in the hierarchy. Customers are connected to the telephone network by a *local* loop which is provided by narrowband twisted pair access lines to the lowest level of the hierarchy, the Class 5 central office telephone switch. A *twisted pair* is composed of two wires that are twisted to minimize impairments in analog transmission such as crosstalk and interference. These twisted pairs are bundled into cables and then aggregated at the central office. Generally, if the distance traversed is greater than a few miles, the local loops are aggregated into larger bandwidths for transmission over microwave, copper, or increasingly optical fiber. Indeed, the DS1 and DS3 multiplex levels were created for this very reason, their use for data occurring by later invention and innovation rather than initial design.

Class 5 switches may be directly connected, or else connect to larger Class 4 tandem switches, which can be connected to even larger tandem switches such as the Class 3 and 2. The final route choice is via Class 1 switches, which are connected at the highest bandwidth aggregate level. All switches still operate only at the single voice channel level, however. In general, switches are larger at higher levels in the hierarchy, even though a Class 5 switch in a large metropolitan area can be quite large.

There is a move toward a flatter, more distributed network of fully interconnected Class 4 or 5 switches. The number of levels have not yet been reduced to one, but in many cases there are only two or three levels in a hierarchical phone network. Again, this parallels the move of most major businesses from a hierarchical organization to a flatter, distributed organization.

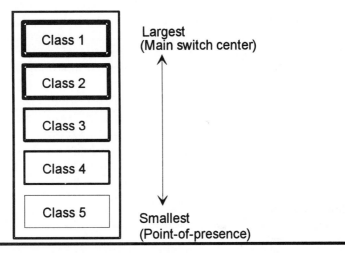

Figure 2.3 Public Telephone Network Hierarchy

Voice channels are time division multiplexed on digital transmission systems. When compared to data, the main difference is that voice calls are circuit switched, whereas data can be either message, packet, cell, or circuit switched as described in Chapter 4. Both voice and data applications require large bandwidths, but for different reasons. Data applications require large bandwidth to support the peak rate of applications, while voice requires larger aggregate bandwidth to multiplex large numbers of individual voice circuits.

2.1.4 Recent History of Data Communications

The challenge corporations face is the flattening of interconnected LAN networks to support the astounding rate of personal computer proliferation. This flattening began the (r)evolution of the corporate data network. What started as a PC for home use has now become a corporate necessity. It was a natural evolutionary choice for the visionary network design engineer to address the users in these islands of information — the distributed LAN interconnecting desktops and the centralized mainframe Management Information Systems (MIS) arena — by creating a common environment using routers to achieve interworking and interconnectivity. This is a key reason for the tremendous success of the router industry.

Just as minicomputers invaded mainframe turf when the cost of minis fell to departmental budget approval levels, bypassing corporate MIS budget approval, so also has the router phenomenon enabled enhanced LAN connectivity due to falling costs. Routers enabled interdepartmental connectivity of diverse computing technologies in a cost-effective manner. Thus, cost and control are more in the hands of the end user than they have ever been before. Many users asked themselves the question; why conform to corporate MIS dictates when you can build your own departmental LAN and handle 90 percent of your data processing needs? When there was a need to interconnect these LANs, the MIS manager and entrepreneurial LAN managers had to work together in order to integrate access from the LANs to the VAX and IBM mainframes. In addition, WAN interconnectivity often went beyond the scope of a single LAN manager and had costs that had to be shared across multiple LANs. The router also found its place here as the gateway to the WAN.

ATM arrives in this era of expanding, even euphoric expectations from users caught up in this new wave of freedom. The high-performance workstation users, for whom even FDDI at 100 Mbps is too slow, are leading the next data communications revolution, with many focused on ATM as the solution. The MIS manager still needs to focus attention on the design and maintenance of the host and front-end processor systems because many mission-critical applications still reside here. The LAN manager needs to focus on the needs of the high-performance workstation user and server, while continuing to enhance the support provided to the users employing legacy technology. Bringing the requirements of these two divergent realms into ATM networking is essential. One should not become caught up in the glamour of a new technology such as ATM until a trail for migrating mission-critical applications is blazed by successful early adopters.

2.2   THE DATA REVOLUTION

The 1990s have yielded a broadband data revolution with a rallying cry of "bandwidth-on-demand!" The expectation of bandwidth-on-demand was created by the LAN where a user shared a high-speed medium with many other users, having access to the full shared medium bandwidth "on demand." This worked well when every user required only a small portion of the shared medium, but has become a significant problem as the power of desktop technology has increased while the Ethernet capability remains at 10 Mbps. There is another interpretation of bandwidth-on-demand advocated by those with a background in circuit switching. This definition is that a user requests bandwidth, and is either granted the full request or is denied access completely (i.e., blocked); this is similar to making a phone call — the call either goes through or you get a busy signal. We shall see that specific applications can make use of one definition or the other, but usually not both.

This section shows how data communications has taken the lead over voice communications in both total volume, growth rate, and the introduction of new services. The discussion then moves to cover how various services best meet application needs in terms of bandwidth, delay, and burstiness. The section concludes with the observation that the data revolution is global, social, and economic in scope.

2.2.1   Data Communications Takes the Lead

Data growth is occurring at a factor of approximately 25 percent per year, far outpacing the average growth of voice, at only 6 to 8 percent per year [2]. The consequence of this differential in growth is plotted in Figure 2.4, which projects the U. S. common carrier voice and data services revenues for the growth rates cited above. If these growth rates remain constant, data network revenue would exceed voice network revenue prior to the year 2000. Of course there are many factors which may cause these growth rates to change, so this is only a possible outcome. However, it makes the point that the era of emerging data communications dominance is near. Groupware, electronic mail, file transfer, local area network interconnection, interactive applications, and emerging computer communications applications represent just a few examples of how data communications is exceeding voice communications in creating the demand for new capabilities.

Billions of Dollars

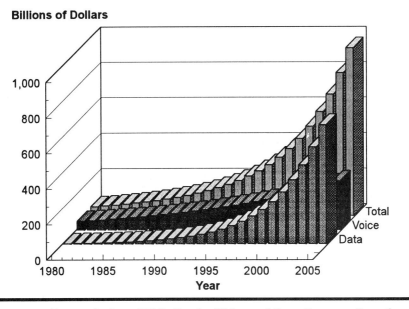

Figure 2.4 Extrapolation of U.S. Carrier Voice and Data Revenue Growth

Data communications not only gives business the competitive edge, it puts them on a leading edge that creates both benefits and risks. Once a business experiences effective data communications, it becomes hooked on its own network. The data and computer communications network quickly becomes the lifeblood of the company. A company needs voice network communications, but a private voice network can only become so large since it is limited by the company's size. Voice communications traffic can be forecast and has predictable characteristics. On the other hand, data communications traffic characteristics can be a different ball game with new rules. Once a company uses a data network, there is no limit to its potential. Key services and switching technologies representing the emerging data market include:

- Private lines
- X.25
- Internet
- N-ISDN
- Circuit switching
- Frame relay
- Switched Multimegabit Data Service (SMDS)
- ATM
- SONET

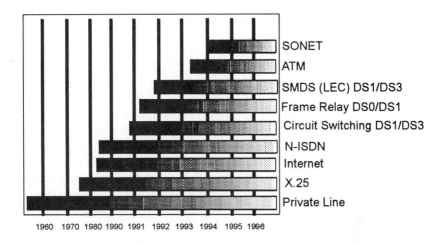

Figure 2.5 Data Services Introduction History

Figure 2.5 shows the time frame for the introduction of the major data services over roughly the past 30 years. Broadband switched data communication is fast becoming the prevalent market in the 1990s — and at the forefront is ATM. This decade may also be called the era of interworking, simply because of the widespread use of these broadband technologies and services and their interaction with LANs.

2.2.2 Business Relies on Data Communications

Many business and corporate managers have little idea of their level of dependence on their data communications networks. This is evident by the small amount of personnel resources usually dedicated to data communications in relation to the funding which pays for networks. Studies show that a major portion of most business office budgets are dedicated to data communications. From this dedication, it is intuitively obvious that data communications networks are fundamental to the successful operation of any business. Computers, terminals, modems, facsimile machines, security systems, and even most telephone systems transit some form of data communications.

Cost also plays a major role in determining network needs, and often is driven more by the cost of services provided by the carrier than by the actual equipment costs. Often services supplied by carriers or vendors are tax-deductible expenses, whereas the purchase of equipment is a depreciable capital expenditure. Equipment expenses often turn out to be a very small part of the total expenses of operating a network. Ongoing support, especially people costs, is a disproportionately lower cost compared to equipment costs.

Many companies planning to offer data communications services to customers must first demonstrate that these services work internally in their own company. A customer is apt to first ask the question: Do you use the same service to transmit your own critical traffic? As is often the case, the company offering the service becomes the test bed for the service before it is sold to a customer, thus becoming in a sense its own best customer.

2.2.3   A New World Order for Data?

The drive is on for corporations to extend their networks into global and international integrated data and voice networks. Corporate global enterprise networks are proliferating at an astonishing rate as international circuit costs decline. Does this point to a new world order for data? One must become part of this new world order of data communications to survive.

Data communications has taken on a global view in many of its facets. This section summarizes some key areas of standardization, global fiber connectivity, and the needs of the multinational corporation.

International standards, of course, are a prerequisite for global connectivity. There is much progress in this area. The latest generation of transmission rates is the same for the first time in history, as shown in Chapter 3. There is increasing cooperation between standards bodies on an international scale, as reviewed in Chapter 6. The entire standards process is changing to better meet the accelerating needs of users; witness the tremendous international success of the ATM Forum. ATM is being standardized on a worldwide basis independent of physical-link speed, as detailed in Part 3.

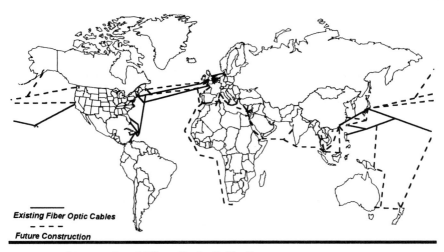

Figure 2.6 International Fiber Optic Cable Systems

International private line connectivity is growing faster than transoceanic cables and fiber can be installed. International switched services are booming, with current product offerings providing a broad range of switched services. The entire world is being connected by fiber optic cable systems. Figure 2.6 illustrates the international fiber optic connectivity that is either planned or already in place.

Recent political, technological, economic, and regulatory changes worldwide have spurred international data network interoperability. New interest is occurring for instance, in the countries of Mexico and Canada through the North American Free Trade Agreement (NAFTA). Growth has primarily been in the United States, Europe, the Pacific Rim, and Southeast Asia. Markets in South America, New Zealand, Australia, Russia, and countries once part of the Eastern Bloc and the now-defunct Soviet Union have been slower to emerge, but are gaining momentum through the development of infrastructures based on state-of-the-art technology. Postal, Telegraphy, and Telephony (PTT) monopolies are realigning with open market competition making worldwide advanced data communications a reality. Many international businesses seek to establish ties and form agreements with the PTTs for purposes of future network planning. These are the communications markets of the 1990s.

The 1990s are also the age of mergers and strategic partnerships. Every day the sun rises on a new international merger or partnership between carriers, hardware vendors, PTTs, governmental agencies, and small companies who fill niche markets. Many joint ventures have sprung up both nationally and internationally. These range from the computer vendors trying to beat out the smaller clone vendors to the large interexchange and international carriers who vie for entrance into foreign markets. One example is the announced plan for a growing international partnership between U.S.-based MCI Communications and the U.K.-based British Telecom (BT), a partnership designed to capture a lion's share of the world data communications market.

2.3   THE DATA NETWORK OF THE 2000s

Have you ever wondered what the architecture of tomorrow's data network will look like? This section takes a look at tomorrow and through the end of the century. Static, predefined private communications networks are migrating to dynamic, virtual networks — networks with ubiquitous access that can interwork past, present, and future protocols. Virtual, public data networks are adding more and more intelligence, in essence enabling the network to become the computer. The corporation is becoming more and more dependent upon virtual data networking to run the day-to-day business, requiring both partitioning and security. The war on technical obsolescence has begun, and is taken to new heights with ATM technology. Will ATM be the Rosetta stone for protocol interworking to allow multiple protocols to be deciphered and understood? Will these networks provide ubiquitous access for all users by public network addressing? Time will tell.

2.3.1   Private and Virtual Private Networks

This section describes and compares two alternative approaches for very high-speed computer networking: dedicated circuits between computers, servers, routers, and workstations (with these systems also performing switching functions) versus shared-access lines connected to network switches which provide virtual, on-demand capacity for workstation-to-computer communications.

Figure 2.7 illustrates the first alternative of a private network that traditionally has been comprised of dedicated circuits. Servers are shown connected to multiplexed remote workstation clusters. Each server/router and remote workstation router/bridge has two links to two separate computers via the network Points of Presence (PoPs) to provide survivability in the event of link failures. When a workstation is not directly connected to the desired server, the intermediate routers must perform switching and routing.

The advantages of this private network approach are full user control and simple, less expensive network technology. The disadvantages are nonproductive redundancy, dedicated link capacity, additional equipment costs to perform switching and routing, and the need to engineer the private line trunks for peak capacity and redundancy.

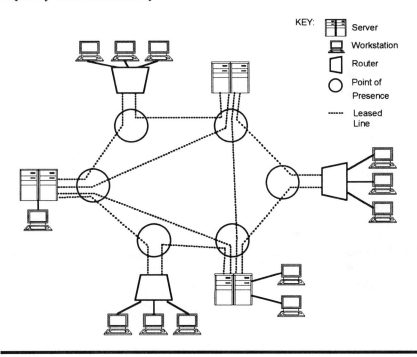

Figure 2.7 Dedicated Circuit Private Network Example

Figure 2.8 illustrates the same set of servers, routers, remote routed/bridged workstation clusters, and network PoPs as in the private network, but as part of a shared, virtual network. A separate virtual video network shares the backbone, as shown in the figure. An ATM switch is placed at each network PoP which has access lines to each user. The ATM switches are highly interconnected by very high-speed links shared across multiple users, providing multiple services by dynamically allocating shared resources. Only a *single* access line is required for each user, thus halving the access cost when compared with the previous example. The reliability of access is identical to that of the dedicated network, assuming that the access circuits in the dedicated network take the same route to the network PoP. The ATM switches perform routing and switching, relieving the routers and servers of this task.

In summary, the advantages of the virtual network alternative are: reduced access line charges, the capability to satisfy high peak demands (particularly during low activity intervals for other services), cost impacts proportional to usage (versus cost proportional to peak rate in the dedicated network alternative), and enhanced reliability. Disadvantages are less predictable peak capacity and less user control.

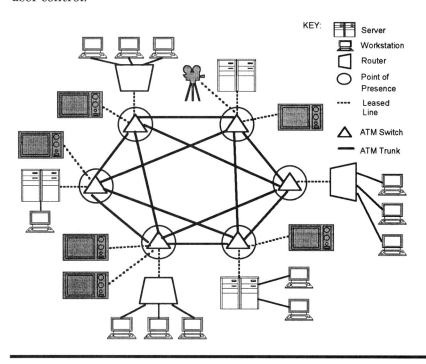

Figure 2.8 Virtual Private Network Example

Virtual private networks are defined as network partitions of shared public network resources between multiple users to form a private network that *appears private to the users* but is still part of the larger public network. Shared network resources are assigned in fair proportion to the bandwidth required by customers.

In a virtual private network a single access circuit from the site to the network is sufficient, because multiple virtual circuits can be provided from multiple users at a site to their destination on the network. For example, each virtual circuit can be allocated a peak rate equal to the access circuit, but have a sum of average rates that is less than the access circuit. Figure 2.9 demonstrates this concept by showing how users A, B, C, and D at site 1 all have a single physical circuit into their premises ATM device which converts these inputs to four ATM virtual circuits (as indicated by the different line styles) and then transmits them over a *single* physical ATM access circuit to the ATM network switch. These individual user virtual circuits are logically switched across the ATM network to the destination premises ATM device where they are delivered to the physical access circuit of the end user, as illustrated in the figure. Later, Part 3 of this book details the various types of physical access circuits and virtual ATM circuits.

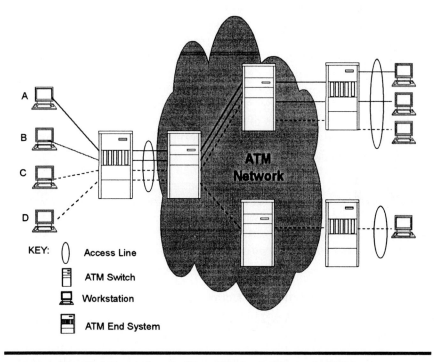

Figure 2.9 Detailed Example of a Virtual Private Network (VPN)

2.3.2   The Intelligent Data Network

Corporations and goverments are moving towards using faster, larger, and more intelligent data communications networks — where the intelligence lies *within* the network, rather than outside it. They are also looking for intelligent network services, rather than simple, traditional private line or circuit switched data services. The term *intelligent network* connotes some level of *value-added* service provided by the network. Examples include address translation versus interpretation, intelligent routing decisions made within the network by route servers rather than predefined routes across the network, and protocol conversion rather than just transparent (to the user) protocol transport. Network intelligence can also mean a service offering based on centralized, intelligent, network-based devices that serve as information servers offering voice, video, and imaging interpretation, routing, and on-line service features. Prime examples are the interactive database services, such as CompuServe and America On-line.

Users want networks that are "smarter," not just faster and better interfaces. Users want access to intelligent public data services so that they can better leverage the intelligence within their own network. Current and emerging data communication services are just now slowly adding this type of intelligence to their networks — intelligence which in the 1980s resided at the premises. Now, network service providers offer alternative network intelligence that is extended to the user premises where significant intelligence may not be practical.

There is thus a trade-off between intelligent networks and intelligent user equipment. Many factors, driven by global industry standardization and the development of technology, will influence decisions on where the network intelligence will reside. The market is both technology-driven and user-driven. For example, international providers want the network intelligence to reside in international gateway nodes. National carriers want intelligence to reside in carrier Point of Presence (PoP). Local exchange carriers (LECs) or PTTs want intelligence mainly in the serving Central Office (CO). Customer Premises Equipment (CPE) vendors want the intelligence to reside in the CPE. The profits of the next century lie in the intelligent functions, not the connectivity function, with all of the aforementioned groups recognizing this fact. The user needs to mix and match all of the above options for the best cost and functional advantage — typically in a "hybrid" networking environment using a mixture of components to meet the needs of the intelligent data communications network of the future. However, the network users and network providers must also work together to ensure that the mix and blend of technology being used meets business goals. This is critical for continued successful business operation. Relating these responsibilities to the strategic business objectives of the company will guarantee on-going success.

2.3.3 Meeting the Needs of the Enterprise

Large enterprises typically have a few large locations that serve as major traffic sources and sinks. Typical applications are large computer centers, large office complexes with many information workers, campuses requiring high-tech communication, data or image repositories, and large-volume data or image sources. These large locations have a significant community of interest between them; however, the enterprise usually also requires a relatively large number of smaller locations needing at least partial, lower performance access to this same information. The smaller locations have fewer users, and generally cannot justify higher cost equipment or networking facilities. Cost generally increases as performance, number of features, and flexibility are increased. ATM-based interworking will initially be useful to the largest locations in the largest enterprises. However, the high performance and flexibility of ATM interworking must somehow also meet the need for connectivity to the many smaller locations that also need to be served.

For the lower-speed access, efficiency is a more significant concern than the flexibility and higher performance of ATM. This is because the cost per bit, per second generally decreases as the public network access speed increases. For example, the approximate ratio of DS1/DS0 and DS3/DS1 tariffs is approximately 10:1, while the speed difference is approximately 25:1. This means that a higher-speed interface can be operated at 40 percent efficiency at the same cost per bit, per second. Conversely, the lower-speed interface costs 2.5 times as much per bit per second, and therefore efficiency can be important.

Figure 2.10 illustrates an ATM-based interworking network cloud connecting a few large sites to many small sites. Such a network is composed of many smaller sites and few larger sites which is typical of large enterprises, such as corporations, governments and other organizations. Principal needs are multiple levels of service characterized by parameters such as throughput, quality, and billing.

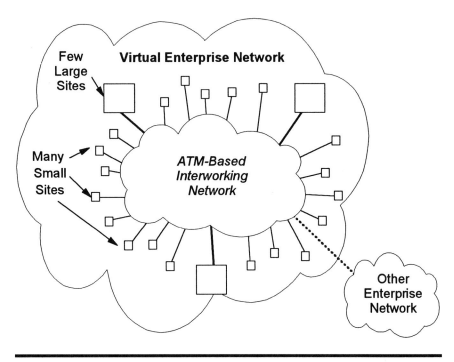

Figure 2.10 Typical Enterprise Network

2.3.4   Virtual Network Partitioning and Security

Data traffic often represents the most private and sensitive information of an enterprise. Therefore, the majority of data traffic is intra-enterprise. In contrast, a large amount of voice traffic is communicated between enterprises over the public telephone network. Data communications in large part still occurs on private networks, or else uses modem communications over the telephone network. The tremendous growth in Internet traffic is an example of the burgeoning demand for public data communication. Even though an increasing amount of data traffic is communicated between enterprises, it is still typically less than 10 percent of the total data traffic. If public data services provide good security and partitioning, then interenterprise traffic could increase in public, shared carrier services. An enterprise, however, needs to control what communication can occur within as well as outside its boundaries. Security and screening can be done in a connection-oriented environment through signaling, or in a connectionless environment through the use of screening tables, or filters.

Corporate, government, and university internetworking have also exploded. When the corporate network of company A is tied to the corporate network of company B, via either private lines or switched services, internetworking takes place as shown in Figure 2.11. This is often the case when two companies need to share vital information such as engineering CAD/CAM files, databases, groupware, and other applications. This connectivity between two or more corporate, government, or university private networks usually occurs through the use of public network services. Internetworking works well as long as security precautions are taken by both the service provider and the end user networks. It is a common practice today for users to install separate routers, with filters as "firewalls" in such configurations.

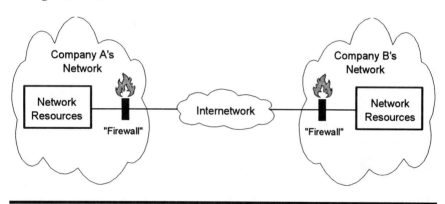

Figure 2.11 Inter-Corporate Internetworking

2.3.5   The War on Technological Obsolescence

Business users are concerned with maximizing their investment in computing and data communications equipment. Rapid advances in computing make a maximum productive lifetime of 3 to 5 years typical. Generally, the most expensive computer equipment is justified for only a small set of mission-critical applications. A similar situation exists in the area of data communications.

Currently, most intraenterprise data communication networks are constructed from Customer Premises Equipment (CPE) interconnected by private lines at DS1/E1 speeds or less. The advent of public data network services, such as frame relay and SMDS, and their attractive associated pricing, has motivated some customers to migrate a portion of their bursty data from private lines to virtual private networks. Burstiness is defined as the ratio of the peak traffic rate to the average traffic rate. Usage-based billing operates by charging for only the transferred data. Virtual networks offer higher peak rates at affordable levels because there is normally no penalty for idle time, or equivalently, a low average rate.

Frame relay usually can be economically installed in most existing CPE with only a software upgrade. In addition to software upgrades, SMDS usually requires new CPE, or an external CSU/DSU. ATM will also require software upgrades, as well as new CPE and/or external CSU/DSUs.

There is an alternative to both SMDS and ATM called the Data eXchange Interface (DXI). This is an HDLC-based protocol that encapsulates and/or maps the SMDS and ATM header functionality between a DTE and a DCE. This means that most CPE can be upgraded via software, at a cost similar to that of frame relay. The DXI protocols are better suited for early, cost-effective, low-speed implementation of SMDS and ATM. There is, of course, additional cost for the CSU/DSU; however, the CSU/DSU can multiplex many lower-speed DXI interfaces onto a higher-speed ATM or SMDS cell interface cost-effectively. The ATM DXI protocol is defined in Chapter 17.

With the large amount of publicity about ATM and B-ISDN, many users are delaying decisions on moving data traffic onto frame relay and SMDS — in order to see what will happen with ATM and because of the perceived incompatibility and competition between these services. One thrust of this book is that customers can economically justify the performance and flexibility of ATM at their largest locations first, while simultaneously achieving connectivity to their many smaller locations using existing, or less expensive, equipment and access speeds enabled by ATM-based protocol interworking. Chapters 8 and 17 provide further detail on these aspects of protocol interworking.

2.3.6   ATM as the Rosetta Stone for Protocol Interworking?

In 1799, Napoleon's army discovered the Rosetta stone in Egypt, which provided the key to translation to the undeciphered Egyptian hieroglyphic and demotic writing. The key was that the Rosetta stone contained the same message in the known Greek. Can ATM provide a similar role in translating the myriad of existing, sometimes arcane data communication languages into the modern language of data communications? Will ATM-based networks be the Rosetta stone for public data network protocol interworking? ATM may provide a similar role in the complex world of protocol interworking along the trail that routers have blazed.

Users want a network where any one user can connect with any type of interface and protocol — talk to any other user on any other network that may have a different interface and protocol. This is the expectation that router-based networks are setting today. This same interconnection via multiple interfaces and protocols provides a vision of future value-added data network services.

Protocol interworking standards are being developed in on-going standards bodies and industry forums. The ITU and ANSI have specified AAL3/4 and AAL5 in detail as the foundation for frame relay and SMDS protocol interworking. AAL1 provides a basis for circuit emulation and interworking with Narrowband ISDN (N-ISDN). IETF RFC 1483 defines what can also be viewed as a service-specific AAL called multiprotocol encapsulation carried over the common part of AAL5. Figure 2.12 illustrates the use of ATM as the basis for interworking between multiple interfaces and protocols. There is an Inter Working Function (IWF) for each native protocol that allows interoperation with an ATM-based end device. Chapters 8, 11, and 17 cover these concepts in detail.

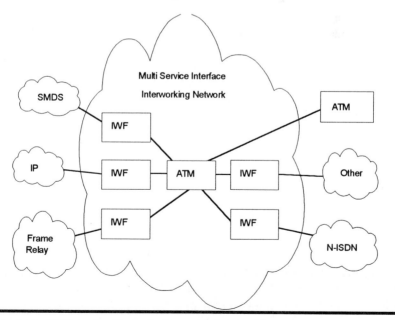

Figure 2.12 ATM as the Basis for Protocol Interworking

Protocol interworking removes the obstacle of connectivity for smaller locations within the enterprise and slows or circumvents technological obsolescence for existing, or lower cost, equipment. Furthermore, protocol interworking facilitates more rapid evolution to ATM by allowing a seamless migration on a site-by-site basis, retaining full connectivity at every step, as traffic and application performance requirements grow. Indeed, as traffic grows at large sites, the older, lower-speed ATM equipment can be migrated down to the medium sized sites and replaced by the latest high-performance equipment, further justifying the need for protocol interworking as a weapon in the war on the high cost of technological obsolescence.

2.3.7 Ubiquitous Access and Addressing

Ubiquitous access to an intelligent data communications network spanning the globe has become the rallying cry of many users. Users want to access the data network as a large "cloud" and thus be able to talk to any other user connected to that cloud without requiring any knowledge of the internals of the network cloud. A prerequisite to this capability is the assignment of a globally-unique address to each user. The public voice network has these characteristics, with several lessons from that domain applicable to data networks. There is also the experience of the Internet, which has different addressing characteristics, that is also a significant factor. If a user cannot reach any other user on the public data network, as is taken for granted in telephony and has become a *fait accompli* on the Internet, then the resulting data service will have little utility.

The public phone network uses an addressing, or numbering, scheme called E.164 that basically has a country code part and then a nationally assigned part for each country, as detailed in Chapter 9. This is the familiar international dialing plan. Internationally, and usually nationally, this is a geographic numbering plan, since the digits taken from left to right hierarchically identify the geographic location of the user. For example, in North America the first two digits identify world zone 01, the next three digits identify the area code, the next three digits identify the exchange switch, and the last four digits identify the user access line on the switch. This convention greatly simplifies the routing decisions in telephone networks and has allowed the goal of universal connectivity to be achieved in the telephone network.

The Internet assigns 32-bit numbers to each user employing an organizational hierarchy. Entire blocks of numbers are assigned to an organization which need not have any geographic meaning whatsoever. A user may move geographically, and it becomes the job of the intelligent network to find him or her using a routing discovery protocol. The organization may structure its block of addresses however it chooses, either geographically, organizationally, or in some other manner. The Internet has also achieved worldwide, nearly ubiquitous access and addressing as well.

Which addressing method will be employed in data networks of the future? For now, both of these methods, and even more, are being pursued by the industry standards groups, as reviewed in Chapters 9 and 11.

2.3.8 As Reliable as the Phone Network

Similar to the expectation of universal connectivity from telephony, data users expect public data networks to be as reliable as the telephone network. Intelligent ATM networks may rely on their fiber transport to be both near error and near outage-free or may detect errors and faults at the ATM level. Redundancy and restoration must be observed at every step in the design, with SONET technology providing this capability in some configurations or ATM-based restoration algorithms in others. Since many applications do not provide error correction or switching to alternate paths, the capability for an ATM network to guarantee nearly error-free transmission and continuous availability is important.

Successful service providers offer services with high availability and low error rates as required by the corporations and government entities that build their enterprise networks on the virtual network. The switch from conventional private lines to broadband switched service is progressing, but there will always be a need for dedicated private lines for specialized applications. The incremental reliability of the public data network, low error rates, and the reduced price of switched data services based on the economies of scale inherent in the carrier frame and cell-based infrastructures make switched data services even more appealing in comparison to dedicated private line services as time progresses.

2.4   OUTSOURCING TRENDS CONTINUE

The choice between outsourcing versus an internally designed and developed network is one that spans every aspect of the business, and is one of the most important decisions of the network manager. An internally designed and developed network involves all stages of planning, building, and maintaining a network for the business. Outsourcing of network needs involves a third party taking over some or all aspects of the corporation's data network.

What is outsourcing? What issues and considerations are associated with an outsourcing contract? What are the benefits and drawbacks of outsourcing? This section summarizes the primary factors to consider before signing an outsourcing agreement. A more detailed analysis of outsourcing can be found in Reference .

2.4.1   Outsourcing Defined

Outsourcing is contracting one or more outside vendors to design, develop, and implement a solution for a company's communication needs. This may include, but is not limited to, planning, designing, installing, managing, owning, leasing, operating, and controlling a communications network. Full responsibility for some portion of the company's network communications assets will be transferred to, or assumed by, the outsourcing vendor.

Good strategic planning and innovative techniques must prevail regardless of whether the design is performed internally or is outsourced. First, evaluate the existing available resources. Compare the pros and cons of obtaining new resources versus contracting out the service. If the analysis determines that outsourcing is not justified, then analyze the systems and technologies available that can optimize the in-house operation. If the decision is to outsource, begin by shopping among multiple vendors. Always remember that portions of the organization may be outsourced, while others could be retained in-house. In fact, this is one method of retaining your key people should the outsourcing deal go sour or terminate unexpectedly.

The requirements gathering process is a critical step before outsourcing can begin. Compare the two major scenarios — *private line* versus *virtual private network* configurations — before making the decision to outsource. Cost factors, of course, are a major consideration. During the planning stages, a balance must be maintained between designing a network to accommodate internal applications and designing the internal applications to accommodate the network. Network applications and the network itself often grow together after the network is built. The capacity-planning process is critical throughout the life cycle of the network. Applications are analyzed on an individual and aggregate basis to determine the best network solution. The network design engineers will have various levels of knowledge at their disposal, ranging from projected traffic bandwidths between sites to a complete protocol profile for each application.

One way of looking at the decision of internally designed networks versus outsourcing is *right-sizing*. Right-sizing is the euphemism, or buzz word, that combines the downsizing of network staff and resources with distributing the computing power closer to the user. All this occurs while expanding the features, functionality, and intelligence of the network. A strong set of network management capabilities is also a key requirement.

2.4.2 Outsourcing Considerations

The decision to outsource is often one of business policy based on many factors, including:

*   Corporate resource availability
*   Sensitivity of the data
*   Return on investment (ROI) analysis
*   Skill set and reliability of the vendor
*   Cost factors of either owning the business elements or leasing them from the outsourcer
*   Retention of control
*   Business charter of the company
*   Deductible expense versus capital investment

Other factors, such as how much support is required for how long, contract stipulations between user and vendor, and the loyalty to existing company employees, also play a major role in the outsourcing decision. Many business aspects should be considered for possible impact by outsourcing, including:

◊   Resources (staff and existing investments)
◊   Questions on skill sets and reliability of the vendors performing the outsourcing function

◊   Cost savings or eventual loss
◊   Control of the hardware
◊   Network monitoring
◊   Future of the company to either continue outsourcing or to bring the business back in-house

As with any business case analysis, all expenses of an outsourcing deal should be analyzed in detail for proof of the validity of cost savings versus expenditures. This is the process of determining what needs to be outsourced. A business considering outsourcing must consider the following:

- Understand what needs to be outsourced
- Plan what to do with current resources (most importantly people!)
- Compare current employee skill sets with those of outsourcing company
- Choose a reliable outsourcing company
- Understand the monetary impacts before and after outsourcing
- Clearly define levels of control and methods of regaining it should that be required
- Define the extent of outsourcing, for example, protocols, interfaces, and locations
- Determine the duration of the outsourcing contract, ideally five years or less
- Fully understand the vendor-user relationship
- Maintain the loyalty of the retained staff, and ensure loyalty of the new staff
- Do not announce your intentions prematurely

If ability to manage or control the network internally is lost, and the external vendor source fails to perform, how difficult is it to rebuild the internal networking department? The loss of skilled people may be difficult and costly to recover. When outsourcing is chosen as the alternative, a strategic plan must be implemented with contingencies for each of these possible scenarios. Turning over the network to the vendor in increments until the vendor's capabilities can be judged, called selective outsourcing, is one answer. Regardless of which solution is chosen, a clear-cut plan must be in place vis-à-vis the vendor for a minimum of two years. Define the vendor's plan for updating technology and workforce. Make sure the outsourcing vendor is able to adapt to your company's business needs, as well as their own, to maintain your future growth and competitiveness. Alternate outsourcing vendors must also remain an option.

2.4.3   Summary of Benefits and Drawbacks

The primary *benefits* of outsourcing are:

+ Uses vendor experience and specialists
+ Allows focus on core business rather than on running networks
+ Outsources network management
+ Taps a good source of quick network resources
+ Reduces costs
+ Augments existing work force with skilled workers
+ Reduces computing and communications staff
+ Combines computing and communications departments
+ Develops applications and business practices

The *drawbacks* of outsourcing are:

− Loss of control
− Possible loss of resources if networking operations are retained
− Loss of in-house expertise
− Alienation of users
− Possible sacrifice of technology flexibility
− Risk of impact to critical systems if vendor fails

These drawbacks can be countered with good planning, smart management, and proper choice of vendor. Unfortunately, these factors are not always controllable.

2.4.4   The Future of Outsourcing

The market for outsourcing continues to grow. More third-party companies are providing outsourcing packages. The complexities of outsourcing grow as users move from bridged and multiplexed to routed environments. Also, as LANs continue to proliferate, the number of disparate protocols increases drastically, and all of them require interconnectivity. Outsourcing is clearly here to stay, and it is having a major impact on the computer and communications industries. There are obviously many short-term benefits from outsourcing, but what about the long-term effects? Many people who disagree with outsourcing say that the long-term expenses outweigh the short-term gains. This underscores the need to first develop an accurate business case. Currently, most companies that outsource do so for purposes of systems integration.

2.5 REVIEW

Throughout this chapter the evolution from voice to data networks is a recurring theme. History has taken us through various reincarnations of the same types and cycles of technologies. Infrastructures developed for more cost-effective voice networks have evolved into support for data networks. A revolution in data is occurring, led by data communications protocols and services that address business dependence on data communications. Customers invest in newer, higher performance equipment at the locations that have business applications that justify the cost. There is a move from private data networks to virtual private data networks offered by public providers, allowing users to meet the needs of the enterprise through partitioning and security. Is ATM the Rosetta Stone for protocol interworking? It is already unlocking the some differences and revealing common needs of multiple protocols as later chapters reveal. ATM presents some new tactics in the war on technological obsolescence, while outsourcing remains a viable strategy to control costs.

2.6 REFERENCES

[1]   D. Spohn, *Data Network Design*, McGraw-Hill, 1993.
[2]   D. Dunn, M. Johnson, "Demand for Data Communications," *IEEE Network*, May 1989.

2

Transmission, Networks, Protocols, and Services for Data Communications

This part describes the transmission, networks, protocols, and services used in modern data communications networks. Chapter 3 introduces the layered OSI Reference Model (OSIRM), and then proceeds to define the basics of the digital transmission hierarchy. The Integrated Services Digital Network (ISDN) is introduced, followed by definitions of connection-oriented versus connectionless data services.

Chapter 4 reviews common network topologies and circuit types, continuing with an explanation of the principles of multiplexing and switching. Chapter 5 reviews the public and private data services of this decade, including X.25 packet switching, frame relay, Switched Multimegabit Data Service (SMDS), and Transmission Control Protocol/Internet Protocol (TCP/IP).

3

Transmission and Protocol Basics and Review

Protocols shape our everyday lives. A protocol is similar to a language, conveying meaning and understanding through some form of communication. Computer communication protocols are defined as sets of rules and message exchanges. Protocols are modeled in a layered fashion with lower layer protocols providing services to the next higher layer. For one computer to talk to another, each must be able to understand the other's protocol. Protocols play an important role in data communications; without them islands of users would be unable to communicate. Protocols are defined through protocol architectures, the most well known being the seven-layer Open Systems Interconnect Reference Model (OSIRM), which will be introduced in this chapter. The concept of layered protocols is largely due to the OSI model; however, the specific protocols of OSI are not widely implemented. This concept of layering has enabled the entire industry of multiprotocol routing.

The physical layer is the lowest layer protocol which provides access to the transmission medium, specifying details concerning physical interface signaling, timing, and pin level configuration. At the physical layer there are three predominant types of transmission applicable to ATM: plesiochronous, SONET/SDH, and basic pulse transmission over fiber or twisted wire pairs. Data link layer protocols allow communications with the physical layer and provide link-by-link error detection/correction, multiplexing, and flow control. Included in the data link layer are both the Logical Link Control (LLC) and the Medium Access Control (MAC) sub-layers in Local Area Network (LAN) communications. Network layer protocols then take over by providing end-to-end addressing, flow control, and integrity checking. The transport layer provides multiplexing onto the network layer, expedited delivery, and further integrity checking. The session layer establishes a connection between end systems. Finally, the presentation layer manipulates data into different forms for the highest and most complicated layer, the application layer.

This chapter also examines the Integrated Services Digital Network (ISDN) protocol reference model and the role it plays in the Broadband-ISDN (B-ISDN) protocol reference model. Next, a comparison of the OSI Connection-Oriented Network Services (CONS) and ConnectionLess Network Services (CLNS) is presented. This chapter concludes with an overview of circuit switching and switched network services.

3.1 BASIC PROTOCOL LAYERING CONCEPTS

Figure 3.1 illustrates the basic concept of protocol layering that is relevant to ATM and ATM Adaptation Layers (AALs), as described in the following discussion.

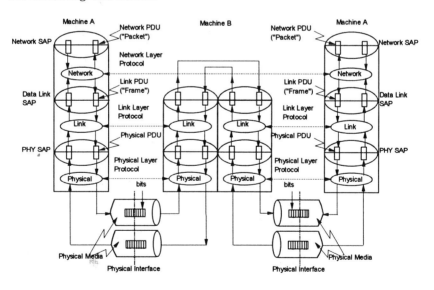

Figure 3.1 Physical, Link, and Network Layers

The term *interface* is used in two ways in different standards bodies. First, primarily in the CCITT/ITU view, physical interfaces provide the physical connection between different types of hardware, with protocols providing rules, conventions, and the intelligence to pass data over these interfaces between peer protocol layers. Normally, the view is that bits flow over physical interfaces. Secondly, primarily in the OSI view, interfaces exist between protocol layers. Normally the view is that *Protocol Data Units* (PDUs), or *messages*, pass over protocol interfaces. These interfaces between layers are called Service Access Points (SAPs) because they are the points where the lower layer provides service to the higher layer. Stated another way, the physical interfaces provide the path for data flow between machines, while protocols manage that data flow across this path using protocol interfaces within the machines. Physical interfaces and protocols must be compatible for accurate data communications. Many network designs now incorporate multiple levels of protocols and interfaces, always starting at the physical layer.

The concepts behind the use of multiple protocol layers are important. The concepts of physical, data link, and network layer protocols can now be defined on a high level. Asynchronous Transfer Mode (ATM) has its roots in the physical, data link, and network layers of the OSI protocol model.

3.2 LAYERED REFERENCE MODELS — THE OSIRM

The Open Systems Interconnection Reference Model (OSIRM) was intended to define the functions and protocols necessary for any computer system to connect to any other computer system, regardless of the manufacturer. This model was developed by the International Organization for Standardization (ISO), beginning in 1977 with the formation of ISO Technical Subcommittee 97 (TC97), Subcommittee 16 (SC16), and was officially documented in 1983 in ISO standard 7498. Figure 3.5 depicts the basic OSI reference model showing a source, intermediate, and destination node and the protocol stack within each. The layers are represented starting from the bottom at the first layer, which has a physical interface to the adjacent node, to the topmost seventh layer, which usually resides on the user end device or a host that interacts with user applications. Each of these seven layers represents one or more protocols that define the functional operation of communications between user and network elements. All protocol communications between layers are peer-to-peer, depicted as horizontal arrows between the layers. Emerging standards span all seven layers of the model, as summarized below. Although OSI has standardized many of these protocols, only a few are in widespread use. The layering concept; however, has been widely adopted by the computer and communications standards bodies.

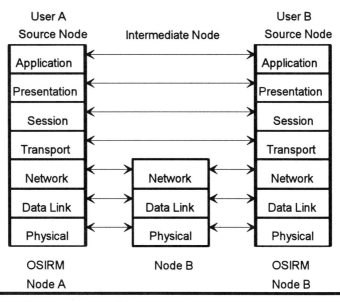

Figure 3.2 OSI Reference Model

Figure 3.3 illustrates the basic elements of every layer in the OSI reference model. This is the portion of the OSIRM that has become widely used to categorize computer and communications protocols according to characteristics from this generic model. Often the correspondence is not exact; for example, ATM is often described as embodying characteristics of both the physical and data link layers. The formal concept of layering is used in the definitions of ATM, ATM Adaptation Layers (AALs), and even higher layer protocols associated with ATM. A layer (N+1) entity communicates with a peer layer (N+1) entity by way of a service supported at layer (N) through a Service Access Point (SAP). The layer (N) SAP provides the primitives to layer (N+1) of request, indicate, confirm, and response. Parameters are associated with each primitive. Protocol Data Units (PDUs) are passed down from layer (N+1) to layer (N) using the request primitive, while PDUs from layer (N) are passed up from layer (N) to layer (N+1) using the indicate primitive. Control and error information utilize the confirm and response primitives.

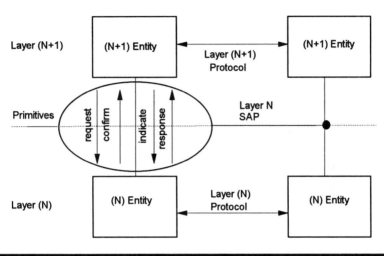

Figure 3.3 Illustration of Layered Protocol Model

The shorthand notation illustrated in Figure 3.4 is used to express this concept graphically in later chapters.

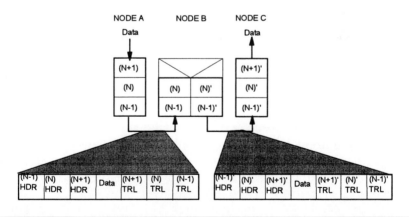

Figure 3.4 Shorthand Protocol Model Notation

Starting at the left-hand side, Node A takes data at layer (N+1), which is connected to Node B by a layer (N-1) protocol. On the link between Nodes A and B we illustrate the resultant enveloping of the layer headers (HDR) and trailers (TRLR) that is carried by the layer

(N-1) protocol. Node B performs a transformation from layer (N) to the corresponding layered, different protocols called layer (N)' and layer (N-1)'. The resultant action of these protocol entities is shown by the layer (N-1)' PDU on the link between nodes B and C. Several examples using this notation are given in Chapters 8 and 17.

3.3 LAYERS OF THE OSI REFERENCE MODEL

Many computer networking architectures can be modeled by the basic structure of the seven-layer Open Systems Interconnect Reference Model . We now cover each layer of the OSIRM in more detail. The OSIRM presents a layered approach to data transmission: seven layers, with each successively higher layer providing a value-added service to the layer above it. Data flows down from layer 7 at the originating end system to layer one, where it is transmitted across a network of intermediate nodes over interconnecting physical media, and back up to the layer 7 of the destination end system. Not all seven levels need be used — this is dependent upon the application and user needs. The specific OSI protocols for each of the seven layers are not well defined in standards or widely adopted in practice, particularly the application, presentation, and session layers. The following sections summarize the generic functions of all seven layers starting with the physical layer, the one closest to the physical transmission medium.

3.3.1 Physical Layer

The first layer encountered is the physical layer (L1), which provides for the transparent transmission of a bit stream across the physical interconnection of network elements. The intelligence managing the data stream and protocols residing above the physical layer are transparently conveyed by the physical layer.

The physical layer connections may be point-to-point, or multipoint. The connection may be operated in full duplex or half duplex mode. Simplex means that transmission is in one direction only, while full duplex means that transmission occurs in both directions simultaneously. Half duplex involves the use of physical layer signaling to change the direction of simplex transmission. The bit stream may be transmitted serially or in parallel.

The physical layer includes specification of electrical voltages and currents, mechanical connector specifications, basic signaling through connections, and signaling conventions. The physical layer can also activate or deactivate the transmission medium and communicate its status through protocol primitives with the data link layer. The physical medium can either be an actual physical transmission medium or a wireless medium.

Examples of the physical layer include EIA-RS-232-C, EIA-RS-449, CCITT X.21/X.21bis, CCITT V.35, IEEE 802 LAN, ISO 9314 FDDI, and the new HSSI interface. One example of a wireless physical interface is a wireless LAN.

The terms Data Termination Equipment (DTE) and Data Communication Equipment (DCE) refer to the hardware on either side of a communications channel interface. DTE equipment is typically a computer or terminal which acts as an end point for transmitted and received data via a physical interface to a DCE. DCE equipment is typically a modem or communication device, which has a different physical interface than that of the DTE. One commonly used type of DCE is called a Channel Service Unit/Data Service Unit (CSU/DSU); it converts from the DTE/DCE interface to a telephony-based interface covered later. Figure 3.5 shows a common end-to-end network configuration where DTE1 talks to DCE1, which in turn formats the transmission for transfer over the network to the end DCE, which then interfaces to the end DTE.

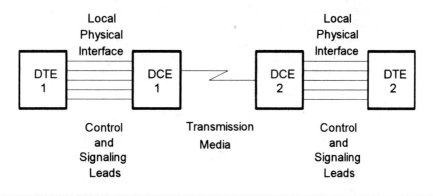

Figure 3.5 DTE to DTE Communications

3.3.2 Data Link Layer

The data link layer is layer 2 (L2) in the seven-layer OSI reference model, and the second layer in most other computer architecture models as well. The primary function of the data link layer is to establish a reliable protocol interface across the physical layer (L1) on behalf of the network layer (L3). This means that the link layer performs error detection and possibly even error correction. Toward this end, the data link control functions establish a peer-to-peer relationship across each physical link between machines. The data link layer entities exchange clearly delimited protocol data units which are commonly called *frames*. The data link layer may use a limited form of addressing such that multiple data link layer protocol interfaces can be multiplexed across a single physical layer interface. There may be a flow control function to control the flow of frames such that a fast sender does not overrun a slow receiver.

Computer communications via local area networks utilizes special functions of the data link layer called the Medium Access Control (MAC) and Logical Link Control (LLC) layers. The MAC layer protocols form the basis of Local Area Network (LAN) and Metropolitan Area Network (MAN) standards used by the IEEE 802.X LAN protocol suite, which includes Ethernet, Token Ring, and Token Bus. Examples of the link layer include ISO 7776, CCITT X.25 link layer, ISDN LAP-D, ISO HDLC, and MAC layer protocols such as the ISO 9314-2 FDDI Token Ring MAC.

Some of the new services, such as frame relay and ATM, can be viewed as using only the first two layers of the OSI reference model, and rely heavily on reducing the link layer services to increase speeds at lower costs because of the resulting protocol simplification. A key difference between frame relay and ATM is that the addresses take on an end-to-end significance, whereas in the OSI link layer addresses are only significant between nodes. Part of the 802.6 Distributed Queue Dual Bus (DQDB) standard is also called a layer 2 protocol.

3.3.3 Network Layer

The third layer (L3) encountered is the network layer. The principal function of the network layer is to provide reliable, in-sequence delivery of protocol data between transport layer entities. In order to do this, the network layer always has an end-to-end addressing capability. A unique network layer address is assigned to each network layer protocol entity. A network layer protocol may communicate with its peer over a route of intermediate machines with physical, data link, and network layers. The determination of this route is called the *routing function*. Network layer PDUs are often called *packets*.

The network layer may also perform end-to-end flow control and the segmentation and reassembly of data. The network layer is the most protocol intensive portion of packet networks. Some examples of protocols used in the network layer are the CCITT X.25 and X.75 packet level and gateway protocols, the Internet Protocol (IP), CCITT/ITU-T Q.931, Q.933, Q.2931, and the OSI CLNP.

The network layer is used to define data call establishment procedures for packet and cell switched networks in ISDN and B-ISDN. B-ISDN signaling utilizes a layer 3 protocol for call setup and disconnection, as covered in Chapter 9. SMDS also employs a layer 3 protocol to provide an end-to-end datagram service using E.164 (i.e., telephone numbers) for addressing.

3.3.4 Transport Layer

The fourth layer encountered is the transport layer. The principal function of the transport layer is to interconnect session layer entities. Historically it has been called the host-host layer. Principal functions that it performs are segmentation, reassembly, and multiplexing over a single network layer interface. The transport layer allows a session layer entity to request a class of service, which must be mapped onto appropriate network layer capabilities. It is the fourth layer's responsibility to manage end-to-end flow control. The transport layer may often perform error detection and correction as well, which has become increasingly more important since it provides a higher level error correction and retransmission protocol for new services such as frame relay, SMDS, and ATM. An example of the Transport layer includes the CCITT X.224 OSI transport protocol TP4. Another widely used example of a transport type of protocol is the Internet Transmission Control Protocol (TCP).

3.3.5 Session Layer

The fifth layer encountered is the session layer. The session layer is essentially the user's interface to the network, which may have some data transformations performed by the presentation layer. Sessions usually provide connections between a user, such as a terminal or LAN workstation, and a central processor or host. So called peer-to-peer session level protocols can directly connect user applications. Session layer protocols are usually rather complex, involving negotiation of parameters and exchange of information about the end user applications. The session layer has addresses that are meaningful to end users. Other session layer functions include flow control, dialog management, control over the direction of data transfer, and transaction support.

Some examples of the session layer are terminal-to-mainframe log-on procedures, transfer of user information, and setting up information and resource allocations. The ISO standard for the session layer is the ISO 8327/CCITT X.225 connection-oriented session protocol.

3.3.6 Presentation Layer

The sixth layer encountered is the presentation layer. The presentation layer determines how data is presented to the user. Official standards are now complete for this layer. Many vendors have also implemented proprietary solutions. One reason for these proprietary solutions is that the use of the presentation layer is very equipment dependent. Some examples of the Presentation layer protocols are video and text display formats, data code conversion between software programs, and peripheral management and control, using protocols such as CCITT X.410, and CCITT X.226 OSI connection-oriented protocol.

3.3.7 Application Layer

The seventh layer encountered is the application layer. The seventh layer manages the program or device generating the data to the network. The application layer is an "equipment-dependent" protocol, and lends itself to proprietary vendor interpretation. Examples of standardized application layer protocols include CCITT X.400, X.420 X.500 - X.520 directory management, ISO 8613/CCITT T.411-419 Office Document Architecture (ODA), and the ISO 10026 distributed Transaction Processing (TP).

3.3.8 Mapping of Generic Devices to OSI Layers

In the past mainframe and minicomputer era, the bottom three layers (network, data link, and physical) were implemented on different equipment than the next three higher layers (presentation, session, and transport). The first three layers were implemented on a Front End Processor (FEP), while the higher three layers were implemented on a host. Current customer premises devices such as bridges, routers, and hubs usually manipulate the protocols of the first three layers: network, data link, and physical. They can often connect dissimilar protocols and interfaces. Many implementations of user software that cover the top three non, application layers (presentation, session, and transport) form together as a single program.

Now, let's get physical!

3.4 DIGITAL TIME DIVISION MULTIPLEXING (TDM)

Public networks developed plesiochronous digital transmission for economic transmission of voice which was then used for data. Recently the Synchronous Optical NETwork (SONET) in North America and the Synchronous Digital Hierarchy (SDH) have been developed internationally in support of higher speed and better quality digital transmission. This section reviews some basics of these technologies.

3.4.1 Plesiochronous Digital Hierarchy

The so-called plesiochronous (which means nearly synchronous) digital hierarchy was developed nearly 40 years ago by Bell Labs to carry digitized voice over twisted wire more efficiently in major urban areas. This evolved first as the North American Digital Hierarchy, depicted in Table 3.1. Each format is called a digital stream (DS) and is assigned a level in the hierarchy. The lower numbered digital streams are multiplexed into the higher numbered digital streams within a certain frequency tolerance. There is no fixed relationship between the data between levels of the hierarchy, except at the lowest level called DS0 at a rate of 64 kbps.

Table 3.1 North American Digital Hierarchy

Signal Name	Rate	Structure	Number of DS0s
DS0	64 kbps	Time Slot	1
DS1	1.544 Mbps	24xDS0	24
DS1c		2xDS1	48
DS2		2xDS1c	96
DS3	44.736 Mbps	7xDS2	672

A transmission repeater system over a four-wire twisted pair was defined and called T1. The term "T1" is often used colloquially to refer to a DS1 signal. There is actually no such thing as a "T3" signal, even though it is often used to colloquially refer to a DS3 signal. The actual interfaces for DS1 and DS3 are called the DSX1 and DSX3 interfaces, respectively, in ANSI standards. The DSX1 is a four-wire interface, while the DSX3 interface is a dual coaxial cable interface.

Closely related hierarchies were also developed in Europe and Japan. These hierarchies are summarized in Table 3.2 [2]. All of these hierarchies have the property that multiplexing is done in successive levels to move between successive speeds, and that the speed of each of these levels is asynchronous with respect to the others within a certain tolerance.

An important consequence of these digital hierarchies on data communications is that only a discrete set of fixed rates is available, namely nxDS0 (where $1 \leq n \leq 24$ in North America and Japan and $1 \leq n \leq 30$ in Europe), and then the next levels in the respective multiplex hierarchies. The details of the DS0 to DS1 and E1 mappings are defined in the next section on ISDN.

Indeed, one of the early ATM proposals [3] emphasized the capability to provide a wide range of very granular speeds as a key advantage of ATM over PDH and SDH for B-ISDN.

Table 3.2 Summary of International Plesiochronous Digital Hierarchies

Digital Multiplexing Level	Number of Voice Channels	Bit Rate (Mbps) North America	Europe	Japan
0	1	0.064	0.064	0.064
1	24	1.544		1.544
	30		2.048	
	48	3.152		3.152
2	96	6.312		6.312
	120		8.448	
3	480		34.368	32.064
	672	44.376		
	1344	91.053		
	1440			97.728
4	1920		139.264	
	4032	274.176		
	5760			397.200
5	7680		565.148	

3.4.2   SONET and the Synchronous Digital Hierarchy (SDH)

The North American defined Synchronous Optical NETwork (SONET) and the closely related international Synchronous Digital Hierarchy (SDH) are the next step in the evolution of Time Division Multiplexing (TDM).   SONET/SDH have two key benefits over Plesiochronous Digital Hierarchy (PDH): rates of higher speeds are defined, and direct multiplexing is possible without intermediate multiplexing stages.   This is accomplished through the use of pointers in the multiplexing overhead that directly identify the position of the payload.   Furthermore, the fiber optic transmission signal transfers a very accurate clock rate along transmission paths all the way to end systems.

The North American SONET signal formats are designated as Synchronous Transfer Signals (STS) at N times the basic STS-1 (51.84 Mbps) building block rate by the term STS-N.   Signals at speeds less than the STS-1 rate are called Virtual Tributaries (VTs). The optical characteristics of the signal that carries SONET payloads is called the Optical Carrier (OC-N).   An STS-N signal can be carried on any OC-M, as long as M≥N.   The standard SONET STS and VT rates are summarized below. Table 3.3 shows the SONET speed hierarchy by OC-level, illustrating the number of DS0s, DS1s, and DS3s equivalents assuming M13 multiplexing.

Table 3.3 SONET STS-N/OC-N Speed Hierarchy

STS-N or OC-N level	Bit Rate (Mbps)	Number of DS0s	Number of DS1s	Number of DS3s
1	51.84	672	28	1
3	155.52	2,016	84	3
6	311.04	4,032	168	6
9	466.56	6,048	252	9
12	622.08	8,064	336	12
18	933.12	12,096	504	18
24	1,244.16	16,128	672	24
36	1,866.24	24,192	1008	36
48	2,488.32	32,256	1344	48
96	4,976.00	64,512	2688	96
192	9,952.00	129,024	5376	192

The CCITT/ITU has developed a similar synchronous multiplex hierarchy with the same advantages using a basic building block called the Synchronous Transfer Module (STM-1) with a rate of 155.52 Mbps that is exactly equal to the STS-3 rate. There is therefore a direct mapping between the SONET STS-3N rates and the CCITT/ITU STM-N rates. An STM-1 frame is equivalent to an STS-3c frame in structure. The overhead byte definitions differ between SONET and the SDH so that direct interconnection is currently not possible. This was done on purpose so that interworking could be achieved more readily. However, there are still some incompatibilities between SONET and SDH in the definition and use of overhead information. A set of lower speed signals, called Virtual Containers (VCs) is also defined in SDH.

Table 3.4 illustrates a similar mapping to that of Table 3.3, comparing the mapping of the North American and CCITT/ITU PDH rates to the corresponding SONET Virtual Tributary (VT) and SDH Virtual Container (VC) rates and terminology. Note that the common 1.5-, 2-, 6- and 44-Mbps rates are mapped consistently. The other common rates are at 155 and 622 Mbps, which is the focus of ATM standardization activity.

Table 3.4 SONET/SDH Equivalent to Plesiochronous Digital Hierar\chy

North American SONET VT	CCITT/ITU SDH VC	SONET Rate (Mbps)	SDH Rate (Mbps)
VT1.5	VC-11	1.544	
VT2.0	VC-12		2.048
VT3.0		3.152	
VT6.0	VC-2	6.312	6.312
	VC-3	44.736	34.368
	VC-4		139.264
STS-1		51.84	
STS-3	STM-1	155.52	155.52
STS-12	STM-4	622.08	622.08

3.4.3 Basic SONET Frame Format

The SONET STS-N frame format is shown in Figure 3.6. Notice that the frame is comprised of multiple overhead elements (section, line, and path) and a Synchronous Payload Envelope (SPE). The frame size for an STS-1 SPE is 9 rows × 90 columns (1 byte per column) for a total of a 783-byte frame (excluding the 27 bytes section and line overhead). The total STS-1 frame of 810 bytes transmitted each 125 µs results in the basic SONET/STM rate of 51.84 Mbps.

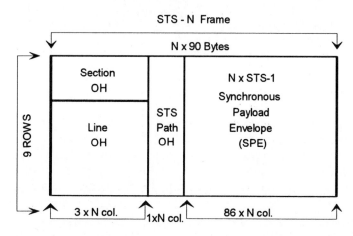

Figure 3.6 STS-N Basic Frame Format

Figure 3.7 shows the breakout of the payload section of the SONET STS-N frame of undetermined size. For frames larger than the STS-1 level, each column is multiplied by N depending on the size of the STS (STS-N).

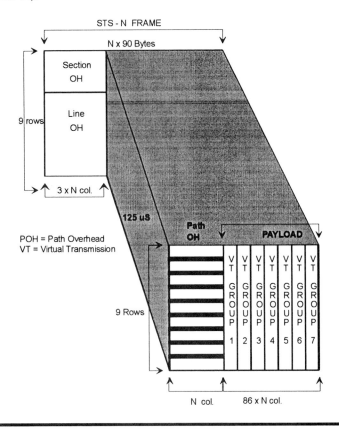

Figure 3.7 STS-N Frame Format Breakout

Figure 3.8 illustrates how VT1.5s are mapped into an STS-1 SONET Synchronous Payload Envelope (SPE). The first column of 9 bytes is the STS-1 path overhead. The next 28 columns are bytes 1 through 9 of the (28) VT1.5 payloads, followed by a column of stuff bytes. Similarly, columns 31 through 58 are bytes 10 through 18 of the (28) VT1.5 payloads, followed by a column of stuff bytes. The last 28 columns are bytes 19 through 27 of the VT1.5 payloads.

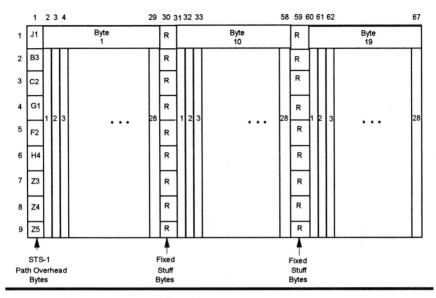

Figure 3.8 VT1.5 Mapping within STS-1 Frame

Figure 3.9 shows the format of an individual VT1.5. Note that there are 27 bytes that are transferred every 125 µs in the SPE as defined above, but that only 24 bytes are necessary to carry the user data. User data byte 25 is included to be able to carry the framing bit transparently. The other two bytes provide a pointer so that the VT can "float" within its allocated bytes and thus allow for timing to be transferred and for the provision of VT level path overhead. The SONET overhead results in a mapping that is 84 percent efficient.

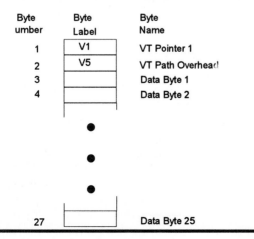

Figure 3.9 Illustration of VT1.5 Format

3.5 BASICS AND HISTORY OF ISDN

To appreciate B-ISDN, one must go back to the original Integrated Services Digital Network (ISDN) standards. First, the ISDN Basic Rate Interface (BRI) and Primary Rate Interface (PRI) services are described. Next, the basic ISDN protocol and framing structure are covered. Subsequent to the definition of Broadband ISDN (B-ISDN), the ISDN standards are now referred to as the Narrowband ISDN (N-ISDN).

3.5.1 ISDN Basics

ISDN was built upon the Time Division Multiplexing (TDM) hierarchy developed for digital telephony as defined in previous sections. The CCITT/ITU has defined and is still defining many standards for Integrated Services Digital Networks (ISDNs). Two standards are defined for the physical interface to ISDN: Basic Rate Interface (BRI), or basic access, as defined in the CCITT ISDN I.430 standard, and the Primary Rate Interface (PRI), as defined in the CCITT ISDN I.431. Both standards define the electrical characteristics, signaling, coding, and frame formats of ISDN communications across the user access interface (S/T) reference point. The physical layer provides transmission capability, activation, and deactivation of Terminal Equipment (TE) and Network Terminations (NTs), Data (D)-channel access for TEs, maintenance functions, and channel status indications. The basic infrastructure for these physical implementations and the definition for the S and T reference points, TE, and NT, are contained in CCITT recommendation I.412.

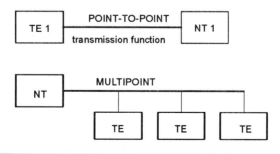

Figure 3.10 ISDN BRI Point-to-Point and Multipoint Configurations

The physical interface in ISDN is one part of the D-channel protocol, and defines a full duplex, synchronous connection between the TE layer 1 terminal side of the basic access interface (TE1, TA, and NT2 functional group aspects) and the NT layer 1 terminal side of the basic access interface (NT1 and NT2 functional group aspects).

Figure 3.10 shows both a point-to-point configuration with one transmitter and one receiver per interchange circuit, as well as a multipoint configuration with multiple TEs, both for BRI. Both bus distances cannot exceed 1000 m, except when using a short passive bus, as opposed to a extended passive bus in multipoint mode, where the limitation is 180 m. The bit rate in both directions is 192 kbps.

3.5.2 BRI and PRI Service and Protocol Structures

The Basic Rate Interface (BRI) and Primary Rate Interface (PRI) service configurations are defined as follows:

- **Basic Rate Interface (BRI):** provides two 64-kbps Bearer (B) channels for the carriage of user data and one 16-kbps control, messaging, and network management data (D) channel. This interface is commonly referred to as 2B+D. The BRI was intended for customer access devices such as an ISDN voice, data, and videophone.

- **Primary Rate Interface (PRI):** provides twenty-three 64-kbps Bearer (B) channels and one 64-kbps Data (D) signaling channel in North America referred to as 23B+D. Internationally 30 B channels are provided in a 30B+D configuration. The PRI was intended for use by higher bandwidth or shared customer devices such as the Private Branch eXchange (PBX), personal computer, and LAN.

The ISDN PRI provides a single 1.544-Mbps DS1 or a 2.048-Mbps E-1 data rate channel over a full duplex synchronous point-to-point channel using the standard Time Division Multiplexing (TDM) hierarchy introduced earlier in this chapter. CCITT Recommendations G.703 and G.704 define the electrical and frame formats of the PRI interface, respectively. The 1.544 Mbps rate is accomplished by sending 8000 frames per second with each frame containing 193 bits. Twenty-four DS0 channels of 64 kbps each comprise the DS1 stream, containing 23 B-channels at 64 kbps each and one D-channel at 64 kbps.

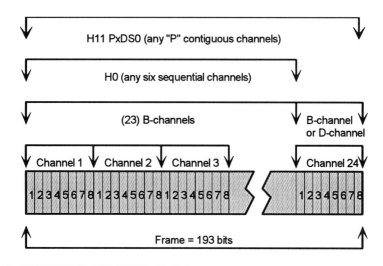

Figure 3.11 1.544 Mbps PRI Frame Structure

Figure 3.11 shows the transmitted framing of the DS1 PRI interface. The CEPT E1-based PRI interface is somewhat different from this, offering 30 B-channels, one D-channel, and a channel reserved for physical layer signaling, framing, and synchronization.

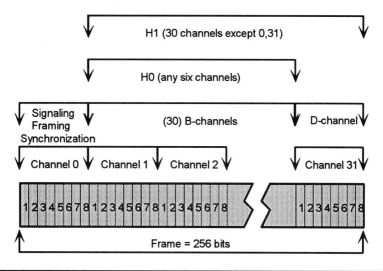

Figure 3.12 CEPT E1 PRI Frame Structure

The frame structure for the CEPT E1 PRI interface is shown in Figure 3.12. A primary attribute distinguishing ISDN service from telephony is the concept of common channel signaling, or out-of-band signaling using the D-channel.

H-channels are used in PRIs. Two types are defined: H_0-channel signals that have a bit rate of 384 kbps and H_1-channels that have a bit rate of 1536 kbps for H_{11}-channels in the U.S., and 1920 kbps for H_{12}-channels in Europe. These channels (except for the H_{12} channel implementation) use B-channel slots on a PRI that is configured as either 24×B or 30×B. Note that this means that the D signaling channel is provided on a separate physical interface. The D-channel and B-channels may share the same physical interface, or the D-channel on one interface may control the B-channels on several physical interfaces.

There is also a capability to establish a nxDS0 bearer service, where n ranges from 1 to 24 (or 30 at the European channel rate) via ISDN signaling. The nxDS0 service uses n contiguous time slots or a bit map specified set of DS0 time slots in the DS1 or E1 frame. This is called the Multi Rate Circuit Mode Bearer Service (MRCMBS). Also, ISDN signaling can establish a Frame Mode Bearer Service (FMBS) or a switched X.25 connection.

3.6 THE INTEGRATED SERVICES DIGITAL NETWORK (ISDN) PROTOCOL MODEL

Integrated Services Digital Network (ISDN) standards were begun by the CCITT in 1972, with the first standards documents published in 1984. The original intent of ISDN was to provide a conversion of telecommunications transmission and switching techniques to a digital architecture, providing end user-to-end user digital service for voice, data, and video. ISDN standards have been used for much more. ISDN standards are also at the root of frame relay standards, as well as the broadband-ISDN/ATM standards which are the focus of this book. The primary ISDN architecture concept consists of multiple devices connecting through an ISDN network termination device (called a TE) into the central office environment where information services are provided. ISDN introduced the notion of multiple planes: the bearer service (or user plane), the control plane, and the management plane. A different OSI layer structured protocol suite for each plane was defined in ISDN as described below.

The ISDN architecture of the user, control, and management planes is shown in Figure 3.13. The user protocol (or bearer service) is layer 1 for circuit mode, layer 2 for frame mode, and layer 3 for packet mode services. Teleservices and value-added services are modeled as higher layers. Intermediate nodes may only provide physical connectivity. Conceptually, another application runs the control, or signaling, plane. The purpose of the control plane protocols is to establish, configure, and release the user plane (bearer) capabilities. Finally, the management plane is responsible for monitoring the status, configuring parameters, and measuring the performance of the user and control planes. This concept of multiple planes — user, control, and management — has been adopted for the B-ISDN/ATM protocol structure. However, the strict alignment with the specific OSI protocol layers has not been entirely retained.

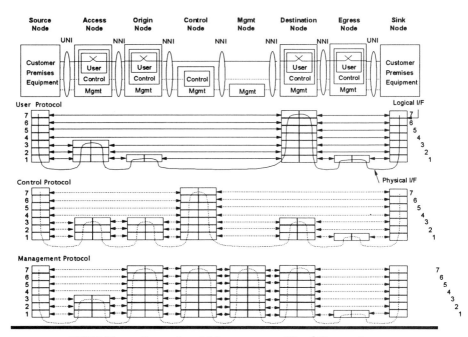

Figure 3.13 ISDN User, Control and Management Plane Protocols

3.7 DEFINITION OF NETWORK SERVICES

Data network services are categorized in the OSI reference model as being either connection-oriented or connectionless. Connection-oriented services involve establishing a connection between physical or logical end points prior to the transfer of data. Examples of

Connection-Oriented Network Services (CONS) are Frame Relay and ATM.

Connectionless services, on the other hand, provide end-to-end logical connectivity and do not require the establishment of a connection prior to data transfer. Examples of Connectionless Network Services (CLNS) are IP and SMDS. Historically, connection-oriented services were generally used in wide area networking, while connectionless services were primarily used in local area networking. Subsequent discussion shows that ATM and its associated adaptation layers provide the capability to support both connection-oriented and connectionless services.

3.7.1   Connection-Oriented Network Services (CONS)

Connection-oriented services require establishment of a connection between the origin and destination before transferring data. Usually the connection is established as a path of links through intermediate nodes in a network. Once established, all data travels over this same path in the network. The requirement that data must arrive at the destination in the same order as sent by the origin is fundamental to all connection-oriented services.

If the connection is established by network management or provisioning actions and is left up indefinitely, then it is called a Permanent Virtual Circuit (PVC). If control signaling of any type is used to establish and take down the connection dynamically, then it is called a Switched Virtual Circuit (SVC). Each of these types of CONS is covered in turn, giving examples for each.

In a PVC service a permanent connection is made between two or more physical or logical interfaces. The connection may be established by physical wiring, equipment configuration commands, service provider provisioning procedures, or combinations of these actions. These actions may take several minutes to several weeks, depending upon exactly what is required. Once the PVC is established data may be transferred over it. Usually PVCs are established, for long periods of time. Examples of physical PVCs are analog private lines, DTE-to-DCE connections, and digital private lines. Examples of logical PVCs are the X.25 PVC, frame relay PVC, and ATM PVC.

In the case of an SVC service, only the access line and address for the origin and each destination point need to be provisioned beforehand. The use of a control signaling protocol plays a central role in SVC services. Via the signaling protocol the origin requests that a connection be made by the network to one or more destinations. The network determines the physical (and logical) location of the destination(s) and attempts to establish the connection through intermediate node(s) to the destination(s). The success or failure of the attempt is indicated back to the originator. There may also be a progress indication to the originator, alerting for the destination, or other handshaking elements of the signaling protocol as well. Often the destination(s) also utilize signaling to allow them to either accept or reject the call. In the case of a failed attempt the signaling protocol usually informs the originator of the reason the attempt failed. Once the connection is established, then data can be transferred over the connection. Usually SVCs are utilized so that resources can be shared by allowing users to dynamically connect and disconnect using the signaling protocol. The signaling protocol usually allows either the originator or destination(s) to initiate a disconnect action. Furthermore a failure in the network or of the originator or destination(s) usually results in an automatic disconnect. Examples of SVCs are telephone calls, ISDN and X.25, frame relay SVCs, and ATM SVCs.

The above description may sound complicated, but it isn't. There is a direct analogy to each of the above terms in the steps of establishing and taking down an SVC connection-oriented service and a normal telephone call, as illustrated in Table 3.5. In fact, much of the complexity of ISDN is introduced by having a more complicated signaling protocol with new names as summarized below. ISDN can support voice calls since the required signaling primitives are part of the signaling protocol.

Table 3.5 General Signaling Comparison to Voice Call

General Signaling Protocol Terminology	Voice Telephone Call Example
Provision Access/Address	Order Service from Phone Company
Handshaking	Dial Tone
Origin Request	Dialing the Number
Successful Attempt Indication	Ringing Tone
Unsuccessful Attempt Indication	Busy Tone
Destination Acceptance	Answering the Phone
Data Transfer	Talking on the Phone
Disconnect Request	Hanging up the Phone

In the case of a switched connection, data is transferred over a connection established through the network, and this same connection is then disconnected, or taken down, after it is no longer needed. The advantage of this additional complexity is that resources can be shared in time, and in some cases the charges for use of a public service can be significantly less for an SVC than they would be for a comparable set of PVCs. In the case of physical SVCs the entire bandwidth of the connection is available to the end points. In the case of logical SVCs the network may be doing some statistical sharing of bandwidth in the switches interior to the network. Connection-oriented services are therefore best used if the required data transfer is intermittent, but lasts quite a bit longer than the time required by the signaling protocol to establish it. Also SVCs can be used to control bandwidth allocation or access to a shared resource, such as a dial-up database service.

Most of the OSI architecture introduced earlier in this chapter is based on connection-oriented services. Private lines and circuit switching are a few examples of connection-oriented services provided by the telephony-oriented carriers, whose primary service is based on connection-oriented services. X.25 PVC and frame relay PVC services are also offered by many carriers.

3.7.2   Connectionless Network Services (CLNS)

In connectionless services, no prior establishment of an end-to-end connection is required for data transmission. Thus, there is no predetermined path that data must take through the network. Therefore, in some connectionless services data may arrive at the destination in a different order than it was sent from the origin. In contrast to connection-oriented services, which may be physical and/or logical, connectionless services are always logical. Another key attribute of many connectionless services is that there is no need for provisioning; you simply plug in the end station equipment, and you are connected! This is often called "plug and play" in data communications.

In connectionless services the originating node transmits packets to the destination via the best path from that node. As other nodes receive these packets, they interpret the address information in the packet and process it based on this node's position in the network and its selection of the best path to the destination as determined by a routing algorithm. The node then switches packets destined for other nodes onto a trunk, and delivers those packets destined for users connected to this node. In a connectionless service the delivery of the packet is not guaranteed; therefore applications must rely on higher level protocols to perform the end-to-end error detection/correction and data integrity checking. Flow control is usually minimal (if any), and the service often does not provide for error detection or correction. Bandwidth efficiency and message sequencing are sacrificed for high performance through fast switching of packets, avoiding the overhead incurred through call establishment and management. Connectionless service is sometimes called datagram service.

The dynamic, automatic determination of which node to select when switching a packet in the above process is, in general, a complicated problem. The generic name for protocols that determine this next hop selection mechanism is routing protocols. Chapter 11 reviews some aspects of routing protocols.

Connectionless service is the predominant mode of communications in Local Area Networks (LANs) because it is well suited to the intermittent, bursty traffic found on applications that use LANs. LANs also take key advantage of the "plug and play" property. Low-level logical LAN connectivity is usually established automatically simply by plugging in the wiring to the LAN adapter interface on the end system. Besides the common LAN technologies such as Token Ring, Ethernet, and FDDI, there is the Internet Protocol (IP) designed for the Wide Area Network (WAN) that has seen great acceptance. By virtue of the routing protocols IP also has the plug and play property that simply attaching (a properly configured) computer to the Internet allows data communication to occur. This is done by the other nodes in the network discovering the added node through routing messages, and adding this to their next hop decision tables.

New connectionless data services are emerging through technologies such as Switched Multimegabit Data Service (SMDS) and the Internet Protocol (IP), both of which can operate over Asynchronous Transfer Mode (ATM) through a specific adaptation layer.

3.8 CIRCUIT SWITCHING

Circuit switching originated in the public telephone network. Let's look at the first telephone usage where each person had a dedicated circuit to every other person, which is essentially a PVC service. This type of connectivity makes sense if you talk to very few people and very few people talk to you. Now let's move forward to the modern day, where the typical person makes calls to hundreds of different destinations for friends and family, for business or pleasure. It is unrealistic to think that in this modern age each of these call origination and destination points would have its own dedicated circuit, since it would be much too expensive. Historically this was how early telephone networks were constructed until the maze of wires began to block out the sun in urban areas before switching was invented by Strowger. Yet a person picks up the phone in one city and calls a person in another city. When this call goes through and both ends begin communicating, they are doing so over a Switched Virtual Circuit (SVC). That circuit is dedicated to the two people until they terminate the call. If they hang up and call back, another circuit is established in the same manner, but not necessarily over the exact same path as before. In this manner, common network resources (circuits) are shared between many users.

Figure 3.14 shows a simplified comparison of two communications networks connecting eight users, labeled A through H, which could be LANs, MANs, PBXs or Hosts. Network (a) shows dedicated private line circuits connecting each user, while network (b) shows circuit switched access to a common, shared network with an access line for each user. In network (a) each user has seven access lines into the network for dedicated circuits connecting to a distant access line for each possible destination. Data or voice is transmitted only via the physical layer. The data is not processed, but instead just passed across the network regardless of the content. The example in the circuit switched network (b) shows User A talking to User H, and User D talking with User E. Any user can communicate with any other user, although not simultaneously, just like in the telephone network.

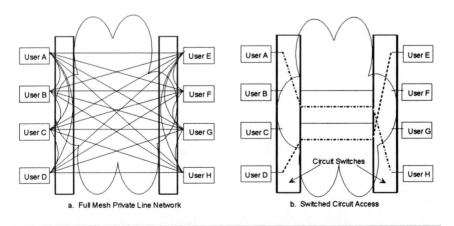

Figure 3.14 Private Line versus Switched Circuit Access

Today, computers need to *talk* to each other in the same manner. Computers can only talk over the telephone network if a modem is used to convert from data signals to signals that can be carried over the circuit switched voice network. Modem is a contraction for modulator/demodulator. The modem presents a DCE interface toward the computer equipment, and outputs a signal that is compatible with a standard phone line. This book does not cover low speed modem communications. See References 4 and 5 for more information on low-speed modem communications.

Data calls which use circuit switching operate in the same manner, but the transmission is not restricted to voice. A point-to-point telephone call is established by a computer whenever data communication is needed and remains up, allowing data to be transmitted. Either the application or the end user disconnects the circuit when they are done. Since circuit switching is a form of connection-oriented service, the same arguments apply. The entire circuit bandwidth is dedicated for the duration of the call. Circuit switching is an ideal technology for traffic which needs constant bandwidth but can tolerate relatively long call establishment and disconnection times.

Figure 3.15 illustrates this process between two users, A and B. Call setup delay associated with the call setup and confirmation is a major consideration with circuit switching for data communications. The call setup time and modem training can often be as large as 10 to 30 seconds. Telephone switch routing usually does not minimize the propagation delay; however, this is not usually a critical factor for data applications operating at the lower speeds of the telephone network.

Circuit switching still remains the most common type of public switched data service. Data circuit switching was much slower to emerge on the market than voice circuit switching, primarily because of the need to digitize switches initially, and it is now limited primarily by whether the access line is digital or analog. Also, the additional complexity of ISDN signaling requires software updates in switches as well as end user equipment.

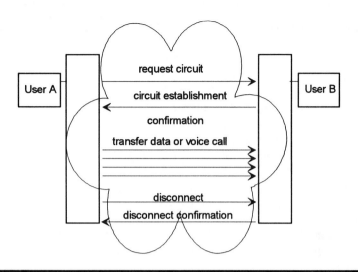

Figure 3.15 Call Establishment and Disconnect

Circuit switching has been used historically as a backup for private line services. It remains the most cost-viable option for private network users, with most switched 56 kbps data services selling at less than 20 cents per minute to less than 5 cents per minute depending on time of day, usage, speed, error-free rate, and other factors. This price is getting very close to that of voice service, since that is basically what it is! This pricing makes it a very cost-effective option to leased-line services if the usage is less than several hours per day, or there are multiple destinations that require dynamic connectivity. The data communications user, however, needs up to three logical types of communication for one call: the data circuit, a signaling capability, and optionally a management capability.

High-speed circuit switching of subrate DS1, DS1, and DS3 speeds is being offered by many LECs and IXCs. Applications which can use high-speed circuit switching as an ideal solution are ones such as bulk data transport and/or those that require all the available bandwidth at predetermined time periods. Circuit switching can provide cost reductions and improve the quality of service in contrast to private lines, depending upon application characteristics.

3.8.1 Switched nx56 kbps and nxDS0

Since Time Division Multiplexing (TDM) has 8000 samples per second per DS0 channel, a difference arises from the fact that 56 kbps uses only seven bits per sample, while 64 kbps uses all eight. The 56-kbps rate resulted from the historical use by the telephone network of one bit per sample in what is called robbed bit signaling.

Switched 56-kbps, or simply switched 56, is a service offered in both the private and public networking environments. Often a Channel Service Unit/Data Service Unit (CSU/DSU) device is used via dedicated 56-kbps or DS0 lines on the CSU side to access a switched 56-kbps service. On the DSU side a standard DCE interface is presented to the computer equipment, as was discussed at the beginning of this chapter.

Another important class of equipment are inverse multiplexers which offer the capability to interface to multiple 56-kbps or DS0 interfaces, as illustrated in Figure 3.16. The inverse multiplexer provides a DCE interface to the computer that aggregates the bandwidth available to the DTE in approximately 56/64-kbps increments. The actual bandwidth provided to the DTE is slightly less than nx56/64 kbps because the inverse multiplexer consumes some bandwidth in its operation. Many of these devices allow 56/64-kbps channels to be dialed-up or disconnected on demand, offering a form of bandwidth on demand. Some router equipment has the intelligence to generate this signaling automatically based upon traffic load. Newer inverse multiplexers can even multiplex multiple DS1 or E1 lines into a DCE interface of approximately nxDS1 or nxE1 speed for very high-speed connectivity.

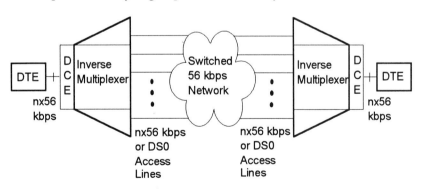

Figure 3.16 Illustration of Inverse Multiplexer Operation

The interface for switched services can be directly from the CPE directly to the IntereXchange Carrier (IXC) Point of Presence (PoP), or via a Local Exchange Carrier's Switched Access service. This is now available in many areas for switched 56-kbps service. Figure 3.17 shows these two types of access. Each type of access has its own merits and drawbacks, such as installation and usage charges, CPE costs, and reliability and availability, which must be traded off against each other.

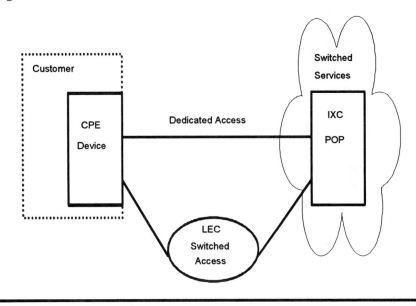

Figure 3.17 Switched Services Interfaces

Many users implement switched 56/64-kbps services as a backup for private lines and to transport non-mission-critical data traffic. Others use it for infrequent high data rate, constant bandwidth data transfers. The typical traffic is long duration, relatively constant bandwidth data transfers, such as batch file transfers, database backups, and highly aggregated, predictable data traffic.

Most switched services are at the 56/64-kbps level, but new services offering higher bandwidth multiples of 56 kbps are now becoming popular, such as switched 384 kbps, DS1, and even DS3. Some carriers offer noncontiguous and contiguous fractional DS1 or nxDS0 reconfigurable or switched services. Reconfigurable services often utilize a computer terminal to rearrange digital cross connects to provide a slower version of nxDS0 switching. The ISDN based version of this service is called the MultiRate Circuit Mode Bearer Service (MRCMBS) which supports switched nxDS0. Videoconferencing is one example, where multiple 56-kbps circuits are combined to form a single high-speed videoconference channel at speeds such as 112 kbps or 224 kbps. Imaging and CAD/CAM file transfer are also good examples of high-bandwidth switched traffic.

Service (MRCMBS) which supports switched nxDS0. Videoconferencing is one example, where multiple 56-kbps circuits are combined to form a single high-speed videoconference channel at speeds such as 112-kbps or 224-kbps. Imaging and CAD/CAM file transfer are also good examples of high-bandwidth switched traffic. Some examples of switched DS1 service traffic include video, imaging, and data center disaster recovery. MCI was the first carrier to offer all of these switched services, but now all the major IXCs offer them.

3.9 REVIEW

This chapter introduced the concept of protocol layering and the notation that is used throughout the book. Next a brief description covered the Open System Interconnection (OSI) reference model and its seven layers: physical, data link, network, transport, session, presentation, and application. Examples of standards and real-world equipment implementing each layer were given. The coverage then moved to digital Time Division Multiplexing (TDM), originally designed to provide more cost-effective telephone calls, and now representing the foundation for the beginnings of high-performance digital data communications. Upon this foundation, construction of the Narrowband Integrated Services Digital Network (N-ISDN) protocol model was begun, but never completed. The concepts of separate user, control, and management protocols from N-ISDN are carried forward into the emerging Broadband ISDN (B-ISDN), but little else. Definitions were given of connection-oriented and connectionless services, and some foreshadowing (as to how ATM is positioned to serve both) was discussed. Finally, some concepts from circuit switching were reviewed.

3.10 REFERENCES

[1]   G. Kessler, *ISDN: Concepts, Facilities and Services*, 2d ed., McGraw-Hill, 1993.
[2]   G. Kessler, D. Train, *MANs*, McGraw-Hill, 1991.
[3]   J. Turner, "Design of an Integrated Services Packet Network," *IEEE Transactions on Communications*, November 1986.
[4]   R. Dayton, *Telecommunications*, McGraw-Hill, 1991.
[5]   D. Spohn, *Data Network Design*, McGraw-Hill, 1993.

4

Networks, Circuits, Multiplexing, and Switching

This chapter provides an overview of common network topologies and circuits, and transmission types, focusing on multiplexing and switching techniques. The five major network topologies defined in this chapter are the basis for most network designs. The presentation then covers the types of circuits that interconnect these topologies. Each circuit type has characteristics, such as the direction of data flow, bit-or-byte oriented transmission, and physical characteristics. Coverage then moves to an explanation of multiplexing techniques — methods of combining and separating units of bandwidth. Finally, the discussion covers the four major methods of switching data: in space, in time, in frequency, or by address. Address switching is the foundation of packet switching and ATM. The reader will then have the background to be able to look at the advanced technologies such as frame relay, DQDB, and ATM that are described in later chapters and are rooted in these multiplexing and switching techniques.

4.1 GENERAL NETWORK TOPOLOGIES

Five commonly used network topologies for computer and data communications networks are: point-to-point, multipoint or common bus, star, ring (or loop), and mesh. The term *node* will be used to designate a network data communications element such as a router, switch, or multiplexer. The term *link* will be used to designate a circuit connection between nodes. A link may be either logical or physical. A concrete example, instead of just an abstract mathematical representation, is given for each topology to illustrate that each of these topologies has a valid data communications application.

4.1.1 Point-to-Point

Point-to-point connectivity is the simplest topology, providing a single link between two nodes. This link can be composed of multiple physical and/or logical circuits. Figure 4.1 shows three examples of point-to-point links. The first example shows a single link between Node A and Node B with a single physical and logical circuit. The second depicts a single link between Node A and Node B with multiple logical circuits riding over a single physical link. The third depicts a single path between Node A and Node B with multiple physical circuits, each carrying multiple logical circuits.

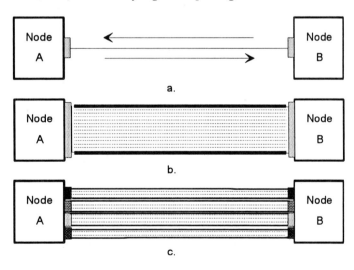

Figure 4.1 Point-to-Point Topology Examples

Point-to-point configurations are the most common method of connectivity. Many data communications applications use point-to-point topologies in Metropolitan and Wide Area Networks (MANs and WANs). For example, almost every user access to the many types of MAN or WAN network services uses some form of point-to-point topology. Examples of the point-to-point topology are private line, circuit switched, and packet switched services.

4.1.2 Multipoint

A common realization of multipoint is the common bus, topology all nodes are physically connected to a common bus structure. Figure 4.2 shows the multipoint common bus topology, where Nodes A through F communicate via a common physical and logical bus. The IEEE 802.4 Token Bus, the IEEE 802.3 Ethernet, and the IEEE 802.6 Distributed Queue Dual Bus (DQDB) all use a common bus topology, as do many other proprietary vendor architectures. The common bus is also called a shared medium topology.

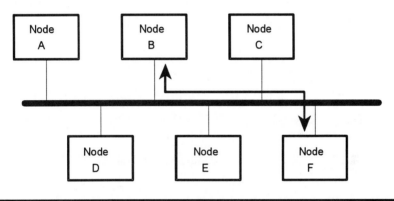

Figure 4.2 Common Bus Multipoint Topology

A multidrop analog line is commonly used for the SNA SDLC loop access. In this example, an analog signal is broadcast from a master station to all slave stations. In the return direction all slave signals are added and returned to the master. See Reference 1 for more information on the SNA SDLC protocol.

Other conceptual examples of the multipoint topology are illustrated in Figure 4.3. Another commonly used multipoint topology is that of broadcast, or point-to-multipoint, which is defined in ATM as the case in which one sender's data is received by many other nodes. Yet another example is that of " incast," or multipoint-to-point, where multiple senders' signals are received at one destination, such as in the slave-to-master direction. In this conceptual illustration note that the multipoint-multipoint (i.e., shared medium, or multicast) topology is effectively the combination of a point-multipoint and multipoint-point topology, as the name implies. The point-to-point topology is also illustrated for comparison purposes.

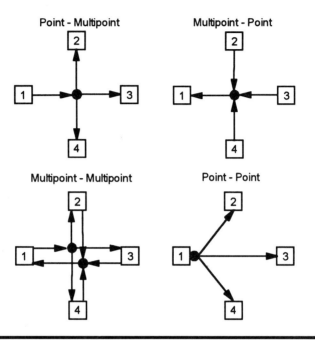

Figure 4.3 Conceptual Illustration of Multipoint Topologies

4.1.3 Star

The star topology developed during the era when most computer communications was centrally controlled by a mainframe. It also has its analogy in the voice world where one central switch is connected to multiple remote switching nodes, each serving hundreds to even thousands of telephones. This network radiates in a star-like fashion from the central switch through the remote switches to the telephones on people's desks. All devices in the network are connected to the central node, which usually performs the processing. Nodes communicate with each other through point-to-point or multidrop links radiating from the central node. The difference between this topology and that of the multipoint topology is that the central node only provides point-to-point connections between any edge node on either a physical or logically switched basis.

Figure 4.4 shows a star topology, where Node A serves as the center of the star and Nodes B through E communicate with each via connections switched through Node A. Another example of a star topology is many remote terminal locations accessing a centralized host processor. These terminals are often called " dumb" terminals since the processing power and intelligence is resident in the host.

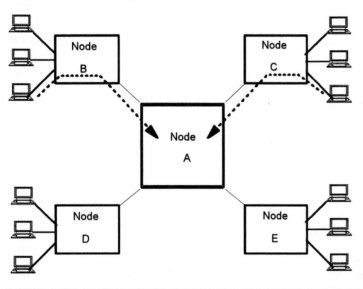

Figure 4.4 Star Topology

The physical star topology is widely used to connect devices to a central hub. The central hub may logically organize the physical star as a logical bus or ring as is commonly done in LAN wiring hubs. The physical star architecture facilitates better network management of the physical interfaces. For example, if a single interface fails in a physical star topology, then the management system can readily disable it.

4.1.4   Ring

The loop, or ring, topology is used for networks in which communications data flow is unidirectional according to a particular protocol. A ring is established, and each device passes information in the direction of the ring.

Figure 4.5 shows a ring network where Node A passes information (frame 1) to Node C via the ring and through Node D. Node C then returns a confirmation (frame 2) to Node A via Node B, at which point Node A removes this data from the ring. There is reuse of capacity in this ring example because the destination removes the information from the ring to make better use of capacity. Examples of the ring topology are the IEEE 802.5 Token Ring and FDDI. Although this topology looks like a special case of a mesh network, we cover it separately because of the different switching action performed at each node. Note that the 802.6 physical topology is often drawn as a ring; however, its operation is logically a bus, as described in Chapter 5. SONET protection rings also use this topology, and are also distinguished from a mesh by the difference in nodal switching action from that of a mesh of circuit switches.

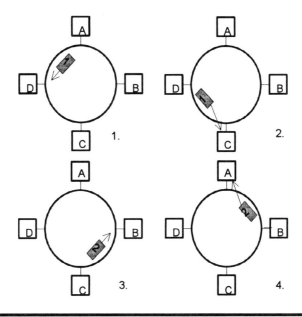

Figure 4.5 Ring or Loop Topology

4.1.5 Mesh

Many switched networks employ some form of mesh architecture. Mesh networks have many nodes which are connected by multiple links. Figures 4.6 and 4.7 show two types of mesh networks. Figure 4.6 shows a partial mesh network where Nodes A, C, F, E, and G have a high degree of connectivity by virtue of having at least three links to any other node, while Nodes A and D have only two links to other nodes. Often the number of links connected to a node is called its degree (of connectivity).

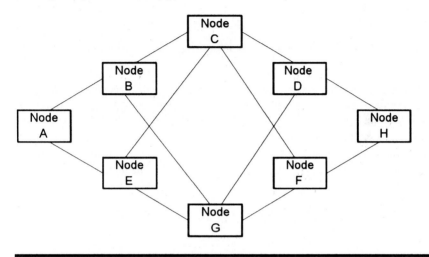

Figure 4.6 Partial Mesh Network

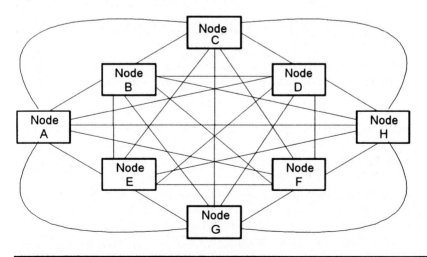

Figure 4.7 Full Mesh Network

Figure 4.7 shows a full mesh network where each node has a link to every other node. Almost every major computer and data communications network uses a mesh topology to give alternate routes for backup and traffic loads, but few use a full mesh topology primarily because of cost factors associated with having a large number of links. This is because a full mesh N-node network has N(N-1)/2 links, which is on the order of N^2. For N greater than 4 to 8 nodes, partial mesh networks are usually employed.

4.2 CIRCUIT TYPES AND SERVICES

This section takes a detailed look at the characteristics of circuits that can be used in network topologies, covering the characteristics of three circuit types: DTE/DCE connections, private lines, and multidrop circuits. These circuit types form the fundamental components of connectivity for most types of data communications, multiplexing, and switching architectures. Additionally, Local Exchange Carriers (LECs), independent telephone companies, and IntereXchange carriers (IXCs) offer private lines and multidrop circuits as a tariffed service.

4.2.1 DTE-to-DCE Connections

DTE-to-DCE connections provide a local, limited distance physical connection between Data Terminal Equipment (DTE) or Terminal Equipment (TE), such as a computer, and Data Communications Equipment (DCE), such as a modem. The physical medium can be two-wire, four-wire, coaxial, fiber optic, or a variety of other interfaces. Figure 4.8 depicts a connection between a DTE and a DCE running simplex, which means that transmission is in a single direction only. Figure 4.9 shows a DTE/DCE connection using half duplex, which means that a transmission in only one direction is allowed at any point in time, as illustrated by the two-headed arrow. The change of transmission direction is accomplished via the control leads between the DTE and DCE at the physical layer. Figure 4.10 depicts full duplex communication, which means that transmission can occur in both directions simultaneously. A separate ground lead is shown for each data signal, indicating a balanced interface which supports higher transmission speeds over longer distances on DTE-to-DCE connections. An unbalanced interface shares a ground lead between multiple signal leads and operates only over shorter distances. All of these examples demonstrate a point-to-point configuration.

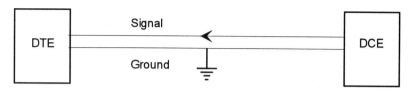

Figure 4.8 Simplex DTE/DCE Connection

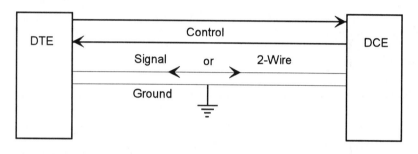

Figure 4.9 Half-Duplex DTE/DCE Connection

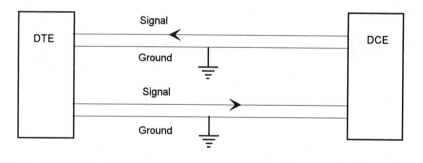

Figure 4.10 Full-Duplex DTE/DCE Connection

4.2.2 Private Lines

A private line, or leased line, is a dedicated circuit leased from a carrier for a predetermined period of time, usually in increments of months. A private line may be upgraded by paying extra for a defined quality of service, such that conditioning is performed to ensure that a better error rate is achieved, which makes a tremendous difference in data communications. As carriers install all-fiber networks, digital private lines are replacing the old voice-grade analog circuits, at the same or even lower costs.

When leased lines are used to access other services, they are called *access lines*. Leased access lines can be purchased through either the local telephone company or, in an increasing number of locations, through alternative access providers or, alternatively, through user-owned access arrangements. Access from these alternate sources is generally less expensive than the local telephone prices. But, of course, the alternative access carrier usually "cream-skims" the lucrative traffic and leaves the "skimmed milk" for the LEC to serve smaller, more remote, or occasional users.

Another form of DS1 rate signal, operating over four wires, is the High-rate Digital Subscriber Line (HDSL) [3]. HDSLs eliminate the cost of repeaters every 2000 ft as in a standard T1 repeater system, and are not affected by bridge taps (i.e., splices). They need to be within 12,000 feet of the serving central office, which covers over 80 percent of the DS1 customers in the United States. Asymmetric Digital Subscriber Lines (ADSLs) are also becoming available and offer higher speeds and better performance. The goal of the ADSL technology is to deliver a video signal and telephone service over a majority of the existing copper, twisted pairs currently connected to homes.

Private lines in Europe and Pacific Rim countries are still very expensive, and transoceanic fiber access is limited. A carrier also must make an agreement with the party at the other side of a fiber to offer the transoceanic service. Prices are dropping, but they require significant investment in small amounts of bandwidth which is often taken for granted in the United States. The high cost of international private lines justifies the cost of sophisticated, statistical multiplexers to utilize the expensive bandwidth as efficiently as possible.

4.2.3 Multidrop

When one user, typically the originator of information, needs to communicate simultaneously with multiple users, a multidrop circuit can be used. Figure 4.11 shows a two-wire multidrop circuit, and Figure 4.12 shows a four-wire multidrop circuit. When using multidrop circuits, there is a master-slave relationship between the master A and slaves B, C, and D. A typical application is where the master A is a cluster controller and B through D are dumb terminals to provide cost-effective access to a centralized host. Note that when dealing with an SDLC loop operating on a two-wire multidrop circuit, only a half duplex connection protocol can be used, as indicated by the two-headed arrow, while full duplex operation can be accommodated on a four-wire circuit, as indicated by the single-headed arrow on the path in each direction. In half duplex operation the master sends out data to the slaves, and polls them for any response. The slave sets a "final" bit in its response, indicating the last frame to be returned. Operation in full duplex is similar, except now the master can send continuously, and uses this channel to poll the slaves.

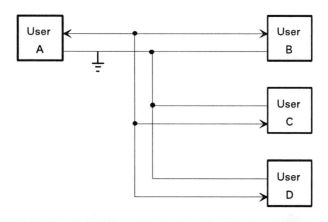

Figure 4.11 Two-Wire Multidrop

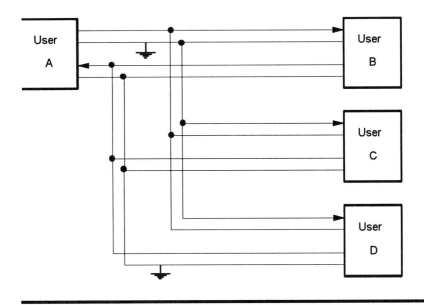

Figure 4.12 Four-Wire Multidrop

4.3 DATA TRANSMISSION METHODS

The digital data transmission method is often characterized as being either asynchronous or synchronous. The terms asynchronous and synchronous are used in different contexts and have entirely different meanings. The meaning not used in this book, but commonly known, of asynchronous versus synchronous character or message transmission is described first. The meaning used in this book of Synchronous versus Asynchronous Transfer Mode (abbreviated as STM and ATM) is then described. These two entirely different meanings of the same term can be confusing, which is why this section presents them together so that the reader can appreciate the differences.

4.3.1 Asynchronous and Synchronous Data Transmission

Asynchronous character transmission has no clock either in or associated with the transmitted digital data stream. Characters are transmitted as a series of bits, with each character identified separately by start and stop bits, as illustrated in the example of ASCII characters in Figure 4.13. There may be a variable amount of time between characters. Analog modem communication employs this method extensively. The baud rate defines a nominal clock rate, which is the maximum asynchronous bit rate. The stop bit can be greater than a baud interval in duration. Since at least 10-baud intervals are required to represent each character, the usable bit rate is no more than 80 percent of the baud rate.

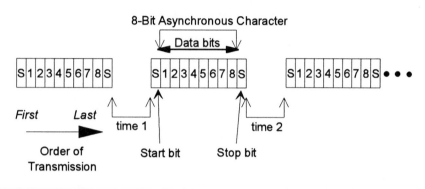

Figure 4.13 Asynchronous Modem Character Transmission

Asynchronous character transmission usually operates at low speeds (typically 9600 bps). Asynchronous interfaces include RS232-C and D, as well as X.21.

Synchronous data transmission clocks the bits at a regular rate by a clocking signal either associated with, or derived from, the transmitted digital data stream. Therefore sender and receiver must have a means to derive a clock within a certain frequency tolerance. On a parallel interface there is often a separate clock lead. Data flows in character streams are called message-framed data. Figure 4.14 shows a typical synchronous data stream. The message begins with two synchronization (SYNC) characters and a start-of-message (SOM) character. The control (C) character(s) denote the type of user data or message following. The data follows next. The cyclic redundancy check (CRC) character checks the data for errors, and the End Of Message (EOM) character signals the end of the transmission stream.

Synchronous data transmission usually operates at speeds of 1200 bps and higher. Synchronous data interfaces include V.35, RS449/RS-442 balanced, RS232-C and D, and X.21.

Synch - Synchronization

SOM - Start of Message

Control - Specifies User Message

Data - User Data Stream

CRC - Cyclic Redundancy Check

EOM - End of Message

Figure 4.14 Synchronous Framed Data Message

4.3.2 Asynchronous versus Synchronous Transfer Mode

Chapter 3 described Synchronous Transfer Mode (STM), or synchronous time division multiplexing, in detail. Asynchronous Transfer Mode (ATM), or asynchronous time division multiplexing, is a different concept that has its roots in packet switching, which Chapter 5 introduces. The following example is a high-level introduction to the basic difference between the STM and ATM multiplexing methods.

Figure 4.15 shows an example of STM and ATM. Figure 4.15a illustrates an STM stream where each time slot represents a reserved piece of bandwidth dedicated to a single channel, such as a DS0 in a DS1. Each frame contains n dedicated time slots per frame; for example, n is 192 in a DS1. Overhead fields identify STM frames that often contain operations information as well, for example, the 193rd bit in a DS1 frame. Thus, if a channel is not transmitting data, the time slot remains reserved and is still transmitted, without any useful payload. In this case, if the other channels have more data to transmit, they have to wait until their reserved, assigned time slot occurs again. Frequent empty time slots result in low line utilization.

ATM uses a completely different approach. A header field prefixes each fixed length payload, as shown in Figure 4.15b. The header identifies the virtual channel. Therefore, the time slots are available to any user who has data ready to transmit. If no users are ready to transmit, then an empty, or idle, cell is sent. Traffic patterns that are not continuous are usually carried much more efficiently by ATM when compared with STM. The current approach is to carry ATM cells over very high-speed STM transmission networks, such as

SONET and SDH. The match between high transmission speeds of SONET and SDH and the flexibility of ATM is a good one.

a. Synchronous Time Division Multiplexing (STM)

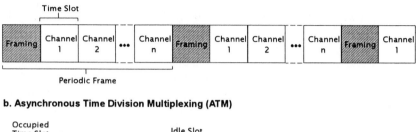

b. Asynchronous Time Division Multiplexing (ATM)

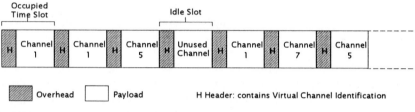

Figure 4.15 Illustration of STM and ATM Multiplexing

4.4   PRINCIPLES OF MULTIPLEXING AND SWITCHING

There is a close relationship between multiplexing and switching. Multiplexing defines the means by which multiple streams of information share a common physical transmission medium. Switching takes multiple instances of a physical transmission medium containing multiplexed information streams, and rearranges the information streams between input and output. In other words, information from a particular physical link in a specific multiplex position is switched to another output physical link usually in a different multiplex position. Multiplexing positions are defined by space, time, frequency, or address. This section first covers multiplexing methods, and then covers point-to-point and point-to-multipoint switching functions.

4.4.1 Multiplexing Methods Summarized

There are four basic multiplexing methods covered: space, frequency, time and address. This is also the historical order in which these were employed in data communications. Space, frequency, and time multiplexing all occur at the physical layer of the OSI reference model. Address switching occurs at higher layers.

4.4.2 Space Division Multiplexing (SDM)

An example of space division multiplexing is where multiple cables interconnect equipment. In other words, space division means physically separate. The original telephone networks, where a pair of wires connected everyone who wished to communicate, is an example of the first use of space division multiplexing. This quickly becomes impractical, as evidenced by old photographs of the sky of major metropolitan cities being blackened by the large number of wire pairs, when only space division multiplexing is used. In early data communications a separate cable was run from every terminal back to the main computer, which is another example of space division multiplexing.

4.4.3 Frequency Division Multiplexing (FDM)

As transmission matured, it was discovered that many analog conversations could be multiplexed onto the same cable, or radio spectrum, by modulating each signal by a carrier frequency. The frequency spectrum of the baseband signal was then placed in separate frequency bands. This yielded a marked increase of efficiency and worked reasonably well for analog signals. The technology was comprised of analog electronics, and suffered problems of noise, distortion, and interference between channels that complicated data communications.

4.4.4 Time Division Multiplexing (TDM)

The next major innovation in multiplexing was motivated by the need over 30 years ago to further increase the multiplexing efficiency in crowded bundles of cables in large cities. This technique made use of the emerging solid-state electronics, and was entirely digital. Analog information was first converted to digital information prior to transmission. The initial cost of this technique was high, but was less than cost of replacing existing cables or digging larger tunnels. Since then, TDM has become the prevalent multiplexing method in all modern telecommunication networks. We now take for granted the fact that our every voice conversation is converted to computer data, transmitted an arbitrary distance, and then converted back to an audible signal. The consequence is that the quality of a voice call carried by digital TDM is now essentially independent of distance. Data communications is more sensitive than digitized voice, but is reaping tremendous benefits from the deployment of TDM infrastructure in public networks. In theory TDM may also be applied to analog signals; however, this application was never widely used.

4.4.5   Address or Label Multiplexing

Address, or label, multiplexing was first invented in the era of poor-quality FDM analog transmission. A more common name for address multiplexing is Asynchronous Time Division Multiplexing (ATDM), which we give as an example later in the chapter. Transmission was expensive, and there was a need to share it among many data users. Each "packet" of information was prefixed by an address that each node interpreted.   Each node decided whether the packet was received correctly, and if not, arranged to have it resent by the prior node until it was received correctly.   SNA, DECNET, and X.25 are early examples of address multiplexing and switching. More recent examples are frame relay and ATM.   The remainder of this book covers the address multiplexing method in great detail.

4.4.6   Point-to-Point Switching Functions

Figure 4.16 illustrates the four basic kinds of point-to-point connection functions that can be performed by a multiplexer or switch.

Space division switching delivers a signal from one physical (i.e., spatial) interface to another physical interface. One example is a copper crosspoint switch.

Time division switching changes the order of time slots within a single spatial data stream, organized by the Time Division Multiplexing (TDM) method.

Address switching changes the address field in data packets, which may be further multiplexed into spatial, time, or frequency signals. This book focuses on this switching method, as applied to packet, frame, and cell switching.

Finally, frequency (or wavelength) switching translates signals from one carrier frequency (wavelength) to another. Wavelength Division Multiplexing (WDM) in optical fibers uses this method.

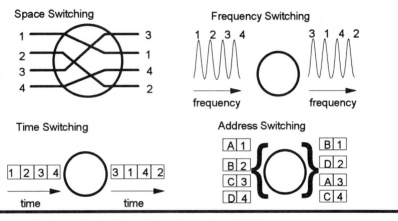

Figure 4.16 Point-to-Point Switching Function Definitions

4.4.7 Point-to-Multipoint Switching Functions

The concept of switching is extended from the case of point-to-point to the broadcast, or point-to-multipoint case, as shown in Figure 4.17. A space division broadcast switch replicates a single input signal on two or more outputs. A simple example is a coaxial television signal splitter. FDM broadcast switching replicates the same signal on multiple output carrier frequencies. TDM broadcast switching fills multiple output time slots with the data from the same input. Address broadcast switching fills multiple packets with different addresses with identical information from the same input packet.

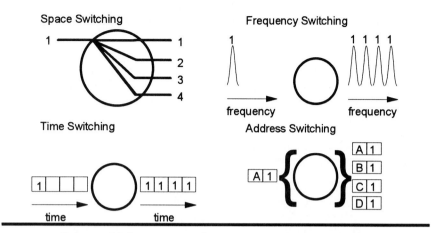

Figure 4.17 Point-to-Multipoint Switching Function Definitions

4.5 EXAMPLES OF MULTIPLEXING

A multiplexer is essentially a very simple switch comprised of a multiplexing function and a demultiplexing function as illustrated in Figure 4.18. The multiplexing function shares the single output between many inputs. The demultiplexing function has one input which must be distributed to many outputs. The multiplexing and demultiplexing functions can be implemented by any of the generic switching functions described in the previous section. Usually a method from the same class is used for both the multiplexing and demultiplexing functions so that the multiplexing method used on each of the interfaces is symmetric in each direction. The overall speed or capacity on the access side interfaces is generally less than that on the trunk side in most multiplexers. For example, this often corresponds to different levels in the Time Division Multiplex (TDM) hierarchy, an example that we will cover in some detail. We will give more detailed examples for each of the generic methods described in the previous section.

The parallelogram symbol with the small end on the side of the single output (called the trunk side) and the large end on the side with multiple interfaces (called the access side) frequently denotes a multiplexer in block diagrams because it graphically illustrates the many-to-one relationship from the large side to the small side, and the one-to-many relationship from the small side to the large side.

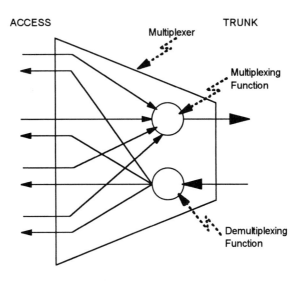

Figure 4.18 Switching Model of a Multiplexer

Multiplexing techniques can be used to share a physical medium between multiple users at two different sites over a private line with each pair of users requiring some or all of the bandwidth at any given time. Some multiplexing techniques statistically assign fixed bandwidth to each user. Other multiplexing methods statistically assign bandwidth to users based upon demand to make more efficient use of the transmission facilities which interface to the network. TDM is often used to reduce the effective cost of a private access line or international private line by combining multiple lower speed users over a single higher speed facility.

4.5.1 Frequency Division Multiplexing (FDM)

Frequency Division Multiplexing (FDM) was widely used as an analog method of aggregating multiple voice channels into larger circuit groups for high-speed transport. FDM multiplexes 12 voice-grade, full duplex channels into a single 48-kHz bandwidth group by translating each voiceband signal's carrier frequency. These groups are then further multiplexed into a mastergroup comprised of 24 groups. Multiple mastergroup analog voice signals are then transmitted over analog microwave systems. A lower frequency analog microwave spectrum was used to frequency division multiplex a DS1 digital data stream in a technique called Data Under Voice (DUV).

Wavelength Division Multiplexing (WDM) on optical fibers is very analogous to FDM in coaxial cable and microwave systems. Optical fiber is *transparent* in two windows centered around wavelengths of 1300 and 1550 nm (10^{-9} m), as shown in the plot of loss versus wavelength in Figure 4.19 [4]. The total bandwidth in these two windows exceeds 30,000 GHz. Assuming 1 bps per hertz (Hz) would result in a bandwidth of over 30 *trillion* bps per fiber!

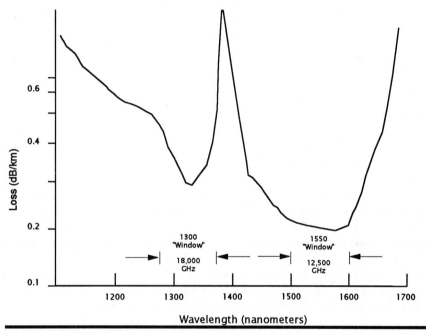

Figure 4.19 Optical Fiber Transfer Characteristic

The carrier frequency at the center of the 1300-nm window is 180 GHz and it is 125-GHz in the 1550-nm window. The sharp attenuation at 1400 nm is due to residual amounts of water (an OH radical) still present in the glass. Continuing improvements in optical fiber manufacturing will likely make even more optical bandwidth accessible in the future. Commercial long-haul fiber optic transmission is now using two wavelengths per fiber, in what is called wideband WDM, in each of these windows. Research and prototype implementations of narrowband WDM, which will allow a much larger number of optical carriers to share a fiber, is nearing commercial deployment.

4.5.2 Time Division Multiplexing (TDM)

Time Division Multiplexing (TDM) was originally developed in the public telephone network in the 1950s to eliminate FDM filtering and noise problems when many signals were to be multiplexed onto the same transmission medium. In the early 1980s, TDM networks using smart multiplexers began to appear in some private data networks, forming the primary method to share costly data transmission facilities among users. In the last decade, time division multiplexers have matured to form the basis of many corporate data transport networks. The premier example of TDM is the DS1 and E1 multiplexing that Chapter 3 described for the ISDN PRI.

4.5.3 Address Multiplexing

A widely used example of address multiplexing is in statistical multiplexers. Statistical Multiplexing, also called Statistical Time Division Multiplexing (STDM), or Asynchronous Time Division Multiplexing (ATDMs), operates similarly to TDM, except it dynamically assigns time slots only to users who need data transmission. Efficiencies of up to 4:1 are gained for voice transmission by utilization of available time slots, rather than wasting them on users who are not speaking. Higher or lower statistical multiplex gains can be obtained for data traffic depending upon the burstiness (peak-to-average statistics) of the data traffic, as covered in Chapter 15. The net effect is an increase in overall throughput for users since time slots are not "reserved" or dedicated to individual users — thus dynamic allocation of bandwidth achieves higher throughput. Figure 4.20 shows a statistical multiplexer which takes multiple low-speed asynchronous and synchronous user inputs for aggregation into a single 56-kbps synchronous bit stream for transmission. The methods used to multiplex the various channels include bit-, character-, block-, and message-oriented multiplexing, each requiring buffering, and more overhead and intelligence than basic time division multiplexing.

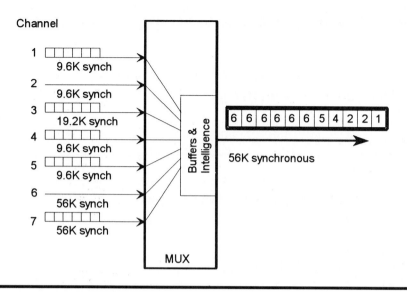

Figure 4.20 Statistical Multiplexer

Figure 4.20 shows an excerpt from a statistical multiplexed data stream. In a statistical multiplexer, the output bandwidth is a 1 to N output-to-input ratio less than the aggregate input bandwidth. This is done on purpose, assuming that not all input channels will be transmitting at the same time when each channel is sampled for transmission. Thus, the output synchronous data stream allocates bandwidth only to users who require it, and does not waste time slots by dedicating bandwidth to users who do not require it. Note in our example that channels 1, 2, 4, 5, and 6 are transmitting, and together utilizing the 128 kbps trunk bandwidth. Using the same example, if channel 3 were to also transmit data at the same instant, the total transmission requirements would exceed the available circuit transmission speed out of the multiplexer, and buffers would begin to store the information until space on the transmission circuit could become available.

Another type of statistical multiplexing is statistical packet multiplexing (SPM). Statistical packet multiplexers combine the packet switching of X.25 with the statistical multiplexing of STDM. SPM operates similarly to STDM in that it still cannot effectively transmit delay-sensitive information such as voice and video. There is still the overhead delay of guaranteed delivery of packets, but efficiencies are gained in dynamic bandwidth allocation and sharing by assigning active bandwidth to the channels which need bandwidth at any given time. Each multiplexer groups the user data into packets passed through the network, multiplexer to multiplexer, similar to packet switching.

4.5.4 Space Division Multiplexing

Space division multiplexing essentially reduces to cable management. This can be facilitated by mechanical patch panels, or increasingly more so by optical and electronic patch panels. To a large extent, space division multiplexing is falling out of favor, and is being replaced by space division switching or other types of multiplexing.

4.6 EXAMPLES OF SWITCHING

This section covers an example for each of the switching techniques: space, time, address, and frequency. The examples chosen for this section define terminology and illustrate concepts that provide background for material in subsequent chapters.

4.6.1 Space Division Switching

Figure 4.21 illustrates a simple two-input, two-output crossbar network, using the crosspoint nodal function. An example connection is shown by the boldface lines and control inputs. Notice that a total of four nodes are required. Classical space division switch fabrics have been built from electro mechanical and electronic elements with the crosspoint function. Future technologies involving optical crosspoint elements with either electronic or optical control are being researched and developed.

Examples of space division switches are DTE matrix switches, supercomputer High Performance Parallel Interface (HPPI) switches, and 3/3 digital cross connects. Many space division switches employ multiple stages of crosspoint networks to yield larger switch sizes.

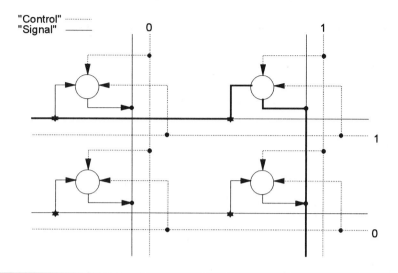

Figure 4.21 Two-Input, Two-Output Crossbar Network

4.6.2   Time Division Switching

The operation of current digital telephone switches may be viewed as being made up of an interconnected network of special purpose computers called Time Division Switches (TDS) [2].

The description of time division switching operation references to Figure 4.22. Each TDM frame has M time slots. The input time slot m, labeled $I(m)$, is stored in the input sample array $x(t)$ in position m. The output address memory $y(t)$ is scanned sequentially by increasing t from 1 to M each frame time. The contents of the address array $y(t)$ identify the index into the input time slot array x that is to be output during time slot t on the output line. In the example of Figure 4.22, $y(n)$ has the value m, which causes input time slot m to be switched to output time slot n. Note that the input sample array must be double buffered in an actual implementation so that time slot phase can be maintained for inputs and outputs with different frame clock phases.

This TDS function is performed for M time slots, that is, once every frame time. This must occur in less than $\tau = 125$ µs (1/8000) for all slots, $n=1,...,M$. The maximum TDS size is therefore determined by the TDS execution rate, I instructions per second (or equivalently I^{-1} seconds per instruction); then the TDS switch size M must satisfy the inequality

$M \le \tau I$

The TDS is effectively a very special purpose computer designed to operate at very high speeds. For I ranging from 100 to 1000 MIPs, the maximum TDS switch size M ranges from 12,500 to 125,000 which is the range of modern single-stage Time Division Switches (TDS). Larger time division switches can be constructed by interconnecting TDS switches via multiple stage crosspoint type networks [2].

Usually, some time slots are reserved in the input frame in order to be able to update the output address memory. In this way, the update rate of the switch is limited by the usage of some slots for scheduling overhead.

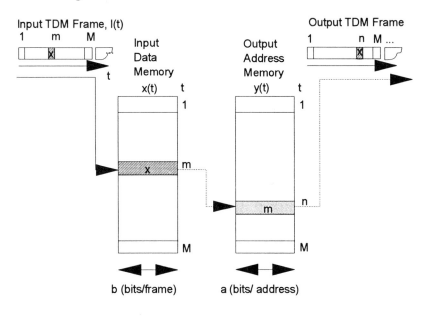

Figure 4.22 Illustration of Time Division Switch Operation

4.6.3 Address Switching

Address switching operates on a data stream in which the data is organized into packets, each with a header and a payload. The header contains address information that is used in switching decisions at each node. The address determines which physical output the packet is output to, along with any translation of the header address value. All possible connection topologies can be implemented within this switching architecture: point-point, point-to-multipoint, multipoint-to-point, and multipoint-to-multipoint. We illustrate these topologies in the following example.

Figure 4.23 illustrates four address switches, each with two inputs and two outputs.

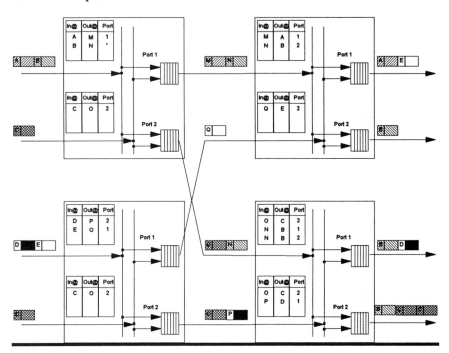

Figure 4.23 Address Switching Example

Packets (either fixed or variable in length) arrive at the inputs as shown on the left-hand side of the figure with addresses indicated by letters in the header symbolized by the white square prior to each shaded payload. The payload shading is carried through the switching operations from left to right to allow the switching result of the address switches to be traced visually. The input address indexes into a table using the column labeled In@, which identifies the address for use on output in the column Out@, and the physical output port on which the packet is sent in the column labeled Port. For example, the input packet addressed as *A* is output on port 1 using address *M*. Conceptually, each switch functions as a pair of busses which connect to the output port buffers. Packets destined for a particular output port are queued for transmission. Buffering is required because contention may occur for the same output port. At the next switch the same process occurs until the packets are output on the right-hand side of the figure.

The packets labeled A, D, and E form point-to-point connections. The packets labeled B form point-to-multipoint connections. The packets labeled C form multipoint-to-point connections. Of course, address switching and multiplexing are at the heart of ATM, which later chapters cover in great detail.

4.6.4 Frequency/Wavelength Switching

A significant amount of research has been conducted recently on all optical networks [5], [6]. The basic concept is a shared media, all-photonic network which interconnects a number of end systems as shown in Figure 4.24.

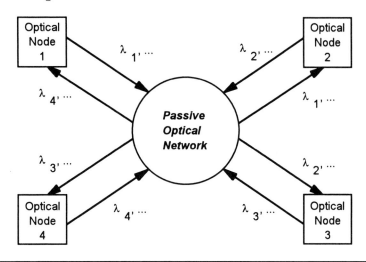

Figure 4. 24 Illustration of Optical WDM Network

The optical end system nodes transmit on at least one wavelength λ and receive on at least one wavelength. The wavelength for transmission and reception may be tunable, currently in a time frame on the order of milliseconds, with an objective of microseconds. The end systems may also be capable of receiving on more than one wavelength. The wavelengths indicated by the subscripts on the character λ are used in the next example of a multiple hop optical network.

If the end system cannot receive all other wavelengths, then some means to provide interconnectivity is required. One early method that has been proposed and implemented is that of multiple hop interconnections. In a multiple hop system, each end system also performs a routing function. If an end system receives a packet that is not destined for it, it forwards it on its transmit wavelength. Eventually the packet reaches the destination, as shown in the trellis drawing of Figure 4.25. For example, in order for station 1 to transmit to station 4, it first sends on wavelength λ_1, which is received by node 2. Node 2 examines the packet header and determines that it is not the destination, and retransmits the packet on wavelength λ_2. Node 3 receives the packet, examines the packet header, and forwards it on λ_3, which is received by the destination, Node 4.

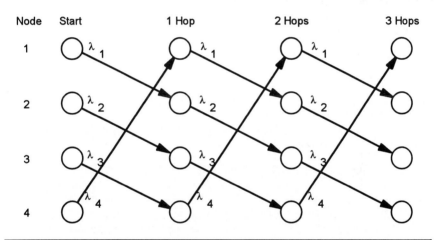

Figure 4.25 Illustration of Multiple Hop WDM Network Trellis

This multiple hop process makes inefficient use of the processing power of each node as the number of nodes grows large, so recent research has focused on single hop designs. In these designs the tunable transmitter and receiver are often employed. There is a need for some means to allocate and share the bandwidth in the optical network. Circuit-based signaling has been tried, but the user feedback is that packet switching, and not circuit switching, is required because the circuit setup time is unacceptable [6]. Fixed allocation of bandwidth in a time-slotted manner is also not desirable. Dynamic scheduling and collision avoidance hold promise for a solution to the demands of very high-speed networking [5].

4.7 REVIEW

This chapter began with a discussion of the five major network topologies: point-to-point, multipoint, star, ring, and mesh. The treatment then moved to a discussion of the relationship between DTE and DCE connections. The multiple uses of the terms asynchronous and synchronous in data communications was clarified. The commonly used definition of asynchronous in modem communications is not what is meant by the word asynchronous in ATM. ATM is a specialized form of address switching derived from packet switching that was described by several examples in this chapter. The major principles of multiplexing and switching were then introduced, followed by a number of examples for the major methods commonly in use today

4.8 REFERENCES

[1]   Ranade, Sackett, *Introduction to SNA Networking*, McGraw-Hill, 1989.
[2]   Keiser, Strange, *Digital Telephony and Network Integration*, Van Nostrand Reinhold, 1985.
[3]   K. Miller, "A Reprieve from WAN's Long Last Mile," *Data Communications*, July 1993.
[4]   Personick, *Fiber Optics Technology and Applications*, Plenum, 1985.
[5]   Jajszczyk, Mouftah, "Photonic Fast Packet Switching," *IEEE Communications*, February 1993.
[6]   Green, "An All-Optical Computer Network: Lessons Learned," *IEEE Network*, March 1992.

5

Review of Modern Data Services

This chapter begins with an introduction to packet switching, articulating the reasons for packet switching, some basic principles, and a discussion on how changes in the telecommunication environment impact the evolution of these packet switching protocols. Next the chapter presents an overview of packet switching and the major public and private data services of the 1990s, including X.25 packet switching, the Internet Protocol (IP), frame relay, and Switched Multimegabit Data Service (SMDS). The aspects of each protocol are described, beginning with the origins of the protocol, and followed by an overview of the packet formats and protocol functions. Next, the operation of the protocol is illustrated through an example. The traffic and congestion control aspects of the various protocols are also surveyed. Finally, aspects of the protocol supported in a public service are covered.

5.1 A BRIEF HISTORY OF PACKET SWITCHING

Packet switched networks have been evolving for over 25 years, and form the basis of many advanced data communications networks today. Packet switching initially provided the network environment needed to handle bursty, terminal-to-host data traffic over noisy analog telephone network facilities. Packet switching has been widely implemented, especially in Europe, where it constitutes the majority of public and private data services.

5.1.1 Early Reasons for Packet Switching

Several factors created the need for packet switching: the need to create standard interfaces between computing devices, and the ability to extend computer communication over noisy transmission facilities, make more efficient use of expensive transmission bandwidth, and enable the interconnection of a large number of computing devices.

The early days of computing saw the development of a new interface and data communication protocol with each computer. Large computer manufacturers, like IBM and DEC, developed protocols that were standardized across their products. One of the goals of the Open System Interconnection (OSI) standardization effort was to enable standard computer communication interfaces and protocols in a multiple vendor environment.

Packets were designed with Cyclic Redundancy Check (CRC) fields that detected bit errors. If a packet was received in error, then a protocol was defined so that it could be resent until it was successfully received. The advent of low bit-error rate fiber optic transmission media means that this function is now cost-effectively done in the end system, since errors rarely occur.

Early packet switching systems were designed for terminal-to-host communications. The typical transaction involved the users typing a few lines, or even just a few characters, and then sending a transaction to the host. The host would then return a few lines, or possibly an entire screen's worth of data. This terminal-host application was very bursty; that is, the peak transmit rate of each terminal was much greater than its average rate. Packet switching allows many such bursty users to be statistically multiplexed onto expensive transmission facilities.

As the number of computers, applications, and people using computers increased, the need for interconnection increased, creating the accelerating need for bandwidth discussed in Chapter 2. Similar to the growth in telephony, it quickly became absurd to have a dedicated circuit to connect every pair of computers that needed to communicate. Packet switching and routing protocols were developed to connect terminals to hosts, and hosts to hosts.

5.1.2   Principles of Packet Switching

Packet switching is a special case of the general address multiplexing and switching method described in Chapter 4.   Packet switching provides a service in which blocks of user data are conveyed over a network.   These blocks of user data are called the payload information.   Packet switching adds overhead to the user data blocks, resulting in a block of data called a *packet*.   All of the protocols studied in this book have this characteristic; including: X.25, IP, frame relay, SMDS, Ethernet, FDDI, Token Ring, and ATM.

The functions that use the packet overhead are either link layer, packet layer, or transport layer from the OSI Reference Model (OSIRM) summarized in Chapter 3.   Older protocols, such as X.25 and IP, perform both the link layer and packet layer function. Newer protocols, such as frame relay, SMDS, and ATM, perform only a subset of link layer functions, but with addressing functions that have a network-wide meaning.

Link layer functions always have a means to indicate the boundaries of the packet, perform error detection, provide for multiplexing of multiple logical connections, and provide some basic network management capability.   Optional link layer functions are: flow control, retransmission, command/response protocol support, and link level establishment procedures.

Network layer functions always have a means to identify a uniquely addressed network station.   Optional network layer functions include: retransmission, flow control, prioritized data flows, automatic routing, and network layer connection establishment procedures.

Packet switching performs the same function as a statistical multiplexer described in Chapter 4, where the packet switch allows multiple logical users to share a single physical network access circuit by providing buffers which are used during times of overflow. Packet switches can control the quality provided to an individual user by allocating bandwidth, allocating buffer space, controlling the traffic offered by users, or by controlling flow.   Part 5 covers the application of these methods to ATM.

Packet switching also extends the concept of statistical multiplexing to an entire network. In order to appreciate the power of packet switching, compare the full mesh network of dedicated circuits in network Figure 5.1a versus that of the packet switched network in Figure 5.1b.   The dedicated network has three lines connected to every host, while the packet network has only one connecting the host to the packet switch. A virtual circuit connects every user through the packet switched network, as shown by dashed lines.

The dedicated circuit network (a) has higher overall throughput, but will not scale well. The packet switched network of (b) requires additional complexity in the packet switches, and has lower throughput, but reduces circuit transmission costs by over 40 percent in this simple example with the nodes placed on the corners of a square. Sharing of network resources allows savings over the cost of many dedicated, low-speed communications channels, each of which is often underutilized the majority of the time. Virtual circuits are a concept that carries through into both frame relay and ATM technology.

a. Dedicated Circuit Network

b. Packet Switched Network

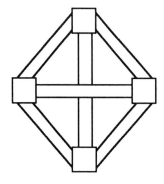 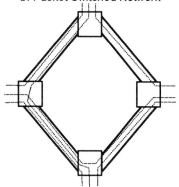

Figure 5.1 The Power of Packet Switching

Packet switching employs queueing to control loss and resolve contention at the expense of added, variable delay. The packet may take longer to get there with packet switching, but the chances of loss are much lower, assuming a reasonable buffer size, during periods of network congestion. There are two types of flow and congestion control used to manage packet data protocols: implicit and explicit congestion notification.

Implicit congestion notification usually involves a layer 4 transport protocol, such as the Internet Transmission Control Protocol (TCP), in either the network device or the user premises equipment. These protocols can adaptively alter the rate of packets sent into the network by estimating loss and delay.

Explicit congestion notification occurs when the protocol notifies the sender and/or receiver of congestion in the network. If the sender or receiver reacts to the explicit indication of congestion quickly enough, then loss can be avoided entirely. Chapter 14 covers the subject of implicit and explicit flow and congestion control in detail.

5.1.3   Darwin's Theory and Packet Switching Evolution

Darwin spent nearly as much time to arrive at his theory of evolution as it has taken for packet switching to be conceived and to sweep the world of data communications. The basic tenets of Darwin's theory of evolution are natural selection, survival of the fittest, and the need to adapt to a changing environment.

This section takes the reader through a brief summary of the genealogy of packet switching with reference to Figure 5.2. The genesis of packet switching began with the proprietary computer communication architectures by IBM, called Systems Network Architecture, and DEC's Digital Network Architecture (DNA). The Synchronous Data Link Control (SDLC) protocol from SNA was refined and standardized upon as the High-level Data Link Control (HDLC) protocol which begat X.25 and Link Access Procedure D (LAP-D) within ISDN. Frame relay evolved as basically a leaner, meaner LAP-D protocol. OSI adopted the X.25 protocol as the first link and packet layer standard.

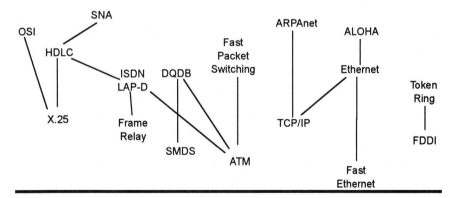

Figure 5.2 Genealogy of Packet Switching

Around the same time that OSI was being developed the U.S. Advanced Research Projects Agency (ARPA) was working on a network with universities and industry that has resulted in the suite of applications and higher level protocols based upon the Internet

Protocol (IP). Ethernet also sprung up at this time as a result of experiments on packet radio communication in Hawaii. Ethernet is evolving into Fast Ethernet. Token Ring was also developed shortly after Ethernet, and has evolved into the higher speed FDDI. Within the past 10 to 15 years the concepts of Fast Packet Switching (FPS) and the Distributed Queue Dual Bus (DQDB) have resulted in the Switched Multimegabit Data Service (SMDS) and ATM.

Packet switching plays an increasingly important role in the rapidly changing environment of distributed processing in the 1990s. Several environmental factors are currently driving the direction of data communications evolution. There is an accelerating need for more bandwidth driven by increasing computing power, increasing need for interconnectivity, and the need to support ever larger networks where any user or application can communicate with any other. The low error rate of modern fiber optic, satellite, and radio communications enables more cost-effective implementation of higher speed data communications. The same technology that increases computer power can be used to increase packet switching performance.

This changing environment creates new opportunities for new species of data communications protocols. The improved quality of transmission facilities alone was a major force in the evolution of frame relay, SMDS, and ATM. These newer protocols are streamlined in that they do not perform error correction by retransmission within the network. The fixed slot and cell size of SMDS and ATM have also enabled cost-effective implementation of switching machines. The increasing capabilities of high-speed electronics are an essential ingredient in SMDS and ATM devices.

5.2   X.25 PACKET SWITCHING

X.25 is the earliest standardized public data network protocol, with initial standards being defined in 1976. This section provides an overview of key features and terminology developed for X.25 that are still used in frame relay and ATM today.

5.2.1   Origins of X.25

In the beginning there were proprietary protocols; then the CCITT standardized upon the first international physical, link, and packet layer protocol, X.25. The CCITT X.25 packet switching standard, along with a number of other "X dot" standards, was developed to provide a reliable system of data transport for computer communications over the noisy, unreliable analog-grade transmission medium.

Speeds have increased beyond 56 kbps, the maximum transmit rate of early packet switches. Many packet switches today can provide networking speeds up to DS1 and E1, with packet sizes up to 4096 bytes.

5.2.2 Structure of X.25

The CCITT set of "X dot" standards for the physical, link, and packet layer protocols shown in Figure 5.3 are known collectively as X.25. The suite of physical, link, and network layer protocols standardized by X.25 has been adopted as part of the OSI Reference Model (OSIRM). These standards define the protocol, services, facilities, packet switching options, and user interfaces for public packet switched networks.

The physical layer is defined by the X.21 and X.21bis standards. X.21 specifies an interface between Data Terminal Equipment (DTE) and Data Circuit terminating Equipment (DCE). X.21 also specifies a simple circuit switching protocol that operates at the physical layer which is widely implemented in the Nordic countries.

The data link layer standard is based upon the High-Level Data Link Control (HDLC) ISO standard. X.25 modified this and initially called it a Link Access Procedure (LAP), subsequently revising it again to align with changes in HDLC resulting in the Link Access Procedure-Balanced (LAP-B).

The packet layer standard is called the X.25 Packet Layer Protocol (PLP). The packet layer defines Permanent Virtual Circuit (PVC) and virtual call, or Switched Virtual Circuit (SVC) message formats and protocols. The concept of PVCs and SVCs from X.25 is also used in frame relay and ATM.

User connectivity to the packet switched network takes on many forms. The suite of "X dot" standards is: X.121, X.21, X.25, X.28, X.29, X.3, and X.32. X.121 defines the network layer numbering plan for X.25.

CCITT Recommendations X.3, X.28, and X.29 define the method for asynchronous DTEs to interface with X.25 networks via a Packet Assembler/Diassembler (PAD) function. A PAD takes strings of asynchronous characters from the DTE, delimited by a carriage return, and assembles these into a synchronous X.25 packet. The PAD delivers a stream of asynchronous characters to the DTE from a received X.25 packet. Recommendation X.3 defines PAD parameters, such as terminal type, line length, break key actions, and speed. Recommendation X.28 defines the terminal-to-PAD interface, while Recommendation X.29 defines the computer-to-PAD interface. Recommendation X.21 defines a dedicated physical interface, and Recommendation X.32 defines a synchronous dial-up capability.

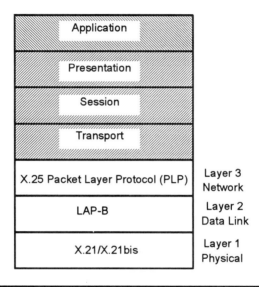

Figure 5.3 X.25 Packet Switching Compared to the OSI Reference Model

5.2.3 X.25 Packet Formats

Each packet to be transferred across the DTE/DCE interface is carried within the LAP-B frame as shown in Figure 5.1. Note that the X.25 layer 3 packet, including packet header and packet data, forms the user data, or information, field of the layer 2 LAP-B frame.

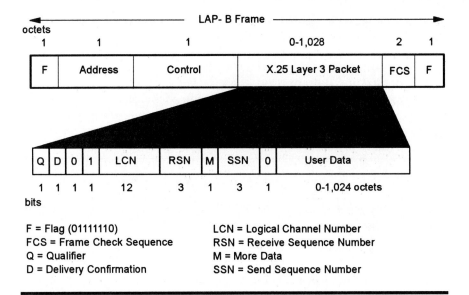

Figure 5.4 LAP-B Frame and X.25 Packet Layer Payload

The address field of the LAP-B frame is primarily used on multidrop lines. The address field also can indicate the direction of transmission and differentiate between commands and responses.

The control field of the LAP-B frame may be one, or optionally two, octets in length in the extended mode. The control field contains the data link layer send and receive sequence numbers which are used in link level flow control and retransmission. The control field also contains a poll/final bit which is also used primarily in multidrop configurtaions.

The Frame Check Sequence (FCS) is computed by the sender, and is recomputed by the receiver to detect if there were any bit errors in the received LAP-B frame. The Flag (F) sequence delimits the beginning and end of the LAP-B frame. An operation of the data link layer is to avoid the occurrence of the flag sequence within user data. If a sequence of six 1s is detected in the data, a zero is inserted prior to the sixth 1 upon transmission. Upon receipt, the inserted zeroes are removed by looking for sequences of five 1s followed by the "stuffed" zero bit.

The X.25 packet layer is composed of a header and a user data, or information, field as shown in Figure 5.4. The Qualifier (Q) bit allows the transport layer to separate control data from user data. The D bit is used in delivery confirmation during call setup. The next two bits indicate the packet type, with 01 indicating a data packet with three-octet header. A four-octet header is also standardized. The X.25 packet layer address has a 4-bit group number and an 8-bit logical channel number, together forming a 12-bit Logical Channel Number (LCN), or virtual circuit number. Virtual circuit zero is reserved, and therefore there can be up to 2^{16} minus 1=4095 virtual circuits. The packet layer Receive and Send Sequence Numbers (RSN and SSN) provide packet layer flow control. The More (M) bit supports segmentation and reassembly by identifying packet segments with a value of 1, with the last segment having a value of zero.

There are also call control packets and control packets defined in X.25. Switched Virtual Circuits (SVCs) are established and cleared using the call control packets. SVCs use the X.121 addressing format. Further information about the details of X.25 can be found in the references at the end of the chapter.

5.2.4   X.25 Functions

There are two types of services defined in the X.25 standard: virtual circuit and datagram. Virtual circuits assure sequence integrity in the delivery of user data, established either administratively as a PVC, or as an SVC through call control procedures. Datagrams are messages unto themselves that do not require call control procedures. Datagrams may be sent in a best-effort mode, or have an explicit receipt notification. X.25 packet switches store and forward each packet to the next node. The packet is deleted from memory only after it is acknowledged.

5.2.5   Example of X.25 Operation

The LAP-B protocol uses a store and forward approach to ensure reliable delivery of packets across noisy, error-prone transmission links. The example of Figure 5.5 illustrates the store and forward approach for recovering from errors between two packet swithing nodes, labelled A and B. Two types of packets are exhanged between the nodes: Link Data (LD) and ACKnowledgments (ACK). Each message has a pair of numbers associated with it: the Receive Sequence Number (RSN) and the Send Sequence Number (SSN). The SSN indicates the number of the packet that is being sent. The RSN indicates the next SSN that is expected by the receiver. The example begins with node A sending user data with send number 1, which is successfully transferred to node B, which acknowledges its receipt with an ACK containing an RSN indicating that the next expected send number is 2. Node B now stores packet 1 and attempts to forward this packet to the next node. Meanwhile node A was sending the second packet; however, it was corrupted due to errors and discarded by node B. The third packet sent by A is received by B, but it is out of order, and B acknowledges this, but indicates that it is still waiting for the second packet. Node A responds by resending packets 2 and 3, which are successfully received and acknowledged by node B, which can now attempt to forward these packets to the next node using the same process.

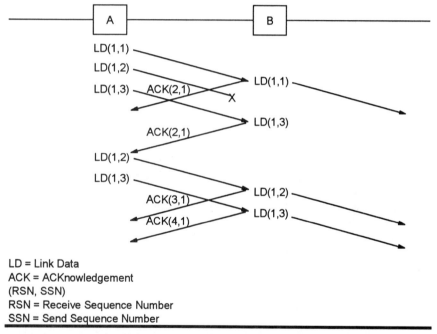

LD = Link Data
ACK = ACKnowledgement
(RSN, SSN)
RSN = Receive Sequence Number
SSN = Send Sequence Number

Figure 5.5 Example of X.25 Store and Forward Operation

This simple example illustrates the complexity involved in store and forward packet switching. This complexity was required when a significant portion of the packets experienced transmission errors. Even more sophisticated schemes were invented in which only the errored packets were retransmitted.

5.2.6 Traffic and Congestion Control Aspects of X.25

The send and receive sequence numbers in the X.25 packet layer are also used to provide flow control between the packet layer source and sink. Figure 5.6 illustrates a simple example of X.25 packet layer flow control. The send sequence number is a sequential number for the current packet. Numbers are incremented modulo the maximum window size. This example uses a modulo of 4, which means that sequence numbers are incremented in the order 0, 1, 2, 3, 0, 1, 2, 3, 0, and so on.

The receive sequence number in the ACKnowledgment indicates the next send sequence number expected in the next packet from the other end for that virtual circuit. Therefore, the receive sequence number acts as an acknowledgment for all packets up to one less than the receive sequence number.

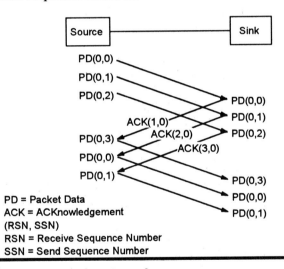

Figure 5.6 Example of X.25 Flow Control

The transmitter can send no more than the modulo minus 1 packets without acknowledgment; otherwise the sender could become confused as to which packets the receiver was acknowledging. As shown in the example, the sender transmits three packets and then waits for the acknowledgment before sending any additional packets. This is called a *sliding window* flow control protocol. This process allows the receiver to control the maximum rate of transmission over a virtual circuit, and is therefore a form of traffic control. This is still an essential function for a slow receiver (such as a printer) to control a fast transmitter (a computer) in many data communications applications today. The receive sequence number acknowledgment can be "piggybacked" in the packet header for a packet headed in the opposite direction on a virtual circuit, or may be sent in a separate acknowledgment packet.

5.2.7 Service Aspects of X.25

X.25 packet switching serves many user communities, especially in Europe, where it constitutes the majority of public and private data services. Over 20 percent of U.S. public data networking is via packet switched networks, whereas in Europe it is just the opposite — over 80 percent of public and private data networking is done through packet switched networks. This is due to the high cost of leased and private lines and the poor quality of transmission facilities in many European countries. The X.25 packet switching market in the United States and the rest of the world continues to grow. Packet switching remains a very popular technology, and will likely continue to be popular through the end of the decade.

5.3   THE INTERNET PROTOCOL SUITE, TCP/IP

The origins of the Internet Protocol (IP) occurred even earlier than X.25. It is also a very important set of protocols with some of its influences seen in SMDS and ATM.

5.3.1   Origins of TCP/IP

The U.S. Advanced Research Projects Agency (ARPA) began development of a packet switched network in 1969, and demonstrated the first packet switching capability in 1972; it was named the ARPAnet. The ARPAnet continued to grow, and in 1983 introduced the Transmission Control Protocol/Internet Protocol (TCP/IP), replacing the earlier Network Control Protocol (NCP) and Interface Message Processor (IMP) Protocol. Also in 1983 the ARPAnet was split into a military network, and a nonmilitary research network, that was the origin of the Internet. In 1990 the National Science Foundation (NSF) embarked upon a program to upgrade the Internet backbone to DS3 speeds (45 Mbps) for supercomputer interconnection. In 1994 the NSF plans to upgrade the Internet backbone for supercomputer communication to OC-3 speeds (150 Mbps). The Internet has its own standards body, called the Internet Engineering Task Force (IETF), which Chapter 6 describes.

5.3.2 Structure of TCP/IP

Figure 5.7 illustrates the Internet protocol suite built atop the Internet Protocol (IP) and the layered Internet protocol architecture. The User Datagram Protocol (UDP), Internet Control Message Protocol (ICMP), routing control protocols, and the Transmission Control Protocol (TCP) interface directly with IP, corresponding to the transport layer in the Internet architecture. This section focuses on TCP and IP.

Both TCP and UDP provide the capability for the host to distinguish between multiple applications through port numbers. TCP provides a reliable, sequenced delivery of data to applications. TCP also provide adaptive flow control, segmentation and reassembly, and prioritized data flows. UDP only provides an unacknowledged datagram capability.

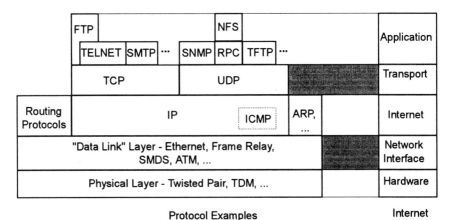

Figure 5.7 Internet Protocol (IP) Suite

A number of applications interface to TCP and UDP as shown in the figure. The File Transfer Protocol (FTP) application provides for security log-in, directory manipulation, and file transfers. TELNET provides a remote terminal log-in capability. The Simple Network Management Protocol (SNMP) supports configuration setting, data retrieval, and alarms, which Chapter 19 describes in more detail. The Trivial FTP (TFTP) protocol provides a simplified version of FTP which is intended to reduce implementation complexity. The Remote Procedure Call (RPC) and Network File Server (NFS) capabilities allow applications to interact over IP.

Routers send error and control messages to other routers using ICMP. ICMP also provide a function in which a user can send a *ping* to verify connectivity to an IP-addressed host.

The Address Resolution Protocol (ARP) directly interfaces to the data link layer, for example Ethernet. The purpose of ARP is to map a physical address (e.g., an Ethernet MAC address) to an IP address.

Routing protocols exchange routing table and topology information with other routers in the network. Examples of routing protocols are the Routing Information Protocol (RIP), Edge Gateway Protocol (EGP), and Open Shortest Path First (OSPF).

5.3.3 TCP/IP Packet Formats

Figure 5.8 illustrates the format of the IP packet. The version field specifies the IP protocol version. The header length field specifies the datagram header length in units of 32-bit words, the most common length being 4 words, or 20 octets. The type of service field specifies a 3-bit precedence of 1 to 7, 1-bit to indicate delay sensitivity, 1-bit to indicate high throughput, and 1-bit to indicate a request for high reliability, and 2 unused bits. The total length field specifies the total IP datagram length for the header and the user data. The identification, flags, and fragment offset fields control fragementation (or segmentation) and reassembly of IP datagrams. The Time To Live (TTL) field specifies how many seconds the packet can remain in the Internet before it is declared "dead." Intermediate nodes decrement TTL, and when it reaches zero, the packet is discarded. The protocol field identifies the higher level protocol type (e.g., TCP or UDP), which identifies the format of the data field.

The header checksum ensures integrity of the header fields through a calculation that is easy to implement in software. Source and destination IP addresses are required, and the user data is placed in the data field. The nonmandatory fields for options and padding can specify routing and time-stamp information.

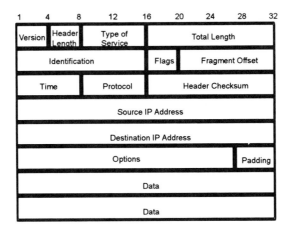

Figure 5.8 IP Datagram Format

The Internet is a huge worldwide network that uses 32-bit IP addresses as a global network addressing scheme. Each user, or host in Internet parlance, is assigned a unique IP address of 32 bits, or 4 octets, represented in the following dotted decimal notation:

XXX.XXX.XXX.XXX

where XXX ranges from 0 to 255 decimal, corresponding to the range of 00000000 to 11111111 in binary. There are 2^{32}, or over 4 billion, IP addresses.

IP addresses are grouped into classes A, B, and C as shown in Figure 5.9, where the class determines the maximum network size, measured in number of hosts. A class A address supports a network of up to 16 million host addresses (2^{24}), a class B address supports up to 64,000 (2^{16}) hosts, while a class C address supports up to 256 (2^8) host addresses. Internet network addresses are assigned and managed by a central authority, the Internet Assigned Numbers Authority (IANA), to ensure that they are unique. A network may assign the host addresses however it wishes, as long as the assignment is unique. If unique addresses are not maintained, problems arise when networks are interconnected.

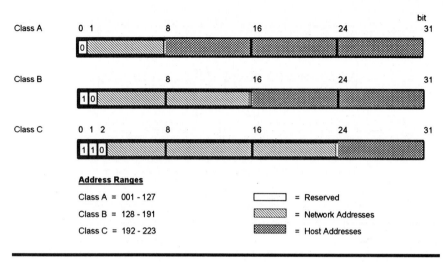

Figure 5.9 Internet Assigned Address Classes

Figure 5.10 illustrates the TCP frame format. The source and destination TCP port numbers identify a specific application program in the source and destination hosts. The sequence number field identifies the position of the sender's byte stream in the data field. The acknowledgment number field identifies the number of the next octet to be received. The HLEN provides the length of the header. The code bits field determines the use of the segment contents (e.g., SYN for synchronize sequence numbers and RST for reset connection). The window field tells the amount of data the application is willing to accept. The checksum is applied across the TCP header and the user data and is used to detect errors. The urgent pointer field specifies the position in the data segment where the urgent data begins if the code bits indicate that this segment contains urgent data. The options and padding fields are not mandatory.

TCP is a connection-oriented protocol and therefore has additional, specific messages and a protocol for an application to request a distant connection, and a means for a destination to identify that it is ready to receive incoming connection requests.

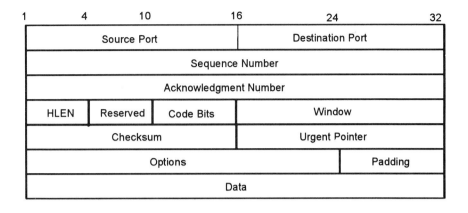

Figure 5.10 TCP Segment Format

5.3.4 TCP/IP Functions

IP provides a connectionless datagram delivery service to the transport layer. IP does not provide end-to-end reliable delivery, error control, retransmission, or flow control; it relies on TCP to provide these functions.

A major function of IP is in the routing protocols, which provide the means for devices to discover the topology of the network, as well as detect changes of state in nodes, links, and hosts. Thus, IP routes packets through available paths around points of failure. IP has no notion of reserving bandwidth; it only finds an available path. Most of the routing algorithms will minimize a routing cost. Further description of routing is covered in Chapter 11.

The design of an IP network based upon the addressing format is a critical issue. A network may have a large number of hosts. If every router in a network needed to know the location of every other host, then the routing tables could become quite large. A key concept used in routing is that of subsetting the host address portion of the IP address, which is called subnetting. Effectively, the host address space is broken down into multiple subnetworks by masking the bits in the host address field to create a separate subnetwork for each physical network. This means that a router need only look at a portion of the address, which dramatically reduces network routing table size.

5.3.5 Example of TCP/IP Operation

Figure 5.11 shows an example of a TCP/IP network transferring data from a workstation client to a server. Both TCP and IP are built upon the principle that the underlying network is a connectionless datagram network that can deliver packets out of order, or even deliver duplicate packets. TCP handles this by segmentation and reassembly using the sequence number in the TCP header, while IP does this using the fragment control fields in the IP header. Either method, or both may be used. A user's data 'ABCD' is segmented into four TCP segments on the left-hand side of the figure. A Router is initially routing this traffic via an X.25 network and sends datagram A via that route. The router then becomes aware of a direct connection to the destination router, and routes the remaining datagrams via the direct route. This routing action causes the datagrams to arrive at the destination server out of order, with datagram "A" traversing the X.25 network and arriving significantly later. TCP resequences the datagrams and delivers the block of data to the destination in the original order. IP performs a very similar process using fragmentation and reassembly.

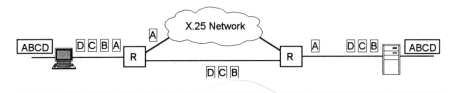

Figure 5.11 Example of Data Transfer Using TCP/IP

This operation by TCP/IP of accepting datagrams out of order, and being able to operate over an unreliable underlying network, makes it quite robust. None of the other standard modern data communication protocols has this attribute.

5.3.6 Traffic and Congestion Control Aspects of TCP/IP

TCP works over IP to achieve end-to-end reliable transmission of data across a network. TCP flow control uses a sliding window flow control protocol, like X.25; however, the window is of a variable size, instead of the fixed window size used by X.25. Figure 5.12 illustrates a simplified example of key concepts in the dynamic TCP window flow control protocol between workstation and a server. The sender starts with a window size equal to that of one TCP segment. The IP datagrams are delivered to the destination workstation, resulting in a delivered segment which is acknowledged. The sender then increases the window size to two segments. When these two segments are received they are both acknowledged, and the sender increases the window size to three segments. The network has become congested at this point, and the third segment is lost. The sender detects this by starting a timer whenever a segment is sent. If the timer expires, then the segment is resent. Upon such a retransmission timeout, the sender resets its window size to one segment and repeats the above process.

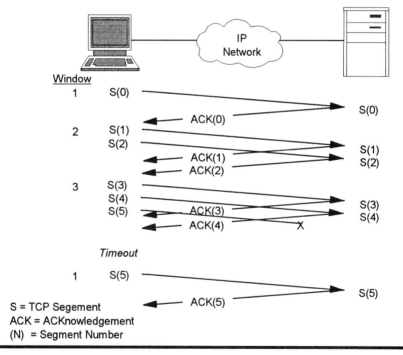

Figure 5.12 Example of TCP Dynamic Windowing Flow Control

The tuning and refinement of the TCP dynamic window flow control protocol has been the subject of a great deal of research. The above operation is often called the Van Jacobson "Slow Start" TCP protocol in the literature.

5.3.7   Service Aspects of TCP/IP

TCP/IP implementations typically constitute a router, TCP/IP workstation and server software, and network management. TCP/IP protocol implementations span UNIX, DOS, VM, and MVS environments. A majority of UNIX users employ TCP/IP for internetworking. Many Network Operating System (NOS) vendors are now integrating TCP/IP into their NOS platforms. Examples include Novell Netware and Banyan Vines.

There are currently many international TCP/IP interests, such as the commercial TCP/IP service offerings of Finland (Lanlink) and Sweden (Swipnet).

Operation of IP over a number of network, data link, and physical layer services is defined. At the network layer, IP operation over X.25 and SMDS is defined. At the data link layer, IP operation over frame relay, Ethernet, and ATM is defined. IP operation over circuit switched and dedicated physical layer facilities is also defined.

5.4   FRAME RELAY

Frame relay led the way in a minimalist trend in data communications, essentially being HDLC on a diet. This section presents a fairly detailed treatment of frame relay, highlighting concepts that have been adopted in ATM.

5.4.1   Origins of Frame Relay

X.25 packet switching and private networks using TDM or statistical multiplexing dominated the data communications marketplace in the 1980s. In order to keep pace with the increased performance of computing and the connectivity required by businesses, a new data communications technology was needed to provide higher throughput at lower costs. Standards work derived from the ISDN Link Access Procedure for the D channel (LAP-D) in CCITT Recommendation Q.921 led to the I.233, Q.922, and Q.933 Recommendations, which standardized the service description, protocol, and status signaling for frame relay. Frame relay also saw the founding of a separate industry group, the Frame Relay Forum, chartered with developing implementation agreements in order to facilitate interoperability. Over 150 manufacturers and service providers have announced support for the frame relay protocol.

5.4.2 Structure of Frame Relay

Figure 5.13 illustrates the structure of the protocols that support frame relay. The user plane of frame relay implements a subset of the OSIRM data link layer functions as specified in ITU-T Recommendation Q.922. Frame relay also has a control plane which is involved with reporting on the status of PVCs, or the establishment of SVCs. Recommendation Q.933 defines the status signaling for PVCs and the call control protocol for SVCs. IT may operate over the frame relay protocols (Q.922), or be signaled via the ISDN protocol (Q.921). The ITU-T signaling standards for ISDN (Q.931), frame relay (Q.933), and B-ISDN (Q.2931) have a common philosophy, message structure, and approach.

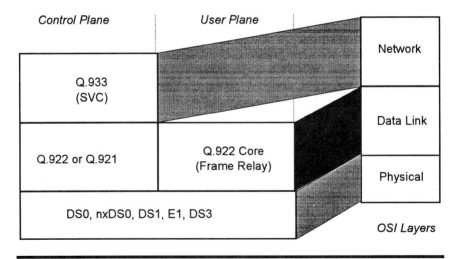

Figure 5.13 Frame Relay Protocol Structure

5.4.3 Frame Format

The frame format used by frame relay services is a derivative of the ISDN Link Access Protocol D-channel (LAP-D) framing structure.

Figure 5.14 shows the standard frame relay frame structure with a two-octet address based upon the CCITT Recommendation Q.922.

The first and last one-octet flag fields, labeled F, are HDLC flags. HDLC zero stuffing, identical to that described in the section above on X.25, is performed to avoid mistaking user data for a flag. Although Figure 5.14 depicts a two-octet address field, there are also three- and four-octet address formats.   The Frame Relay Service Specific Convergence Sublayer (FR-SSCS) defined in ITU-T Recommendation I.365.1 is identical to the frame relay frame without FCS, flags and HDLC zero insertion.   Chapter 17 summarizes FR-SSCS and presents the three- and four-octet frame relay address formats. The user data field can be up to 8188 octets (8192 minus two bytes each for the address and FCS field). The standard states that all implementations must support a maximum frame size of at least 1024 octets. The Frame Check Sequence (FCS) field is two octets long, and is the same used in HDLC, X.25, and a number of other protocols. Now for a detailed review of the address field.

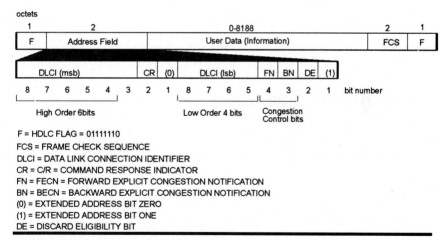

Figure 5.14 Q.922 Frame Mode Bearer Service Frame Structure

The Data Link Connection Identifier (DLCI) is split into two fields, together forming a 10-bit DLCI that identifies up to 1024 virtual circuits per interface. The DLCI only has local significance on an access line. A point-to-point frame relay virtual circuit may have different DLCIs for the access line on each end. Each user CPE device must have a separate DLCI for each destination. This limits the size of a frame relay network to approximately 1000 nodes.

The Command/Response (C/R) bit is not used at this time, being reserved for future functions.

The Forward Explicit Congestion Notification (FECN) and Backward Explicit Congestion Notification (BECN) bits indicate to the receiver and sender, respectively, the presence of congestion in the network. FECN is set in frames traversing the network from sender to receiver which encounter congestion. FECN can be used in receiver-based flow control protocols, such as DECnet. BECN is set in frames traversing the network from receiver to sender when there is congestion in the opposite direction. BECN can be used by a sender which can dynamically change its rate. An increase in the frequency of FECN and BECN bits received is a good indication that congestion is occurring within the network.

The Discard Eligibility (DE) bit, when set to 1, indicates that the frame should be discarded during congestion conditions as opposed to discarding other frames with a higher priority, those with DE bit set at 0. This bit can be set either by the user or by the network to determine what data is discarded when the total traffic exceeds the Committed Information Rate (CIR) and congestion occurs.

The Extended Address (EA) bits act as address field delimiters, set at 0 and 1, respectively. These bits are used to extend the DLCI addressing range to three- and four-octet formats.

5.4.4 Frame Relay Functions

Frame relay provides an upgrade to existing packet switch technology, by supporting speeds up to DS1 (1.544 Mbps) — with speeds up to 45 Mbps expected by the mid-1990s. Frame relay supports Permanent Virtual Circuits (PVCs) and Switched Virtual Circuits. Frame relay is a service that delivers frames in order with high probability. Frames which have detected errors are simply dropped, and therefore, frame relay can operate effectively only on low-error-rate transmission media.

Frame relay virtual circuits may be point-to-point, point-to-multipoint, or multipoint-to-multipoint (called multicast) as defined in the network topologies in Chapter 4. Frame relay virtual circuits may be arranged into closed user groups for security purposes.

Frame relay does not perform the overhead-intensive functions of error management and flow control that X.25 does, and consequently operates at higher speeds. End systems must implement these error management and flow control functions, usually at the transport layer, for example, TCP.

Although the primary use of frame relay is as an interface to a public data service, frame relay can also be used as a backbone protocol. Frame relay is not a switching technique.

A Permanent Virtual Circuit (PVC) is managed by status signaling defined in ITU-T Recommendation Q.933, Annex A, and ANSI Standard T1.617, Annex D. Status signaling provides for status enquiry and reporting of the physical link, or PVCs, as identified by the DLCI value. The status response may be for selected PVCs, or a Full Status (FS) for all PVCs. PVC status is defined by three bits: new, delete, and active. The new bit is set to 1 for a new PVC, otherwise it is set to 0. The delete bit indicates whether the DLCI is configured (1) or not (0). The active bit indicates whether the PVC is operational or not. Status signaling occurs on DLCI 1023.

The maximum number of usable DLCI codes to identify PVCs is limited by the maximum frame relay frame size for the Full Status (FS) message.

5.4.5 Example of Frame Relay Operation

Frame relay has been called "X.25 on a diet" because it is a much simpler protocol. Figure 5.15 illustrates a key aspect of the simplification that frame relay provides. Node A on the left-hand side of the figure is transmitting frames to a destination through an intermediate node B. There are no sequence numbers in the frame, with the ones in the example being shown for illustrative purposes only. Node B simply relays the frames to the outgoing link toward the destination. Node B may begin relaying the frame even before the entire frame is received, in a technique called *fast packet switching* that was popularized by Stratacom. If a frame is received in error, as shown for the fifth frame in the example, the intermediate node simply does not relay it. This simplified relay only protocol allows very simple, fast intermediate nodes to be built; however, it requires more intelligence in the higher layer protocols; such as X.25 or TCP, in the end systems to recover from lost frames. Frames can be lost due to transmission errors or congestion. When frame relay operates over modern transmission media, such as fiber optic systems, frames lost due to errors become a rare occurrence. Loss then occurs primarily due to congestion in frame relay networks. Since loss occurs infrequently, the end systems seldom need to invoke error recovery procedures, which were usually in place anyway regardless of the performance of the underlying network.

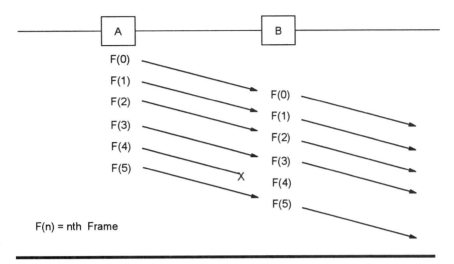

Figure 5.15 Frame Relay Interface, Switching, and Network Service

5.4.6 Traffic and Congestion Control Aspects of Frame Relay

Frame relay defines a Committed Information Rate (CIR) which is the minimum throughput achievable under all network conditions. However, the term "committed" is somewhat ambiguous in that it does not specifically state the frame loss or delay objective. Transmission errors can cause frames to be lost, and finite buffer space in networks which have overbooking can result in lost frames due to momentary, statistical overloads.

The CIR is computed as the number of bits in a committed burst size, Bc, that can arrive during an averaging interval T such that CIR=Bc/T. If the number of bits that arrive during the interval T exceeds Bc, but is less than an excess threshold, Bc+Be, then the subsequent frames are marked as Delete Eligible (DE). The Excess Information Rate (EIR) is defined as (Bc+Be)/T. The bits that arrive during the interval T in excess of Bc+Be are discarded by the access FR node. However, this discarding situation can be avoided for a given access rate, R_A, if Be is set to $R_A \times T$ minus Bc.

At present, there is no uniform method for setting the interval T. If T is set too small, such that Bc is less than the length of a single frame, then every frame will be marked DE. If T is set too large, the buffer capacity in the FR access node required to police CIR may not be practical. Setting T to a value on the order of the round trip delay is a good guideline to achieve good TCP/IP throughput over frame relay.

The Discard Eligible (DE) bit may be set by either the customer or by the network. If a network node becomes congested, it should discard the frames with the DE bit set first. Beware, however, of the danger of discarding frames marked with the DE bit during long periods of congestion. The applications will react by retransmitting lost frames and congestion will intensify.

Congestion notification is provided in the frame relay address field by the FECN and BECN bits. The FECN bit is set by the FR network nodes when they become congested to inform the receiver flow controlled protocols of the congestion situation, while the BECN bit is set in FR frames headed in the downstream direction to inform transmitter flow controlled protocols of the congestion situation. An increase in the frequency of FECN and BECN bits received is a good indication of network congestion. At present, very little use is being made of this technique in end and intermediate systems. One concern is that by the time the FECN/BECN arrives at the controlling end, the congested state may no longer exist in the FR node that sent the FECN/BECN notice. The main concern is that the FECN/BECN notice is delivered to the CPE router, which is not the primary source of flow control. Currently, no technique exists for the CPE router to convey the FECN/BECN message to TCP or the application that could provide flow control.

Figure 5.16 depicts a frame relay network connecting a host in Dallas with many remote user devices in Charleston via a network. In the process of downloading massive files from the mainframe in Dallas to the users in Charleston via a PVC (shown as the dashed line), congestion occurs on the Atlanta-to-Raleigh trunk. The FECN and BECN bit setting is performed on all DLCIs traversing the Atlanta-to-Raleigh trunk to notify them of the congestion condition. The Atlanta node sets the FECN bit to 1 and notifies the Charleston PVC user receiving traffic from Dallas of impending congestion. The Atlanta node also sets the BECN bit to 1 on frames destined from Charleston to Dallas, informing the Dallas PVC user of the same congestion condition. Either the Dallas user could throttle back, or the Charleston user could flow control the Dallas sender using a higher layer protocol. Either action that reduces the frame rate will eventually cause the congestion condition to abate and the FECN and BECN to clear.

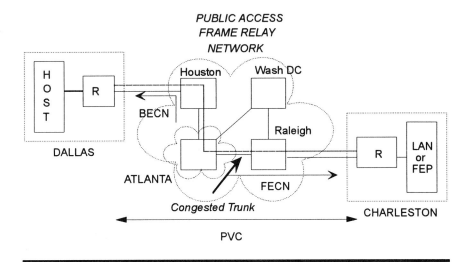

Figure 5.16 Congestion Control Example

Another form of congestion management defined by ANSI T1.618 is the Consolidated Link Layer Management (CLLM) function. CLLM reserves one of the DLCI addresses on a frame relay interface for transmitting control messages to user devices when there are no frames to transmit, yet a congestion notification still needs to be sent. The CLLM message is a contingency for notifying users of congestion activity outside the conventional framing structure, since there is no provision in the standards for empty frames which contain only congestion control information. The CLLM can contain a list of DLCIs that correspond to the congested frame relay bearer connections — all users affected are then notified of congestion. Multiple CLLM messages can be transmitted in a network with many DLCIs that require congestion notification.

5.4.7 Service Aspects of Frame Relay

Frame Relay (FR) is offered as a frame-based public data service that allows access lines of speeds up to 1.544 Mbps from a Customer Premises Equipment (CPE) router, bridge, or Frame Relay Access Device (FRAD) into the public frame relay network. FR network trunks can be of speeds from DS0 (56 kbps) up to and including DS3 (45 Mbps). Some carriers even provide sub-DS0 access speeds. Dial-up access to frame relay is defined in ANSI T1.617.

Frame relay service has emerged full force on the public data network market. Providers of FR hardware include a broad range from CPE vendors to high-end multiplex and switch vendors. Interexchange Carriers (IXCs), Local Exchange Carriers (LECs)), and enhanced service providers all provide FR services, but only the IXCs can provide inter-LATA service. Figure 5.17 shows the connectivity provided by the IXCs across LATAs and outside the LATAs (alternate access scenario). Most of the IXC, and LECs in the U.S. offer some form of frame relay service, along with other packet switch providers, such as Telenet and CompuServe. Frame relay is also being offered in a number of areas in Europe and Asia.

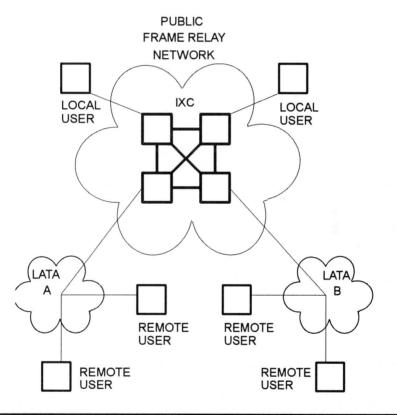

Figure 5.17 Public Frame Relay Network Interconnection

Frame relay also provides some basic public network security in that data originating and terminating through an access line is limited to connectivity established by the PVCs for that access line.

Network providers offer different interpretations of the Committed Information Rate (CIR) and Discard Eligible (DE) functions according to different pricing plans. For example, CIR may be available only in certain increments, or DE traffic may be billed at a substantial discount. Customer network management, performance reporting, and usage-based billing are valued-added services offered by some providers. It pays to shop around carefully.

There are two different philosophies in the setting of the CIR. We call the first regular booking, where the sum of the CIRs is no more than the access rate, and the second is called overbooking, or oversubscription, where the sum of CIRs exceeds the access line rate.

In regular booking the CIRs can be added up across trunk routes, resulting in predictable, deterministic performance, at the expense of a PVC being limited to its CIR during congested intervals. In overbooking, the trunking performance becomes statistical; however, each individual PVC can be provided with a higher CIR rate than in the regular booking case, since the sum of the PVC CIRs can exceed the access line rate.

5.5 SWITCHED MULTIMEGABIT DATA SERVICE (SMDS)

SMDS began a number of trends that have been adopted in ATM. This section covers the service aspects of SMDS and specifics of the Distributed Queue Dual Bus (DQDB) protocol defined in the IEEE 802.6 standard.

5.5.1 Origins of SMDS

SMDS was created as a Metropolitan Area Network (MAN) service by Bellcore. In the purest sense, SMDS is a service definition and not a protocol. The first realization of SMDS was defined using the Distributed Queue Dual Bus (DQDB) technology, as specified in the IEEE 802.6 standard. The IEEE 802.6 DQDB standard defines a. connectionless data transport service using 53-byte slots to provide integrated data, video, and voice services over a Metropolitan Area Network (MAN) which is typically a geographic area of diameter less than 150 km (90 miles). The SMDS implementations based upon the IEEE 802.6 standard were the first public services to use ATM-like technology. Although the IEEE 802.6 standard also defines connection-oriented isochronous services, SMDS supports only a connectionless datagram service primarily targeted for LAN interconnection.

5.5.2   Structure of SMDS/IEEE 802.6

SMDS and the IEEE 802.6 DQDB protocol have a one-to-one mapping to each other as illustrated in Figure 5.18. The SMDS Interface Protocol (SIP) has Protocol Data Units (PDUs) at layers 2 and 3. The layer 2 SIP PDU corresponds to the Distributed Queue Dual Bus (DQDB) Media Access Control (MAC) PDU of the IEEE 802.6 standard. The layer 3 SIP PDU is treated as the upper layers in IEEE 802.6. There is also a strong correspondence between these layers and the OSI reference model as shown in the figure.

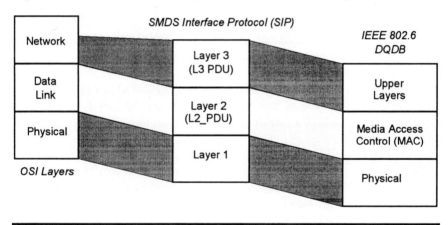

Figure 5.18 Protocol Structure of IEEE 802.6 DQDB and SMDS

5.5.3   SMDS/802.6 Protocol Data Unit (PDU) Formats

Figure 5.19 illustrates the relationship between the user data, the layer 3 SMDS PDU, and the layer 2 SMDS PDU. The user data field may be up to 9188 octets in length. The layer 3 protocol adds a header and trailer fields, padding the overall length to be on a 4-octet boundary.   Layer 2 performs a segmentation and reassembly function, transporting the layer 3 payload in 44-octet segments. The layer 2 PDU has a 7-octet header and 2-octet trailer resulting in a

53-octet slot length, the same length as an ATM cell. The layer 2 header identifies each slot as being either the Beginning, Continuation, or End Of Message (BOM, COM, or EOM).

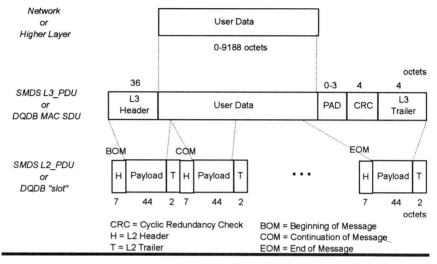

Figure 5.19 SMDS/802.6 Layer 2 and 3 PDU Relationships

Figure 5.20 illustrates the SMDS layer 3 PDU (L3_PDU) format. The first two octets and last two octets of the SMDS L3_PDU are identical to the AAL3/4 Common Part Convergence Sublayer (CPCS) described in Chapter 8. The SMDS L3_PDU header contains the SMDS Source and Destination Addresses (SA and DA) and a number of other fields. Most of these other fields are included for alignment with the IEEE 802.6 protocol and are not used in the SMDS service. When the SMDS layer 3 PDU is segmented by layer 2, all information needed to switch the cell is carried in an SSM or BOM slot. This means that an SMDS switch need only examine the first slot to make a switching decision.

The addressing plan for SMDS and CBDS is based upon the International Telecommunications Union (ITU) Recommendation E.164, and in the United States the proposed format is similar to the North American Numbering Plan (NANP) used for telephone service. As the SMDS E.164 address is globally unique, SMDS provides the capability for ubiquitous connectivity. The IEEE 802.6 standard also allows the option for 48-bit IEEE Media Access Control (MAC) addresses to be employed in the DA and SA fields.

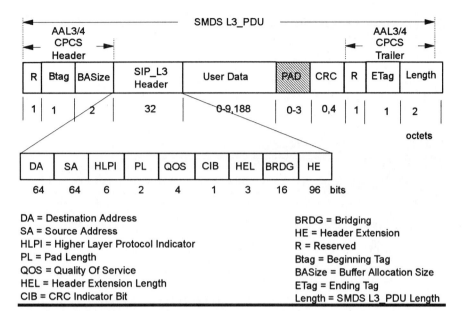

Figure 5.20 SMDS Layer 3 PDU Format

Figure 5.21 illustrates the 48-octet SMDS Layer 2 PDU format encapsulated in a 53-octet DQDB slot. The other four octets of SMDS layer 2 overhead in the DQDB payload are used for the SMDS Segment Type (ST), Message IDentifier (MID), payload length, and a Cyclical Redundancy Check (CRC) on the 44-octet payload. The SMDS layer 2 overhead and function are identical to the ATM AAL3/4 SAR described in Chapter 8. The ST field identifies either a Single Segment Message (SSM), Beginning Of Message (BOM), Continuation Of Message (COM), or End Of Message (EOM) slot. The MID field associates the BOM with any subsequent COM and EOM segments that made up an SMDS L3_PDU. When an SMDS switch receives an SSM or BOM segment, the destination address determines the outgoing link on which the slots are transmitted.

The DQDB Access Control Field (ACF) and header are used to provide a distributed queue for multiple stations on a bus, provide self-healing of the physical network, provide isochronous support, and control management functions. The next sections describe the distributed queueing and self-healing properties of DQDB. Note that the DQDB ACF and header taken together are exactly five bytes, exactly the same size as the ATM cell header. This choice was made intentionally to make the design of a device that converted between DQDB slots and ATM cells simpler.

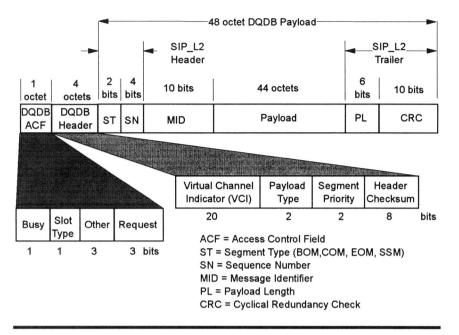

Figure 5.21 DQDB Slot Structure

5.5.4 DQDB and SMDS Functions

This section describes the distributed queueing and self-healing ring properties of the IEEE 802.6 DQDB protocol with reference to Figure 5.22. There are two unidirectional buses A and B that interconnect a number of nodes, often configured in a physical ring. Even though the physical configuration may be a ring, logical operation is bus-oriented. Nodes read from both buses, usually passing along any data onto the next node in the bus. Each node may become the Head Of Bus (HOB) or End Of Bus (EOB) as indicated in the figure. The HOB generates 53-octet slots in a framing structure to which the other nodes synchronize. The EOB node simply terminates the bus. If a node fails or is powered down, it is designed so that it passively passes data. Therefore, each node effectively has four ports, two for each bus. Normally one node would be the HOB for both buses, as shown for node C. However, in the event of a failure of one of the buses connecting a pair of nodes, the DQDB protocol ensures that the nodes on either side of the break become the new HOB within a short period of time.

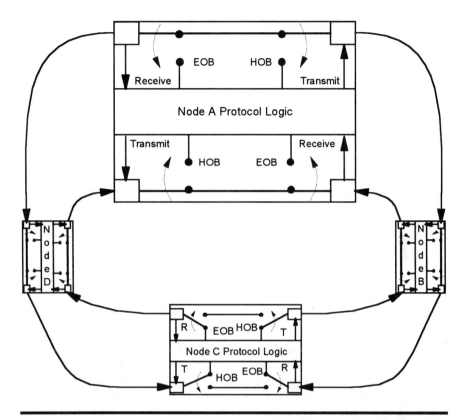

Figure 5.22 Dual DQDB Architecture

The Busy and Request bits in the DQDB Access Control Field
(ACF) implement a distributed queue. Each node has two counters:
one for requests and the other as a countdown for transmission. If a
slot passes through the node with the request bit set, then the
request counter is incremented; otherwise the request counter is
decremented.   Thus, the request counter reflects how many
upstream nodes have slots to send. When a node has data to send, it
writes a logical 1 to the request bit in the first slot received which
has the request bit equal to 0 and loads the count down register with
the value of the request counter.   The countdown timer is
decremented each time a slot passes by in the opposite direction.
Therefore, when the countdown counter reaches 0, the slot can be
sent because all of the upstream nodes have already sent the slots
that were reserved in the opposite direction.

This elegantly simple protocol has several problems that
complicate the IEEE 802.6 standard. First, the nodes which are
closer to the head end of the bus have first access to the request bits,

and can dominate the traffic on the bus. Secondly, provisions must be made for stations to join and leave the bus and handle bit errors. The IEEE 802.6 standard defines procedures to handle all of these cases.

SMDS offers either a point-to-point datagram delivery service, or a point-to-multipoint service, defined as a group multicast address. SMDS service operates on the E.164 source and destination addresses. The SMDS source address is screened by the network to ensure that it is valid for the source SMDS access line. SMDS customers can screen incoming data and only accept data from specific source SMDS addresses or block data from specific SMDS addresses. SMDS users can also limit the destination SMDS addresses that can be reached from their SMDS access lines. SMDS customers can have ubiquitous connectivity, or they can use these screening tools to achieve tightly controlled closed user groups.

5.5.5 Example of SMDS over DQDB Operation

Figure 5.23 illustrates an example of SMDS over DQDB operation. Three DQDB buses, configured as physical rings, are interconnected as shown. A series of slots from a node on the far left are generated with Destination Address (DA) of a node on the far right. The nodes use the DQDB protocol to queue and transmit the slots corresponding to this datagram. The result is that a reassembled datagram is delivered to the destination.

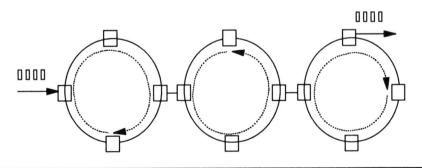

Figure 5.23 Example of SMDS/DQDB Operation

5.5.6   Traffic and Congestion Control Aspects of SMDS

SMDS has an open loop flow control mechanism called Sustained Information Rate (SIR). SMDS SIR is based on the aggregate of all data originating on the SMDS access line regardless of it's destination. For DS3 access lines, five access classes are defined in Table 5.1.

The 34 Mbps for Access Class 5 is the maximum throughput achievable on a 45-Mbps access line after Physical Layer Convergence Protocol (PLCP), and L2_PDU or ATM cell overhead are taken into account. SIR uses a credit manager, or leaky bucket, type of rate enforcement method. Basically, no more than M out of N cells may contain non-idle slots or cells. For class 5 M=N, while for the lower numbered classes the relationship is that SIR=M*34/N Mbps. The value of M controls the number of consecutive slots/cells that can be sent at the DS3 rate.

Table 5.1 SMDS Access Classes

Access Classes	SIR (Mbps)
1	4 Mbps
2	10 Mbps
3	16 Mbps
4	25 Mbps
5	34 Mbps

Data arriving at a rate higher that the SIR rate is discarded at the originating SMDS switch. Note that access classes 1 through 3 line up with standard LAN speeds, so that traffic from a single LAN cannot experience loss due to the SIR credit manager operation. SMDS has a L3_PDU loss objective of 10^{-4}, which is several orders of magnitude greater than that caused by transmission errors. This is consistent with the character of the SMDS service, emulating that of a LAN but providing MAN or WAN coverage.

5.5.7   Service Aspects of SMDS

Switched Multi-megabit Data Service (SMDS) is a combination packet- and cell-based public data service that supports DS1, E1, E3, and DS3 access speeds through a DQDB Subscriber-to-Network Interface (SNI) . SMDS is also supported through a Data eXchange

Interface (DXI) at speeds from 56 kbps up to and including 45 Mbps. SMDS is a connectionless switched data service with many of the characteristics of Local Area Networks (LANs). SMDS can provide the subscriber with the capability to connect diverse LAN protocols and leased lines into a true switched public network solution. SMDS has many public network level security features.

SMDS has also been billed as the gap filler between frame relay and ATM; however, SMDS now finds its unique position in time diminishing. Central office switch vendors, such as Siemens, AT&T, Fujitsu, and NEC, have been the primary players for the first version of cell switching to hit the telecommunications market. Public SMDS services based on the IEEE 802.6 standard architecture have been slow to emerge; however, they are rolling out in 1994 on a worldwide basis. Internationally a close relative of SMDS is the Connectionless Broadband Data Service (CBDS) as defined by the European Telecommunications Standards Institute (ETSI). A North American SMDS Interest Group (SIG) and European SIG have been formed to develop detailed implementation specifications to facilitate interoperable implementations of SMDS and CBDS.

5.6 REVIEW

This chapter began with an introduction to the reasons for packet switching, basic principles of packet switching, and a review of packet switching's relatively short history. This chapter then covered the predominant data services of the 1990s: X.25, IP, frame relay, and SMDS in terms of the origins, protocol formats, functions, operation, traffic aspects, and commercial service aspects. The need to perform store and forward packet switching and windowed flow control was described in coverage of the X.25 protocol. The alternative to OSI, the Internet protocol suite, was introduced. The concepts of resequencing and handling unreliable underlying networks, central to TCP/IP, were covered through an example. The modern notion of a dynamically adjusted window flow control protocol implemented widely in TCP was described along with an example. The concept of reducing protocol function to improve performance was covered in the treatment of frame relay. Finally, the novel concepts of a distributed queue and a self-healing network were covered in the treatment of SMDS and DQDB. This chapter provides the reader with a basic background of the data communication services and concepts that reappear within ATM and related protocols.

5.7   REFERENCES

[1]   D. Spohn, *Data Network Design*, McGraw-Hill, 1993.

[2]   CCITT Recommendation E.164, Numbering Plan for the ISDN Era, Vol.II, Fas.II.2, Blue Book, ITU, 1988.

[3]   CCITT Draft Recommendation I.370, Congestion Management for the Frame Relay Bearing Service.

[4]   CCITT Recommendation Q.921, ISDN User-Network Interface Data Link Layer Specification, Blue Book, ITU, 1988.

[5]   CCITT Draft Recommendation Q.922, ISDN Data Link Layer Specifications for Frame Mode Bearer Services, April 1991.

[6]   ANSI T1.606-1990, ISDN - Architectural Framework and Service Description for Frame Relaying Bearer Service, 1990.

[7]   ANSI T1.607 - Digital Subscriber Signaling System No. 1 - Layer 3 Signaling Specifications for Switched Bearer Service, 1990.

[8]   ANSI T1.617-1991, ISDN - DSS1 Signaling Specification for Frame Relay Bearer Service, 1991.

[9]   ANSI T1.618-1991, ISDN - Core Aspects of Frame Protocol for Use with Frame Relay Bearer Service, May 1991.

3

B-ISDN/ATM
Standards, Basics,
Protocol, and Structure

Part 3 introduces Broadband-ISDN and Asynchronous Transfer Mode (ATM). Chapter 6 covers the foundation of B-ISDN and ATM — the bodies that define the standards and specifications. The major bodies are described along with the important documents developed to date. References are given to subsequent chapters in the book where particular aspects of these standards are covered. The B-ISDN protocol model, organized as user, control, and management planes, is introduced.

Chapter 7 provides a high-level introduction to ATM. Descriptive analogies and several examples explain the concept of addressing and switching fixed length cells.

Chapter 8 covers the three primary layers of the protocol reference model: the Physical layer, the ATM layer where the cell structure occurs, and the ATM Adaptation Layer (AAL) that provide support for higher layer services, such as circuit emulation, frame relay, IP, and SMDS.

Chapter 9 is an overview of the user, control, and management planes. This chapter also covers the specifics of the control plane AAL and higher layers.

6

ATM and B-ISDN Standards and Specifications

This chapter covers ATM and B-ISDN standards and specifications. Beginning with a summary of the major groups of standards and specification bodies currently active in the areas of B-ISDN and ATM, the chapter moves to a discussion of the players involved and the process followed in the production of standards and specifications. The exponential rate of technology development and technological advance presents an ever increasing challenge to the players, and tests the standards and specifications process. The basic standards model of B-ISDN is compared and contrasted with the OSI and ISDN standards models. The major standards and specifications produced to date are listed and briefly summarized. Finally, the current direction of standards and specifications activities is summarized. The conclusion identifies some open issues and areas that readers should follow in the future.

6.1 STANDARDS AND SPECIFICATION BODIES

There are currently two classes of standardization and specification bodies actively involved in B-ISDN and ATM: formal standards bodies and industry forums.

The formal international standards body is the International Telecommunications Union-Telecommunications standardization sector (ITU-T), formerly called the International Telegraph and Telephone Consultative Committee (CCITT). The premier formal B-ISDN/ATM standards organization in the United States is the American National Standards Institute (ANSI). The premier formal B-ISDN/ATM standards organization in Europe is the European Telecommunications Standards Institute (ETSI).

There are four major industry forums currently active in the B-ISDN/ATM specification area: the ATM Forum, the Internet Engineering Task Force (IETF), the Frame Relay Forum, and the SMDS Interest Group (SIG).

The formal standards bodies defined the concept of Broadband-ISDN (B-ISDN) in the late 1980s and chose the ATM technology as the basis for future standards. The formal standards bodies also created the concept of the Integrated Services Digital Network (ISDN) in the early 1970s, which is now called the Narrowband ISDN (N-ISDN), as the predecessor to B-ISDN. Originally it was anticipated that the standards for B-ISDN would take many years to develop. However, new procedures have now been adopted in many of the formal standards bodies in an effort to accelerate the pace of standardization to that required by the industry and users. In fact, the formal standards bodies such as ANSI sometimes take inputs from the industry forums such as the ATM Forum, communicating via liaison letters and representatives that attend meetings of these different groups.

A new style of "jump starting" the standards process is emerging in the 1990s in the formation of industry forums. These are not formal standards committees, but independent groups formed by vendors, users, and industry experts who want to ensure standards for interoperability, but who do not want to add further implementation details to standards. Instead specifications, or implementation agreements, are published by selecting an interoperable subset of requirements from standards, clarifying ambiguities, or in some cases, specifying certain aspects in advance of standards. Sometimes these forums provide valuable contributions to the formal standards organizations (often as already implemented, and hence proven and not theoretical, approaches), speeding along the acceptance of an interface, protocol, or other aspect of B-ISDN and ATM technology. Often, more than writing and publishing the standards is required for success. The multiple vendor and provider agreements developed in these industry forums are often essential to a standard's success.

6.1.1   International Telecommunications Union (ITU)

The International Telecommunications Union (ITU) was founded in 1948 to produce telegraphy and telephone technical, operating, and tariff issue recommendations. An ITU committee formerly known as the CCITT was renamed as the Telecommunications standardization sector, referred to as the ITU-T. The ITU-T is a United Nations sponsored treaty organization. The U.S. voting member in the ITU-T is a representative of the U.S. Department of State, and includes technical advisors through the U.S. National Committee for the ITU-T. Only members may attend meetings. The standards produced by the ITU-T are identified as ITU-T Recommendations in this book.

Up until 1988, the ITU-T published approved recommendations once every four years in the form of a set of books which were often referred to by the "color" of their covers — red, yellow, blue, etc. The first B-ISDN/ATM standards were published in the 1988 blue books. After 1988 an accelerated standards process was adopted, where all subsequent recommendations are published when completed. These are now called "white books."

During a study period, which is now typically two years instead of four, a number of questions are assigned to a study group. The study group then organizes into lower level committees and produces working documents and draft recommendations. These study groups were once referred to by roman numerals in the CCITT days, but in the new modernized ITU-T, the study groups are now referred to by decimal numbers. Study group 1, for example, is involved with B-ISDN services aspects; study group 11 is responsible for signaling protocols; and study group 13 is responsible for defining the ATM-related functions. The address and phone number to obtain further information about the ITU-T are:

International Telecommunications Union-Telecommunications
Place Des Nations
Rue De Varembe
Geneve 20, Switzerland 1211
+41 22 730 5111

6.1.2   American National Standards Institute (ANSI)

The American National Standards Institute (ANSI) is structured into several committees, which cover a broad spectrum of areas. The

ANSI committee T1 is primarily involved in the standardization of B-ISDN and ATM for the United States. These standards are developed in close coordination with the ITU-T, and address characteristics of technology that are unique to North America. Particular ANSI T1 subcommittees are involved with different aspects of standardization, such as T1E1 covering the physical interface aspects of ATM, T1M1 covering the maintenance aspects of ATM, T1A1 covering performance aspects, and T1S1.5 covering the network, service, signaling, ATM layer, interfaces, and AAL aspects of ATM. ANSI committee T1 recommends positions on technical issues that the U.S. Department of State presents at ITU-T meetings. The address and phone number to obtain further information about ANSI are:

American National Standards Institute (ANSI)
11 West 42nd Street, 13th Floor
New York, NY 10036
USA
+01 212 302 1286

6.1.3 European Telecommunications Standards Institute (ETSI)

The European Telecommunications Standards Institute (ETSI) is primarily involved in the standardization of European telecommunications. The ITU-T develops recommendations for worldwide use, while the role of regional bodies, such as ETSI in Europe and ANSI in America, is to generate, on the basis of global standards, more detailed specifications adapted to the unique historical, technical, and regulatory situation of each region. The address and phone number to obtain further information about ETSI are:

European Telecommunications Standards Institute (ETSI)
Route de Lucioles
06291 Sophia Antipolis
Cedex, Valbonne
France
+33 92 944200

6.1.4 ATM Forum

The ATM Forum was formed in October, 1991 by four companies: Northern Telecom, Sprint, SUN Microsystems, and Digital Equipment Corporation (DEC). In January 1992, the membership was opened to the industry. There are currently three categories of membership: principal, auditing, and user. Only principal members can participate in technical and marketing committee meetings. Auditing members receive copies of the technical and marketing committee documents, but cannot participate in the meetings. Only user members may participate in End User Roundtable (ENR) meetings.

There are three types of committees in the ATM Forum: technical, market awareness, and end user. The technical committee produces implementation specifications and is organized into a number of technical "subject matter expert" subcommittees. The accomplishments to date and work in progress of the technical committee are covered later in this chapter.

The Market Awareness and Education (MA&E) committee produces tutorials, presentations, press releases, newsletters, and other informative material. There are branches of this committee in North America and Europe currently, with expansion planned to Asia in 1994. The End User Roundtable (ENR) user group was formed in August 1993, with the goal to collect higher level requirements and provide these to the technical and MA&E committees.

As of January 1994, the ATM Forum had approximately 150 principal members, 300 auditing members, and 25 user members. See [2] for an article on the ATM Forum. The address, phone number, and Internet address to obtain information about the ATM Forum are:

The ATM Forum
303 Vintage Park Drive
Foster City, CA 94404
+1 415 578 6860
info@atmforum.com

6.1.5   Internet Engineering Task Force (IETF)

The Internet Activities Board (IAB) was formed in 1983 by DARPA. By 1989 the Internet had grown so large that the IAB was reorganized, and the principal work of developing specifications to achieve interoperability was assigned to an Internet Engineering Task Force (IETF), split into eight areas, each with an area director. The initial objective of the IAB/IETF was to define the necessary specifications required for interoperable implementations using the Internet Protocol (IP) suite. Specifications are drafted in documents called Request For Comments (RFC). These RFCs pass through a draft stage and a proposed stage prior to becoming an approved standard. Another possible outcome of a draft or proposed standard is that it is archived as an experimental RFC. Out-of-date RFCs are archived as historical standards. The archival of all approved as well as historical or experimental RFCs has created a storehouse of protocol and networking knowledge that is available to the world.

There are a number of good introductory books on the Internet, for example [3], which is the best way to get electronic access to the Internet. The address, phone number, and most importantly, Internet address to obtain more information about the IETF and the Internet are:

Internet Society
1985 Preston White Drive, Suite 100
Reston, VA 22091
+01 703 620 8990
isoc@nri.reston.va.us

6.1.6   Frame Relay Forum and SMDS Interest Group

Many aspects of frame relay are similar to those of ATM. They are both connection-oriented protocols, involve ISDN-based signaling, and require similar network management functions. Indeed, the ATM Forum and the Frame Relay Forum corroborated closely in the production of Frame Relay/ATM interworking specifications.

The SMDS Interest Group (SIG) is also working closely with the ATM Forum to specify access to SMDS service over an ATM UNI interface. Chapter 17 summarizes the status of the interworking of frame relay and SMDS with ATM networks.

6.2 CREATING STANDARDS — THE PLAYERS

Perhaps the single most important driving factor for successful standards and industry specifications is responsiveness to real user needs. Standards created for things with no real user need are rarely successful. Some of the most important questions a user can present to a vendor are "Does it conform to industry standards, which ones, and how?" Standards play a critical role in an age where standardized national and international interoperability is a key to success in data communications.

6.2.1 Vendors

Standards are a two-edged sword for vendors: on the one hand they must consider the standards, while on the other hand they must consider developing something proprietary to differentiate their products. The proprietary feature may increase the cost but add value, or remove some noncritical portion of the standard to achieve lower cost. Vendors are usually very active in the standards process. In the emerging era of ATM, vendors are becoming even more concerned with meeting industry standards. Vendors who remain completely proprietary, or try to dictate the standards with their offerings, are confronted by users unwilling to risk their future business plans on proprietary systems.

Vendors can also drive standards, either by de facto industry standardization, through formal standards bodies, or through industry forums. De facto standardization can occur when a vendor is either an entrepreneur or the dominant supplier in the industry who wants to associate the new technology with their name, such as IBM with SNA. De facto standards in high-technology areas, however, do not last forever. Sometimes the dominant vendor is not the only one in the market with a product, but their product quality or market share makes them the de facto standard around which other vendors must design. B-ISDN and ATM are still too new of a technology for de facto standards to have had a significant effect.

6.2.2   Users

Users do better when they purchase equipment conforming to industry standards rather than non-standards-based products because they can competitively shop for products and services and be assured that there will be some level of interoperability. A certain comfort level exists in knowing that the equipment a company stakes its business communications on has the ability to interface with equipment from other vendors. Especially in the context of international interconnectivity, standards are of paramount importance. Also, users play a key role in developing standards since the use of standard equipment (as well as vendor acceptance) determines the success or failure of the standard. Ubiquitous deployment is often required for success. Vendors say: "We will provide it when customers sign up." Customers say: "We will sign up when it is universally available at the right price, unless we see something else better and less expensive." For example, take the ISDN standard. ISDN in North America has not become available ubiquitously, and hence has not been successful. ISDN is more widely available and has less competition in other countries and has seen greater success. Users usually do not play a very active part in the standardization and specification process. Instead they signal their approval with their purchases — they vote with their money.

6.2.3   Network Service Providers

Network service providers also actively participate in the standard-making process. In a sense they are also users. Service providers are often driven by vendors, but service providers often select vendors that adhere to industry standards that still provide some (usually nonstandard) capability for differentiation. This does not lock them into one vendor's proprietary implementation, and can allow the existence of a multiple vendor environment. Providers must not only make multiple vendor implementations interoperate within their networks, but they must also ensure the availability of industry standard interfaces to provide value-added services to users.

6.3   CREATING STANDARDS — THE PROCESS

In this section the general standards and specification process is reviewed. Figure 6.1 illustrates the generic process of standardization and specification. The process begins with a plan to work on a certain area, which is progressed through written contributions in technical meetings. The result is usually a document which is drafted and updated by the editor in response to contributions and agreements achieved in the meetings. The group reviews the drafts of this document, often progressing through several stages of voting and approval — eventually resulting in a final standard or specification. The standards process can be hindered by business and politics, with the final measure of success being user acceptance and interoperable implementations.

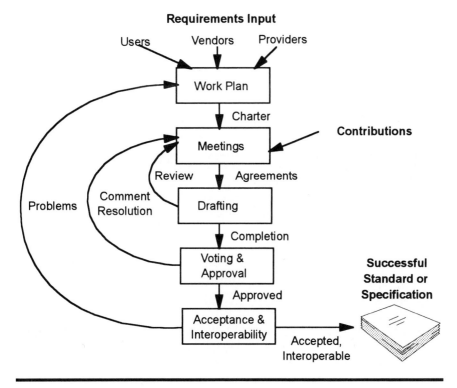

Figure 6.1 Generic Standardization and Specification Process

6.3.1 Work Plan

Most standards and specifications groups first agree on a work plan. A work plan defines the topics to be worked on, a charter for the activity, an organization for performing the work, and usually a very high-level set of objectives. User input and involvement most likely occur at this stage, either indirectly or sometimes even through direct participation. This is the time that vendors and service providers often voice their high-level requirements. The work plan for updating an existing standard almost always includes some changes resulting from user feedback or interoperability issues. Often an approximate time frame is set for completion of the standard or specification.

6.3.2 Meetings and Contributions

The majority of the work occurs at technical meetings, which can last from several days to several weeks. Participants submit written contributions which propose specific text and drawings for the standard, present background information, present arguments for or against a particular approach, or serve as liaisons from other standards or specification bodies. Usually these contributions are discussed, and if the contribution proposes adding text to a baseline document, then a process is employed to determine whether the proposal is accepted, amended, or rejected. In formal standards bodies, there is usually an attempt to achieve consensus before agreeing to include a contribution's input. In some industry forums a straw vote is taken to determine whether the proposal is accepted.

If there is a large committee structure, then a meeting normally begins and ends with a plenary session where representatives from all sub-committees attend. After the plenary meeting, multiple subcommittee meetings usually occur in parallel. The sub-committees are granted some autonomy, however, since they usually must review major changes or key decisions in the plenary session. Meetings also are used to resolve issues that arise from the drafting, review, voting, or approval process described below.

6.3.3   Drafting and Review

A key individual in the development of a standard or specification is the editor. The editor drafts text based upon the contributions, as amended in the meeting. The editor is often trusted to research related standards and specifications and align the document accordingly. A key part of any standards or specification technical activity is the ongoing review, correction, and improvement of the "working" document or baseline text. "Working" documents therefore provide a major input to meetings and become the basis for contributions for the next meeting that will further define the requirements in the document.

6.3.4   Voting and Approval

Once a particular document has reached a "draft" status, it is usually distributed for a preliminary vote. Comments that members believe must be addressed in order to approve the document as a standard or specification are often addressed via a comment resolution process at meetings, resulting in more drafting for the editor. The voting step of the process differs in various bodies in the number of members required to approve a change. If complete concurrence is the objective, then the process can be quite lengthy; if only a majority vote is required, then progress may be more rapid, but possibly increases risk. Once the comment resolution process is completed, the standard or specification then goes to a final vote. Again, depending upon the rules governing the standards or specification body, anything from a simple majority to a certain percentage, or unanimous approval is required for the body to release the document as an approved standard or specification. Often there will also be a supervisory board which will review the proposed standard for consistency with the format, style, scope, and quality required by that body in the final approval stage.

6.3.5   User Acceptance and Interoperability

Since customers have business problems today that sometimes can only be solved by proprietary implementations *prior* to standards, waiting until the perfect standard is designed and approved may put them out of business. Therefore, the user is caught in the dilemma of adopting an emerging standard now or waiting for it to become more mature. Users primarily determine the success of standards by creating the demand for specific capabilities and even technology, and by purchasing implementations from vendors and carriers supporting that standard.

The key technical measure of a standard, or specification's success is whether implementations from multiple vendors or carriers interoperate according to the details of the documentation. The documents should specify a minimum subset of interfaces, function, and protocol to achieve this goal. In support of interoperability, additional documentation, testing, and industry interoperability forums may be required, such as those established for the Internet (IETF), FDDI (ISO), and N-ISDN (CCITT).

If customers do not accept a standard, or if significant interoperability issues arise, then this feedback is provided back into the standards process for future consideration. Acceptance by the vendor community also plays a key role in the success or failure of standards, since if no implementation of the standard is built, no user can buy it!

6.3.6   Business and Politics

Standards organizations and industry forums have had increased participation and scope in recent years. With this increased number of people working on a plethora of problems there comes the inevitable burden of bureaucracy. Service providers, vendors, and to some extent, users view the chance to participate in the standard-setting process as an opportunity to express and impress their views upon the industry. This is a double-edged sword: while participation is necessary, biases are brought to the committees which can tie up decision making and bog down the process of making standards. One example in BISDN is that of Generic Flow Control (GFC), covered in Chapter 8, where the attempt to achieve agreement on a shared medium solution for ATM was never reached for this reason. The impact of this type of situation depends on whether the committee operates on a complete consensus basis or some form of majority rule. All too often a consensus-based approach ends up being a compromise with multiple, incompatible options stated in the standard. There is then a need to further subset a standard as an interoperability specification to reduce the number of choices, and to translate the ambiguities of the standard into specific equipment requirements.

Standards can also have omissions or *holes* that are undefined and left to vendor interpretation because they simply weren't conceived as issues. These *holes* may exist because no agreement could be reached on how the requirement should be standardized. Standards usually identify these items as for further study (ffs) just to point out that there is an awareness of a need for a function or element that isn't standardized yet, but could be in the future.

Vendors can play a game of supporting their proprietary solution to make it a standard before their competitor's proprietary solution becomes a standard. This alone can delay and draw out a standards process for many months or even years. While standards organizations take their time to publish standards, some vendors try to take the lead and build equipment designed around a proposed standard, or a partially issued standard, and then promise compliance with the standard once it is finally published. If they guess right, they can be well ahead of the pack; if they miss the mark, a significant investment could be lost.

6.4   STANDARDS AND SPECIFICATIONS — RESULTS TO DATE

This section summarizes the standards and specifications that have been approved by the CCITT/ITU-T, ANSI, the ATM Forum, the IETF, and the Frame Relay Forum as of early 1994.

6.4.1   CCITT/ ITU-T Standards

The current ITU-T standards relating to B-ISDN are listed in Table 6.1. Starting in 1988, the CCITT published Recommendations I.113 and I.121 which defined the vocabulary, terms, principles, and basic objectives for broadband aspects of ISDN, called B-ISDN. These recommendations were revised and approved in November 1990, and published in 1991. Also in 1991, eleven more recommendations I.150, I.211, I.311, I.321, I.327, I.361, I.362, I.363, I.413, I.432, and I.610 were published, further detailing the functions, service aspects, protocol layer functions, Operations, Administration, and Maintenance (OAM), and user-to-network and network-to-network interfaces. In 1992, the following additional recommendations were approved: I.364, I.371, and I.580. In 1993, I.113 and I.321 were revised, and new Recommendations I.356, a new section of I.363 for AAL5, I.365, and I.555 were approved. A number of signaling B-ISDN signaling standards are up for approval in 1994, as summarized in Chapter 9. A brief description of each standard follows:

Table 6.1 CITT/ITU-T B-ISDN Standards

Number	Title
I.113	Vocabulary for B-ISDN
I.121	Broadband Aspects of ISDN
I.150	B-ISDN Asynchronous Transfer Mode Functional Characteristics
I.211	General Service Aspects of B-ISDN
I.311	B-ISDN General Network Aspects
I.321	B-ISDN Protocol Reference Model and Its Application
I.327	B-ISDN Functional Architecture
I.350	General Aspects of Quality of Service and Network Performance in Digital Networks, including ISDN
I.356	B-ISDN ATM Layer Cell Transfer Performance
I.361	B-ISDN ATM Layer Specification
I.362	B-ISDN ATM Adaptation Layer (AAL) Functional Description
I.363	B-ISDN ATM Adaptation Layer (AAL) Specification
I.364	Support of Connectionless Data Service on a B-ISDN
I.365.1	Frame Relaying Bearer Service Specific Convergence Sublayer (FR-SSCS)
I.371	Traffic Control and Congestion Control in B-ISDN
I.413	B-ISDN User-Network Interface
I.432	B-ISDN User-Network Interface - Physical Layer Specification
I.555	Frame Relay Bearer Service Interworking
I.580	General Arrangements for Interworking between B-ISDN and 64 kb/s ISDN
I.610	B-ISDN OAM Principles and Functions
G.804	ATM Cell Mapping Into Plesiochronous Digital Hierarchy (PDH)

Recommendation I.113, Vocabulary for B-ISDN — is a glossary of terms and acronyms used in B-ISDN and ATM.

Recommendation I.121 CCITT, Broadband Aspects of ISDN — defines basic principles and characteristics of B-ISDN, and how it can be evolved from TDM and ISDN networks.

Recommendation I.150 CCITT, **B-ISDN Asynchronous Transfer Mode Functional Characteristics** — defines functional characteristics of ATM, such as the establishment of and signaling for virtual paths and channels, cell level multiplexing, per virtual connection Quality of Service (QoS) and Generic Flow Control (GFC).

Recommendation I.211 CCITT, **General Service Aspects of B-ISDN** — defines interactive and distribution service classes, the types of information needing support, example applications, and some possible attributes such as bit rate, QoS, synchronization, responsiveness, and source characteristics. These were covered in Chapters 1 and 2 as business and technology drivers.

Recommendation I.311 CCITT, **B-ISDN General Network Aspects** — peals back the multilayered ATM onion and begins to detail the concepts behind ATM, such as the physical and ATM sublayers and the way Virtual Path (VP) and Virtual Channel (VC) connections are constructed from smaller links, and defines the notions of VC switching and VP cross-connects using a number of examples. It then covers the control and management of B-ISDN, including the physical network management architecture and general principles of signaling. Chapters 7, 8, and 9 summarize many of these concepts.

Recommendation I.321 CCITT, **B-ISDN Protocol Reference Model and Its Application** — defines the layered model which is used as the road map for the remainder of this book. A summary of I.321 is covered later in this chapter.

Recommendation I.327 CCITT, **B-ISDN Functional Architecture** — defines a basic architectural model for B-ISDN, and how it relates to ISDN and connectionless services. This is covered in the next section.

Recommendation I.350 CCITT, **General Aspects of Quality of Service and Network Performance in Digital Networks, including ISDN** — defines the terms Quality of Service (QoS) as the user's perception and Network Performance (NP) as the network operator's observation. It defines specific performance parameters in terms of a number of generic categories. Chapters 12 and 21 cover these topics.

Recommendation I.356 CCITT, **B-ISDN ATM Layer Cell Transfer Performance** — defines the reference events, the definitions, and how the detailed ATM layer performance parameters identified in

I.350 can be theoretically calculated. Chapters 12 and 21 cover these topics.

Recommendation I.361 CCITT, **B-ISDN ATM Layer Specification** — defines the bits and bytes of the ATM cell format, what they mean, and how they are to be used. Chapter 8 covers this subject in detail.

Recommendation I.362 CCITT, **B-ISDN ATM Adaptation Layer (AAL) Functional Description** — defines basic principles and sublayering. It also defines service classification in terms of constant or variable bit rate, timing transfer requirement, and whether the service is connection-oriented or connectionless. Chapter 8 details this subject.

Recommendation I.363 CCITT, **B-ISDN ATM Adaptation Layer (AAL) Specification** — defines the specifics of the AALs 1, 2, 3/4, and 5. AAL1 is for connection-oriented, continuous bit rate service that requires timing transfer. AAL2 is for connection-oriented, variable bit-rate service that requires timing transfer. AAL3/4 and AAL5 can be used for connection-oriented or connectionless, variable bit-rate service that does not require timing transfer. Chapter 8 covers the details of AALs.

Recommendation I.365.1 CCITT, **Frame Relaying Bearer Service Specific Convergence Sublayer (FR-SSCS)** — defines the specifics for interworking frame relay and ATM. Chapter 17 covers this subject.

Recommendation I.364 CCITT, **Support of Connectionless Data Service on a B-ISDN** — defines an approach for support of connectionless services, such as 802.6/DQDB, over AAL3/4.

Recommendation I.371 CCITT, **Traffic Control and Congestion Control in B-ISDN** — defines terminology for traffic parameters, a traffic contract, conformance checking, resource management, connection admission control, prioritization, and implementation tolerances. These complex topics are covered in layman's terms using examples in Chapters 12, 13, and 14.

Recommendation I.413 CCITT, **B-ISDN User-Network Interface** — defines the reference configurations and terminology used in the B-ISDN standards. The simpler mapping used by the ATM Forum in this book is explained in Chapter 8 with relation to this standard.

Recommendation I.432 CCITT, **B-ISDN User-Network Interface - Physical Layer Specification** — defines the details of how ATM cells are mapped into the Synchronous Digital Hierarchy (SDH) TDM structure, how the ATM Header Error Control (HEC) is generated, and how bit errors impact HEC and cell delineation time. Chapter 8 covers this at a high level.

Recommendation I.555 CCITT, **Frame Relay Bearer Service Interworking** — defines how frame relay interworks with a number of other services, including B-ISDN. Chapter 17 covers this subject.

Recommendation I.580 CCITT, **General Arrangements for Interworking between B-ISDN and 64 kb/s ISDN** — defines in general terms how the narrowband ISDN can be interworked with the Broadband-ISDN in support of user data transfer, control and management.

Recommendation I.610 CCITT, **B-ISDN OAM Principles and Functions** — covers the initial principals and functions required for Operation, Administration, and Maintenance (OAM) of primarily the ATM layer. We cover this important topic in Chapters 19 through 21.

Recommendation G.804, ATM Cell Mapping Into Plesiochronous Digital Hierarchy (PDH) — defines how ATM cells are mapped into various TDM structures, such as E1, DS1, E3, and DS3. Chapter 8 covers this subject.

6.4.2   ANSI Standards

ANSI committee T1 adapts CCITT/ITU-T standards to the competitive environment and the unique physical layer transmission requirements of North America. The standards approved to date are listed and briefly described below.

T1.624-1993, BISDN UNI: Rates and Formats Specification — defines the mappings of ATM cells into DS3 and SONET payloads and how fault management is performed.

T1.627-1993, BISDN ATM Functionality and Specification — defines the ATM layer following I.361, adding additional explanations of the protocol model, further interpretations of traffic management, and some examples describing functions required in an implementation as annexes.

T1.629-1993, BISDN ATM Adaptation Layer 3/4 Common Part Functionality and Specification — defines the AAL3/4 functionality based on I.363, expanding on the protocol model and giving an example state machine in an annex. Chapter 8 presents AAL3/4.

T1.630-1993, BISDN - Adaptation Layer for Constant Bit Rate Services Functionality and Specification — defines AAL1 based upon I.363, but in considerably more detail through more explanatory, detailed requirements and a number of good technical annexes. T1.630 also defines the specifics of emulating the North American DS1 circuit function, interface, and alarm management. Chapter 8 summarizes AAL1.

T1.633, Frame Relay Bearer Service Interworking — is based very closely on recommendation I.555. Chapter 17 covers this subject.

T1.634, Frame Relay Service Specific Convergence Sublayer (FR - SSCS) — is based upon a draft CCITT recommendation that will likely be part of I.365 in the future. Chapter 17 covers this subject.

T1.635, BISDN ATM Adaptation Layer Type 5 — defines AAL5 based precisely upon I.363. Chapter 8 covers this subject.

6.4.3 ATM Forum Specifications

The ATM Forum has produced several implementation specifications which are summarized below.

ATM User-Network Interface (UNI) Specification Version 2.0 — defined a PVC capability for the ATM UNI, added physical layers based upon FDDI and fiber channel technology for the local area, defined an SNMP-based Interim Local Management Interface (ILMI), and adopted a subset of standardized ATM OAM functions. Chapter 19 describes the ILMI.

ATM User Network Interface (UNI) Specification Version 3.0 — supersedes version 2.0, correcting errors and adding new functions. The major new functions were definition of an initial signaling protocol defined as a subset of the ITU-T standard, definition of traffic control beyond the peak rate in an unambiguous manner, and the specification of scrambling for the DS3 rate in order to allow operation over current transmission systems based upon implementation experience. Chapters 12 and 13 describe extensions to traffic management defined by the ATM Forum.

ATM Data eXchange Interface (DXI) Specification Version 1.0 — defines a frame-based interface that allows a DTE to pass the ATM VPI/VCI in the frame address. It is similar in nature to the SMDS DXI specification. Chapter 17 summarizes this specification.

ATM Broadband-InterCarrier Interface (B-ICI) Specification Version 1.0 — defined characteristics of service for PVC connection between carrier networks. The OAM functions, frame relay network interworking, and transport of SMDS-ICI over ATM are defined in great detail. Chapter 17 summarizes this specification.

6.4.4 IETF RFCs Related to ATM

The IETF has completed several documents to date in support of ATM, with several others in progress.

RFC 1483 MultiProtocol Encapsulation Over ATM — defines how higher layer protocols, such as IP, are encapsulated for bridging and routing over an ATM network. Interworking at the protocol encapsulation level with frame relay networks is also defined. Chapter 17 summarizes this standard.

RFC 1577 Classical IP Over ATM — defines how the current Internet Protocol (IP) can utilize the ATM Switched Virtual Connection (SVC) capability. Chapter 11 provides a detailed example of this standard.

6.5 B-ISDN PROTOCOL MODEL AND ARCHITECTURE

This section describes the protocol model for B-ISDN that is built on ATM. The standards vision of how B-ISDN interconnects with ISDN, SS7, and OSI is then presented.

6.5.1 B-ISDN Protocol Reference Model

Figure 6.2 depicts the B-ISDN protocol reference model from ITU-T Recommendation I.321, which is used to structure the remaining recommendations. A significant portion of the architecture exists in this protocol dimension. This subsection introduces the B-ISDN protocol reference model. The remainder of the book steps through each facet of this model in detail. The shaded version of the model in Figure 6.2 illustrates what each chapter or part covers.

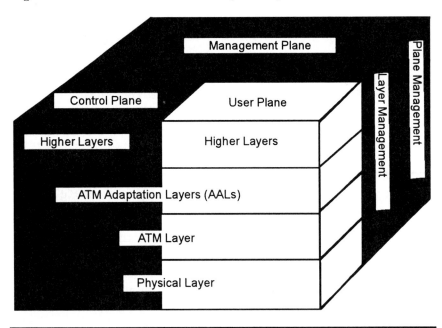

Figure 6.2 B-ISDN Protocol Model

The top of the cube illustrates the planes, which are defined on the front and side of the cube. The user plane and control plane span through the higher layer, down through the AALs (which can be null), to the ATM layer and physical layer. Therefore the physical layer, ATM layer, and ATM Adaptation Layers (AALs) are the foundation for B-ISDN. The user and control planes may make use of common ATM and physical layer protocols; however, the end purpose differs in the AALs and higher layers.

The management plane is further broken down into layer management and plane management. As shown in the figure, layer management interfaces with each layer in the control and user planes. Plane management has no layered structure and is currently only an abstract concept with little standardization at this point. It can be viewed as a catch-all for the things that do not fit into the other portions of this model, by having the role of overall system management.

6.5.2 B-ISDN Architecture

Figure 6.3 depicts the vision of how B-ISDN should interconnect with ISDN, SS7, and OSI as defined in CCITT Recommendation I.327, which complements the ISDN architecture defined in Recommendation I.324. SS7, ISDN, and B-ISDN are all lower level capabilities that serve to interconnect Terminal Equipment (TE) or service providers through local functional capabilities which represent the point of physical interface. SS7 provides out-of-band interexchange signaling capabilities for telephony and ISDN, while ISDN provides signaling capabilities for TDM-based services, X.25, and frame relay. B-ISDN provides bearer services of various types and signaling for speeds in excess of 34 Mbps. These all act in support of higher layer capabilities above layer 3 in the OSI model.

6.6 CURRENT DIRECTIONS AND OPEN ISSUES

This section concludes with a discussion identifying critical success factors, new approaches, standardization and specification work that is in progress, and an enumeration of some open issues.

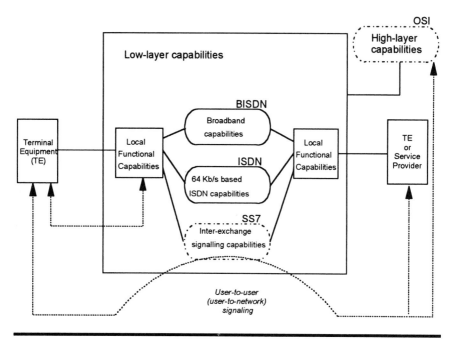

Figure 6.3 Integrated B-ISDN, ISDN, SS7, and OSI Architecture

6.6.1   Measures of Success and New Approaches

It is important for standards-conscious players to keep in mind the mission, regardless of the dangers that may be present. Those bodies that keep the user's application foremost in mind while dealing with the problems and dangers of the standards process will live long and prosper, or at least live longer than those who concentrate primarily on the development of standards.

Some forums now provide training centers, seminars, and interoperability tests labs and strictly adhere to parallel standards development by both national and international standards bodies. User trial communities and university test beds are another method employed by these forums to help speed up the testing and acceptance of these new technologies. The Internet protocols are an example of the work by university and academia in successfully leading standards development.

Standards development seems to have improved by the adoption of this "free market" approach, just as the same "free market" of ideas has stimulated the world.

6.6.2 Future Work Plans

The ITU-T, ANSI, and ETSI continue to refine and expand upon the set of standards introduced earlier. Significant activity is occurring in the areas of the control and management planes. Chapter 9 summarizes the extensive new additional features proposed for the signaling protocol. ATM layer management is also being defined in detail, and network management, in general, is being defined as covered in Chapters 19 through 21.

The ATM Forum has announced that it planned to work on the following nine areas, starting in the fall of 1993, which are elaborated on at the appropriate point in this book [4]. Some results from these ATM Forum activities are expected in 1994 and 1995, even though specific documents and schedules have not been announced. This is a very ambitious charter, and it will likely take years to complete specifications in all of these areas.

Signaling
Traffic Management
System Aspects and Applications (SAA)
   SMDS Access over ATM
   Frame Relay Interworking
   Video Support
   Circuit Emulation
   Application Program Interface (API)
LAN Emulation
Testing and Interoperability
Private Network-Network Interface (NNI)
Broadband InterCarrier Interface (B-ICI)
Physical Layer
Network Management

The Frame Relay Forum and ATM Forum have announced that they will define service interworking with ATM and address signaling interworking in 1994.

The SMDS Interest Group (SIG) and ATM Forum have announced a joint effort to specify how SMDS service can be accessed by users over ATM in 1994.

The IETF has several activities ongoing in the area of ATM, including how routing should be performed, what initial steps can be taken to support classical IP over ATM, and definition of how the IP Maximum Transfer Unit (MTU) can be negotiated, which Chapter 11 covers.

6.6.3 Unaddressed Areas

There are a number of areas that are currently not addressed in any standard, specification, or documentation. Several reasons account for this. A common reason is that product differentiation can occur in this area, and standardization is not required for interoperability. Details of an implementation are often in this category. In other areas, it is a matter of prioritization of the standards and specification work to be performed. Certain areas are simply not as important or as urgent as others. Usually, if these areas are acknowledged by a standards body, they are briefly defined and then identified as being designated for further study (ffs).

6.6.4 Predicting the Future

One critical aspect often overlooked that influences standards acceptance is the development of services and applications in conjunction with and accompanying the standards. An example of this is the ISDN basic (BRI) and primary (PRI) rate interfaces, where the technology is fully developed but applications are scarce. ISDN was basically a telephone company plan to upgrade, digitize, and put the latest technology into the utilities' networks for maximum efficiencies — the end user was an afterthought and not really a significant factor in the standardization equation. Little wonder, then, that end users did not perceive any immediate value.

Does the same fate that obsoleted N-ISDN before await B-ISDN? We predict that it will not. Many of the same players from N-ISDN have a vested interest in the success of B-ISDN and will not repeat the same mistakes, and industry forums are bringing the data communications and telecommunication vendors, providers, and users together in a new process increasingly adopted and moving into the twenty-first century.

6.7 REVIEW

This chapter identified the organizations that are taking an active role in standardizing and specifying ATM and B-ISDN. The CCITT, now called the ITU-T, started it all in 1988. The process and level of detail have come a long way since then, as shown by the number of standards and specifications on B-ISDN and ATM that have been produced. The role of the various players in the standards process, user, vendors, and service providers, was covered. The standards process was then described in terms of how they are developed and finalized. We commented on differences in approach between the various organizations. The CCITT/ITU-T, ANSI, ATM Forum, and IETF standards related to ATM that have been published to date were listed and briefly summarized. An overview of the B-ISDN layered protocol model was presented, describing how it relates to the ISDN, SS7, and OSI protocols that were introduced earlier. This chapter concluded with the current directions for future work in each of the organizations and identified some open issues.

6.8   REFERENCES

[1]   B. Figi, "Broadband Aspects of ISDN," IIR Conference, London, UK, February 1993

[2]   J. Johnson, "The ATM Circus Gets a Ringmaster - The ATM Forum Acts as Coordinator and Catalyst for ATM Standards," *Data Communications*, March 21, 1993

[3]   D. Dern, *The Internet Guide for New Users*, McGraw-Hill, 1993

[4]   ATM Forum, "53 Bytes Newsletter," January 1994, Volume 2, Issue 1

[5]   ITU-T, "Recommendation I.113, Vocabulary for B-ISDN," 1993

[6]   ITU-T, "Recommendation I.121, Broadband Aspects of ISDN," 1991

[7]   ITU-T, "Recommendation I.150, B-ISDN Asynchronous Transfer Mode Functional Characteristics," 1991

[8]   ITU-T, "Recommendation I.211, General Service Aspects of B-ISDN," 1991

[9]   ITU-T, "Recommendation I.311, B-ISDN General Network Aspects," 1991

[10] ITU-T, "Recommendation I.321, B-ISDN Protocol Reference Model and Its Application," 1991

[11] ITU-T, "Recommendation I.327, B-ISDN Functional Architecture," 1991

[12] ITU-T, "I.350 - General Aspects of Quality of Service and Network Performance in Digital Networks, Including ISDN," 1988

[13] ITU-T, "I.356 - B-ISDN ATM Layer Cell Transfer Performance," 1993

[14] ITU-T, "Recommendation I.361, B-ISDN ATM Layer Specification," 1991

[15] ITU-T, "Recommendation I.362, B-ISDN ATM Adaptation Layer (AAL) Functional Description," 1991

[16] ITU-T, "Recommendation I.363, B-ISDN ATM Adaptation Layer (AAL) Specification," 1993

[17] ITU-T, "Recommendation I.364, Support of Connectionless Data Service on a B-ISDN," 1992

[18] ITU-T, "Recommendation I.365.1, Frame Relaying Bearer Service Specific Convergence Sublayer (FR-SSCS)," 1993

[19] ITU-T, "Recommendation I.371, Traffic Control and Congestion Control in B-ISDN," 1992

[20] ITU-T, "Recommendation I.413, B-ISDN User-Network Interface," 1991

[21] ITU-T, "Recommendation I.432, B-ISDN User-Network Interface - Physical Layer Specification," 1991

[22] ITU-T, "Recommendation I.555, Frame Relay Bearer Service Interworking," 1993

[23] ITU-T, "Recommendation I.580, General Arrangements for Interworking between B-ISDN and 64 kb/s ISDN," 1992

[24] ITU-T, "Recommendation I.610, OAM Principles for the B-ISDN Access," Geneva, 1991

[25] ITU-T, "Recommendation G.804, ATM Cell Mapping Into Plesiochronous Digital Hierarchy (PDH)," 1993

[26] ANSI, "T1.624-1993, BISDN UNI: Rates and Formats Specification," 1993

[27] ANSI, "T1.627-1993, BISDN ATM Functionality and Specification," 1993

[28] ANSI, "T1.629-1993, BISDN ATM Adaptation Layer 3/4 Common Part Functionality and Specification," 1993

[29] ANSI, "T1.630-1993, BISDN - Adaptation Layer for Constant Bit Rate Services," 1993

[30] ANSI, "T1.633, Frame Relay Bearer Service Interworking," 1994

[31] ANSI, "T1.634, Frame Relay Service Specific Convergence Sublayer (FR-SSCS)," 1994

[32] ANSI, "T1.635, BISDN ATM Adaptation Layer 5," 1994

[33] ATM Forum, "ATM User-Network Interface Specification Version 2.0," June 1992

[34] ATM Forum, "ATM User-Network Interface Specification Version 3.0," August 1993

[35] ATM Forum, "BISDN Inter Carrier Interface (B-ICI) Specification, Version 1.0," August 1993

[36] ATM Forum, "ATM Data eXchange Interface (DXI) Specification, Version 1.0," August 4, 1993

[37] J. Heinanen, "IETF RFC 1483, Multiprotocol Encapsulation over ATM Adaptation Layer 5," July 1993

[38] M. Laubach , "IETF RFC 1577 - Classical IP and ARP over ATM," January 1994

7

Basic Introduction to ATM

This chapter introduces the reader to the basic principles of ATM. ATM in a most basic sense is a technology, defined by protocols standardized by the ITU-T, ANSI, ETSI, and the ATM Forum introduced in the previous chapter. The coverage begins with the building blocks of ATM — transmission paths, virtual paths, and virtual channels. Next a look is taken at the ATM cell and its transmission through a series of simple examples. The fact that the 53-octet cell size turned out to be a compromise between a smaller cell size optimized for voice and a larger cell size optimized for data is presented. This chapter concludes with a discussion of how ATM means many things to many people, such as a method of integrated access, a public virtual data service, a hardware and software implementation, or a network infrastructure.

7.1   OBJECTIVES OF ATM

Asynchronous Transfer Mode (ATM) is a cell-based switching and multiplexing technology designed to be a general-purpose, connection-oriented transfer mode for a wide range of services. ATM is also being applied to the LAN and private network technologies as specified by the ATM Forum.

ATM handles both connection-oriented traffic directly or through adaptation layers, or connectionless traffic through the use of adaptation layers. ATM virtual connections may operate at either a Constant Bit Rate (CBR) or a Variable Bit Rate (VBR). Each ATM cell sent into the network contains addressing information that establishes a virtual connection from origination to destination. All cells are then transferred, in sequence, over this virtual connection. ATM provides either Permanent or Switched Virtual Connections (PVCs or SVCs). ATM is asynchronous because the transmitted cells need not be periodic as time slots of data are in Synchronous Transfer Mode (STM), as described in Chapter 4. Subsequent chapters define these terms and concepts in detail.

ATM offers the potential to standardize on one network architecture defining the multiplexing and switching method, with SONET/STM providing the basis for the physical transmission standard for very high-speed rates. ATM also supports multiple Quality of Service (QoS) classes for differing application requirements on delay and loss performance. Thus, the vision of ATM is that an entire network can be constructed using ATM and ATM Application Layers (AALs) switching and multiplexing principles to support a wide range of all services, such as:

★ Voice
★ Packet data (SMDS, IP, FR)
★ Video
★ Imaging
★ Circuit emulation

ATM provides bandwidth-on-demand through the use of SVCs, and also support LAN-like access to available bandwidth.

7.2   THE ATM CELL AND TRANSMISSION

The primary unit in ATM is the *cell*. This section defines the basics of the ATM cell; the next chapter presents a detailed explanation. Several examples are used to illustrate the basic concepts of ATM.

7.2.1   ATM Cell

ATM standards define a fixed-size cell with a length of 53 octets (or bytes) comprised of a 5-octet header and a 48-octet payload as shown in Figure 7.1. The bits in the cells are transmitted over the transmission path from left to right in a continuous stream. Cells are mapped into a physical transmission path, such as the North American DS1, DS3, or SONET; European, E1, E3 and E4; or ITU-T STM standards; and various local fiber and electrical transmission payloads as defined in Chapter 8. This is only a brief introduction to the ATM cell format, followed up by a detailed exposition in Chapter 8.

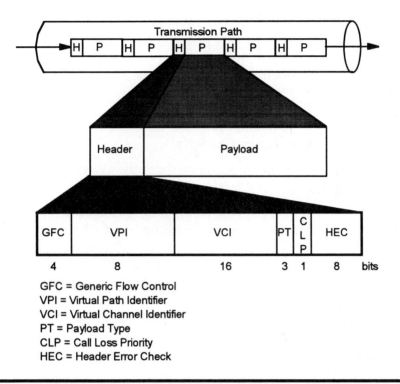

GFC = Generic Flow Control
VPI = Virtual Path Identifier
VCI = Virtual Channel Identifier
PT = Payload Type
CLP = Call Loss Priority
HEC = Header Error Check

Figure 7.1 ATM Cell Transmission and Format

All information is switched and multiplexed in an ATM network in these fixed-length cells. The cell header identifies the destination, cell type, and priority. The Virtual Path Identifier (VPI) and Virtual Channel Identifier (VCI) hold local significance only, and identify the destination. The Generic Flow Control (GFC) field allows a multiplexer to control the rate of an ATM terminal. The Payload Type (PT) indicates whether the cell contains user data, signaling data, or maintenance information. The Cell Loss Priority (CLP) bit indicates the relative priority of the cell. Lower priority cells are discarded before higher priority cells during congested intervals.

Because of its critical nature, the cell Header Error Check (HEC) detects and corrects errors in the header. The payload field is passed through the network intact, with no error checking or correction. ATM relies on higher layer protocols to perform error checking and correction on the payload. The fixed cell size simplifies the implementation of ATM switches and multiplexers and enables implementations at very high speeds.

When using ATM, longer packets cannot delay shorter packets as in other packet switched implementations because long packets are chopped up into many cells. This enables ATM to carry Constant Bit Rate (CBR) traffic such as voice and video in conjunction with Variable Bit-Rate (VBR) data traffic, potentially having very long packets within the same network.

7.2.2 Cell Segmentation Example

ATM switches take a user's data, voice, and video and chops it up into fixed length cells, and multiplex it into a single bit stream which is transmitted across a physical medium. An example of multimedia application is that of a person needing to send an important manuscript for a book to his or her publisher. Along with the letter, this person would like to show his or her joy at receiving a contract to publish the book.

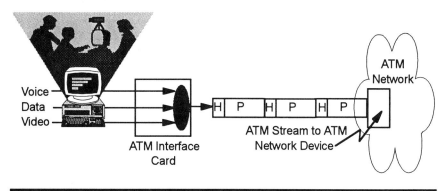

Figure 7.2 Multimedia Communications Example Using ATM

Figure 7.2 illustrates the role of ATM in this real-life example, where Jeanne is sitting at her workstation. Jeanne's workstation has an ATM interface card, sound board with microphone, and videocamera. The workstation is connected to a local ATM switch, which in turn is attached to a public ATM-based wide area network service to which the publisher is also connected.

Jeanne places a multimedia call to the publisher, begins transmitting the data for her manuscript, and begins a conversation with the publisher, with Jeanne and the publisher able to see each other's faces — providing text, voice, and video traffic, respectively, in real time. The publisher is looking through the manuscript at its workstation all the while and having an interactive dialogue with Jeanne. Let's take this scenario one piece at a time.

Video and voice are very time-sensitive; the information cannot be delayed for more than a blink of the eye, and the delay cannot have significant variations. Disruption in the video image of Jeanne's face or distortion of the voice destroy the interactive, near real-life quality of this multimedia application. Data can be sent in either connection-oriented or connectionless mode. In either case, the data is not nearly as delay-sensitive as voice or video traffic. Data traffic, however, is very sensitive to loss. Therefore, ATM must discriminate between voice, video, and data traffic, giving voice and video traffic priority and guaranteed, bounded delay, simultaneously assuring that data traffic has very low loss.

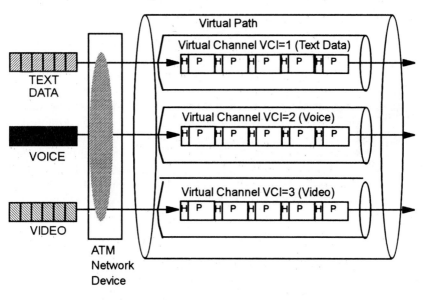

Figure 7.3 Virtual Channels Supporting Multiple Applications

Examining this example in further detail, a virtual path is established between Jeanne and the publisher, and over that virtual path three virtual circuits are defined for text data, voice, and video. Figure 7.3 shows how all three types of traffic are combined over a single ATM Virtual Path (VP), with Virtual Circuits (VCs) being assigned to the text data (VCI=1), voice (VCI=2), and video (VCI=3).

7.3   THEORY OF OPERATION

This section presents two examples of how user traffic is segmented into ATM cells, switched through a network, and processed by the receiving user.

7.3.1   A Simple ATM Example

Let's look at the last example, where Jeanne is simultaneously transmitting text, voice, and video data traffic from her workstation, in more detail. The workstation contains an ATM interface card, where the chopper "slices and dices" the data streams into 48-octet data segments as shown in Figure 7.4. In the next step the postman addresses the payload by prefixing it with the VPI, VCI, and the remaining fields of the 5-octet header. The result is a stream of 53-octet ATM cells from each source: voice, video, and text data. These cells are generated independently by each source, such that there may be contention for cell slot times on the interface connected to the workstation. The text, voice, and video are each assigned a VCC: VCI=1 for text data, VCI=2 for voice, and VCI=3 for video, all on VPI=0. This example has been greatly simplified, as there would normally be many more than just three active VCI values on a single VPI.

Figure 7.4 shows an example of how Jeanne's terminal sends the combined voice, video, and text data. A gatekeeper in her terminal shapes the transmitted data in intervals of eight cells (about 80 μs at the DS3 rate), normally allowing one voice cell, then five video cells, and finally what is left — two text data cells — to be transmitted. This corresponds to about 4 Mbps for high-fidelity audio, 24 Mbps for video, and 9 Mbps for text data. All data sources (text, voice, and video) contend for the bandwidth each shaping interval of eight cell times, with the voice, video, and then text data being sent in the above proportion. Cells are retained in the buffer by the gatekeeper in case all of the cell slot times were full in the shaping interval. Chapter 13 covers a more detailed example of traffic shaping. A much larger shaping interval is used in practice to provide greater granularity in bandwidth allocation.

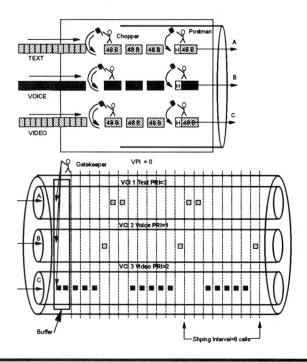

Figure 7.4 Asynchronous Transfer Mode Example

7.3.2 An ATM Switch Example

An illustration of an ATM switch is shown in Figure 7.5. A continuous video source is shown as input to a *packetizing* function, with logical destination VPI/VCI address D. The continuous bit stream is broken up into fixed-length cells comprised of a header and a payload field (indicated by the shading). The rate of the video source is greater than the continuous DS3 bit stream with logical destination address A, and the high-speed computer directly packetized input addressed to B. These sources are shown *time division multiplexed* over a transmission path, such as SONET or DS3.

The initial function of the ATM switch is to *translate* the logical address to a physical outgoing switch port address and to an outgoing logical VPI/VCI address. This additional ATM switch header is prefixed to every input ATM cell as shown previously. There are three point-to-point vitual connections in the figure. The DS3 has address A which is translated into C destined for physical port 1. The video source has address D which is translated into address E, destined for port 2. The computer source has address B which is translated to address F, destined for port 1.

The ATM switch utilizes the physical destination address field to deliver the ATM cells to appropriate physical switch port and associated transmission link. Chapter 10 describes examples of ATM switches that perform this function.

At the output of the ATM switch, the physical address is removed by a *reduce* function. The logically addressed ATM cells are then time division multiplexed onto the outgoing transmission links. Next these streams are demultiplexed to the appropriate devices. The Continuous Bit Rate (CBR) connections (i.e., video and the DS3) then have the logical addresses removed, and are reclocked to the information sink via the *serialize* function. Devices, such as workstations, can receive ATM cells directly.

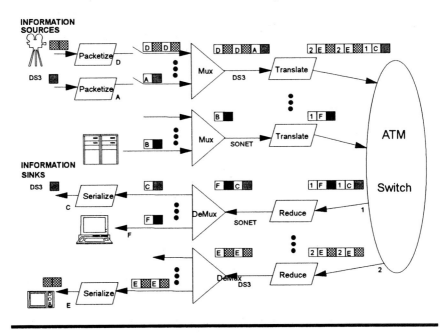

Figure 7.5 Asynchronous Transfer Mode Example

7.4 CHOICE OF PAYLOAD SIZE

When a standard cell size was under discussion by the ATM Forum, there was a raging debate between a 32-octet versus a 64-octet payload size. The decision on the 48-byte payload size was the compromise between these positions. The choice of the 5-octet header size was a separate tradeoff between a 3-octet header and an 8-octet header.

There is a basic tradeoff between efficiency and packetization delay versus cell size illustrated in Figure 7.6. Efficiency is computed for a 5-octet cell header. Packetization delay is the amount of time required to fill the cell at a rate of 64 kbps, that is, the rate to fill the cell with digitized voice samples. Ideally high efficiency and low delay are both desirable, but cannot be achieved simultaneously. As seen from the figure, better efficiency occurs at large cell sizes at the expense of increased packetization delay. In order to carry voice over ATM and interwork with two-wire analog telephone sets, the total delay should be less than about 12 ms, otherwise echo cancellation must be used. Two TDM to ATM conversions are required in the round-trip echo path. Allowing 4 ms for propagation delay and two ATM conversions, a cell size of 32 octets avoids the need for echo cancellation. Thus, the ITU-T adopted the fixed-length 48-octet cell payload as a compromise between a long cell sizes for time-insensitive traffic (64 octets) and smaller cell sizes for time-sensitive traffic (32 octets).

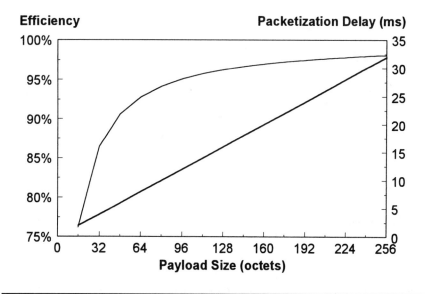

Figure 7.6 Delay versus Cell Size Tradeoff

7.5   ATM NETWORKING BASICS

Three major concepts in ATM are: the transmission path, the Virtual Path (VP), and, optionally, the Virtual Channel (VC). These form the basic building blocks of ATM.

7.5.1   Transmission Path, Virtual Path, and Virtual Channel Analogy

Let us look at a simple example of these concepts in relation to vehicle traffic patterns. These analogies are not intended to be exact, but to introduce some concepts that later chapters elaborate on. Think of cells as vehicles, transmission paths as roads, virtual paths as a set of directions, and virtual channels as a lane discipline on the route defined by the virtual path. Figure 7.7 illustrates the example described in this section.

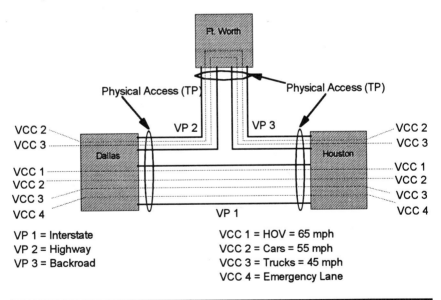

Figure 7.7 Transportation Example of ATM Principles

Three transmission paths form the set of roads between three cities: Dallas, Fort Worth, and Houston. There are many interstates, highways, and back roads between the two cities which create many possibilities for different routes — but the primary routes, or virtual paths, are the interstate (VP1) from Dallas to Houston, the highway from Dallas to Fort Worth (VP2), and a back road (VP3) from Fort Worth to Houston. Thus, a car (cell) can travel from Dallas to Houston either over the highway to Fort Worth and then the back road to Houston, or take the direct interstate. If the car chooses the interstate (VP1), it has the choice of three lanes: car pool or High Occupancy Vehicle (HOV) (VCC1), car lane (VCC2), or the truck lane (VCC3). These three lanes have speed limits of 65 mph, 55 mph, and 45 mph, respectively, which will cause different amounts of delay in reaching the destination. In our analogy, vehicles strictly obey this lane discipline (unlike on real highways).

In our example, the interstate carries high-speed traffic: tractor trailers, buses, tourists, and business commuters. The highway can carry car and truck traffic, but at a lower speed. The back roads carry locals and traffic avoiding backups on the interstate (spillover traffic), but at an even slower speed.

Note that our example of automotive traffic (cells) has many opportunities for missequencing. Vehicles may decide to pass each other, there can be detours, and road hazards (like stalled cars in Texas!) may cause some vehicles (cells) to arrive out of sequence or vary in their delay. This is evident in normal transportation when you always seem to leave on time, but traffic causes you to be delayed. Automotive traffic must employ an Orwellian discipline where everyone follows the traffic routes exactly (unlike any real traffic) in order for the analogy to apply.

The routes also have different quality. When you get a route map from the American Automobile Association (AAA), you have a route selected based on many criteria: least driving (routing) time, most scenic route, least cost (avoids most toll roads), and avoid known busy hours. The same principles apply to ATM.

Now, let's give each of the road types (VPs) and lanes (VCCs) a route choice. A commuter from Dallas to Houston in a hurry would first choose the VP1, the interstate. A sightseer would choose the highway to Fort Worth (VP2) to see the old cow town, and then the back road to Houston (VP3) to take in Waxahachie and Waco on the way. When commuters enter the interstate toward Houston, they immediately enter the HOV lane (VCC1) and speed toward their destination.

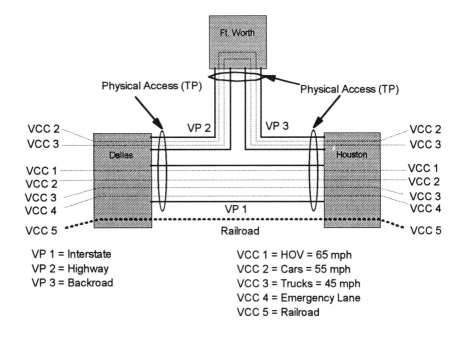

Figure 7.8 Transportation Example — STM versus ATM

Figure 7.8 adds a railroad (VCC5) running from Dallas to Houston along the same interstate route (VP1) in the previous example. Assuming no stops between Dallas and Houston, the railroad maintains the same speed from start to finish, with one railroad train running after another according to a fixed schedule. This is like the Synchronous Transfer Mode (STM) or Time Division Multiplexing (TDM) discussed in Chapter 4. Imagine there are passengers and cargo going between Dallas and Houston, each having to catch scheduled trains. The arriving passengers and cargo shipments originating at Dallas must wait for the next train. Trains travel regardless of whether there is any passenger or cargo present. If there are too many passengers or cargo for the train's capacity, the excess must wait for the next train. If you were a commuter, would you want to rely on the train always having capacity, or would you prefer to have a car and statistically have a better chance of making it to Houston in an even shorter time period using ATM?

Studying this analogy, observe that the private vehicles (and their passengers) traveling over VCC1, VCC2, or VCC3 have much more flexibility (ATM) than trains (STM) in handling the spontaneous needs of travel. The trains are efficient only when the demand is accurately scheduled and very directed, such as during the rush hour between suburbs and the inner city.

Note that the priorities, or choice, of each VCC can vary throughout the day, as can priorities between VPs in ATM. An additional VCC can be configured on a moment's notice (VCC) and assigned a higher priority, as in the case of an ambulance attempting to travel down the median during a traffic jam to get to the scene of an accident.

7.5.2 Transmission Path, Virtual Path, and Virtual Channels

Bringing our last analogy forward into ATM transmission terms, Figure 7.9 depicts graphically the relationship between the physical transmission path, Virtual Path (VP), and Virtual Channel (VC). A transmission path contains one or more virtual paths, while each virtual path contains one or more virtual channels. Thus, multiple virtual channels can be trunked a single virtual path. Switching can be performed on either a transmission path, virtual path, or virtual circuit level.

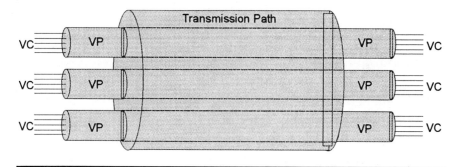

Figure 7.9 Relationship of VC, VP, and Transmission Path

This capability to switch down to a virtual channel level is similar to the operation of a Private or Public Branch eXchange (PBX) or telephone switch in the telephone world. In the PBX/switch, each channel within a trunk group (path) can be switched. Figure 7.10 illustrates this analogy. In the literature devices which perform VC connections are commonly called VC switches because of this analogy with telephone switches. Transmission networks use a cross-connect, which is basically a space division switch, or effectively an electronic patch panel. ATM devices which connect VPs are commonly often called VP cross-connects in the literature by analogy with the transmission network.

These analogies are useful for those familiar with TDM/STM and telephony to understand ATM, but should not be taken literally. There is little reason for an ATM cell switching machine to restrict switching to only VCs and cross-connection to only VPs.

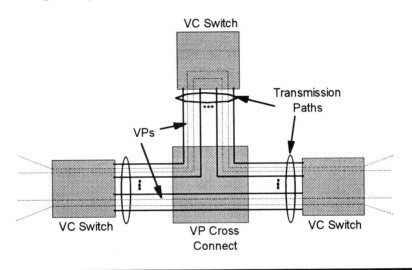

Figure 7.10 Switch and Cross-Connect Analogy

7.5.3 Virtual Path Connections (VPCs) and Virtual Channel Connections (VCCs)

At the ATM layer, users are provided a choice of either a VPC or a VCC, defined as follows:

Virtual Path Connections (VPCs) are switched based upon the Virtual Path Identifier (VPI) value only. The users of the VPC may assign the VCCs within that VPI transparently since they follow the same route.

Virtual Channel Connections (VCCs) are switched upon the combined VPI and Virtual Channel Identifier (VCI) value.

Both VPIs and VCIs are used to route cells through the network. Note that VPI and VCI values must be unique on a specific transmission path (TP). Thus, each TP between two network devices (such as ATM switches) uses VPIs and VCIs independently. This is demonstrated in Figure 7.11. Each switch maps an incoming VPI and VCI to an outgoing VPI and VCI. In this example, switch 1 and switch 2 have a single transmission path (TP) between them. Over this TP there are multiple virtual paths (VPs). At the ATM UNI, the input device to switch 1 provides a video channel over Virtual Path 1 (VPI 1) and Virtual Channel 6 (VCI 6). Switch 1 then assigns the VCI 6 to an outgoing VCI 15, and the incoming VPI 1 to outgoing VPI 12. Thus, on VPI 12 switch 2 specifically operates on virtual channel (VC) number 15 (VCI 15). This channel is then routed from switch 2 to switch 3 over a different path and channel (VPI 16 and VCI 8). Thus, VPIs and VCIs are tied onto each individual link across the network. This is similar to frame relay, where Data Link Connection Identifiers (DLCIs) address a virtual circuit (VC) at each end of a link. Finally, switch 3 translates VPI 16 into VPI 1, and VCI 8 on VP 16 to VCI 6 on VP 1 at the destination UNI. The destination VPI and VCI need not be the same as at the origin. The sequence of VPI/VCI translation across the switches can be viewed as a network address in an extrapolation of the OSI layer 3 model.

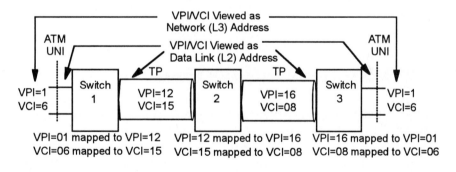

Figure 7.11 Illustration of VPI/VCI Usage on Link and End-to-End Basis

7.6 ATM — ARCHITECTURE, TECHNOLOGY, OR SERVICE?

ATM technology takes on many forms and means many different things to different people, from providing software and hardware multiplexing, switching, and cross-connect functions and platforms, to serving as an economical, integrated network access method, to becoming the core of a network infrastructure, to the much-touted ATM service. Let's now explore each.

7.6.1 As an Interface and Protocol

Asynchronous Transfer Mode (ATM) is defined as an interface and protocol designed to switch variable bit-rate and constant bit-rate traffic over a common transmission medium. The entire B-ISDN protocol stack is often referred to as ATM.

7.6.2 As a Technology

ATM is often referred to as a technology, comprised of hardware and software conforming to ATM protocol standards which can provide a multiplexing, cross-connect, and switching function in a network. ATM technology takes the form of a network interface card, multiplexer, cross-connect, or even a full switch. Chapter 10 covers ATM switching systems, while Chapter 11 covers ATM end systems and applications.

7.6.3 As Economical, Integrated Access

Public ATM service providers offering ATM-based services are now appearing on the scene, enabling users to capitalize on a basic advantage of ATM — integrated access to reduce cost. The development of circuit emulation technology based upon ATM will make this benefit available to users who already have a large number of TDM access lines today. The TDM access lines can be multiplexed onto an E3, DS3, or even SONET access line, leaving large amounts of bandwidth available for ATM applications at little incremental cost.

7.6.4 As an Infrastructure

Where ATM technology can also have an advantage is as the core of a network infrastructure. ATM hardware and associated software together can provide the backbone technology for an advanced communications network. In fact, many experts view an ATM-based architecture as the future platform for data and eventually voice. ATM also provides a very scalable infrastructure, from the campus environment to the central office. Scalability occurs in the dimensions of interface speed, switch size, network size, and addressing.

7.6.5 As a Service

ATM is not a service, but services can be offered over an ATM architecture. The Cell Relay Service (CRS) involves the direct delivery of ATM cells. Other services involve ATM Adaptation Layers, and include private line emulation service as defined using AAL 1, variable-rate video as defined using AAL 2, Switched Multimegabit Data Service as defined using AAL 3/4, and frame relay as one of the service-specific connection-oriented services defined for AAL 5. Chapter 8 explains the above ATM Adaptation Layers in detail.

7.7 REVIEW

The chapter began with definitions of ATM terminology and concepts. Examples were presented throughout the chapter in an attempt to compare ATM concepts to everyday life examples. The presentation then moved on to define the basic building block of ATM — the cell — and the method of constructing cells by assigning VPI and VCI addresses to a header field which prefixed the user payload. With these basics in hand, the chapter progressed through two examples of ATM protocol operation. Next the story moved to a little ATM history, describing the process of standardizing on a 53-Byte cell size and the tradeoffs that were involved. Next the building blocks of ATM networking, the transmission path, the Virtual Path (VP), and the Virtual Channel (VC), were covered. The concept of both Virtual Path Connections (VPCs) and Virtual Channel Connections (VCCs) was then introduced. ATM has many facets, and is referred to as an interface, a technology, integrated access, a network infrastructure, and even a service. Each of these aspects was summarized to set the stage for the rest of the book.

8

Physical, ATM, and AAL Layers

This chapter explores in detail the foundation of the entire ATM-based B-ISDN protocol stack. The three lowest protocol layers are introduced, first defining what functions they perform and then how they interface. It is logical to start at the bottom with the physical (PHY) layer, and then move to the ATM layer, which defines virtual paths and virtual channels, and finish with the ATM Adaptation Layer (AAL). Many of these concepts were introduced in the last chapter and are covered in greater detail in this chapter.

The primary layers of the B-ISDN protocol reference model are: the PHYsical layer, the ATM Layer where the cell structure occurs, and the ATM Adaptation Layer (AAL) that provides support for higher layer services such as circuit emulation, frame relay, and SMDS. The PHY layer corresponds to OSI Reference Model (OSIRM) layer 1, the ATM layer and part of the AAL correspond to OSIRM layer 2, and higher layers correspond to OSI layer 3 and above.

First the description covers the various physical interfaces and media currently defined and specified. A detailed discussion of definitions and concepts of the ATM layer, defining the cell structure for both the User-to-Network Interface (UNI) and the Network Node Interface (NNI), follows. At a lower level, a description of the meanings of all of the cell header fields, payload types, and generic functions that they support is provided. Lastly, the next higher layer in the protocol stack — the ATM Adaptation Layer (AAL) —is covered. An in-depth study of ATM Adaptation Layers (AALs) 1 through 5 relating them to the ITU-T, defined service classes A through D, is also provided.

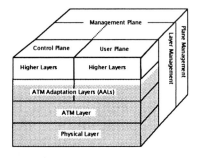

Throughout the remainder of the book you will see figures like the one shown on the left. They depict the B-ISDN protocol model from I.321 introduced in Chapter 6, with a portion of the B-ISDN/ATM protocol model shaded out to represent the subject matter of that particular section. This figure shows what this chapter covers: the physical layer, the ATM layer, and the AALs that are common between the user and control planes.

8.1   THE PLANE-LAYER TRUTH — AN OVERVIEW

If the front and right sides of the B-ISDN protocol cube were unfolded, they would yield a two-dimensional layered model like that shown in Figure 8.1.

Layer Name			Functions Performed	
Higher Layers			Higher Layer Functions	L a y e r M a n a g e m e n t
A A L	Convergence Sublayer (CS)		Common Part (CP)	
			Service Specific (SS)	
	SAR Sublayer		Segmentation And Reassembly'	
ATM			Generic Flow Control Call Header Generation/Extraction Cell VCI/VPI Translation Cell Multiplexing/Demultiplexing	
P h y s i c a l	Transmission Convergence (TC) Sublayer		Cell Rate Decoupling Cell Delineation Transmission Frame Adaptation Transmission Frame Generation/Recovery	
	Physical Medium (PM)		Bit Timing Physical Medium	

Figure 8.1 B-ISDN/ATM Layer and Sublayer Model

Figure 8.1 lists the functions of the four B-ISDN/ATM layers along with the sublayer structure of the AAL and PHYsical (PHY) layer as defined in ITU-T Recommendation I.321. Starting from the bottom, the Physical layer has two sublayers: Transmission Convergence (TC) and Physical Medium (PM). The PM sublayer interfaces with the actual physical medium and passes the recovered bit stream to the TC sublayer. The TC sublayer extracts and inserts ATM cells within the Plesiochronous or Synchronous (PDH or SDH) Time Division Multiplexed (TDM) frame and passes these to and from the ATM layer, respectively. The ATM layer performs multiplexing, switching, and control actions based upon information in the ATM cell header and passes cells to, and accepts cells from, the ATM Adaptation Layer (AAL). The AAL has two sublayers: Segmentation And Reassembly (SAR) and Convergence Sublayer (CS). The CS is further broken down into Common Part (CP) and Service-Specific (SS) components. The AAL passes Protocol Data Units (PDUs) to and accepts PDUs from higher layers. PDUs may be of variable length, or may be of fixed length different from the ATM cell length.

The Physical layer corresponds to layer 1 in the OSI model. The ATM layer and AAL correspond to parts of OSI layer 2, but the address field of the ATM cell header has a network-wide connotation that is like OSI layer 3. A precise alignment with the OSI layers is not necessary, however. The B-ISDN and ATM protocols and interfaces make extensive use of the OSI concepts of layering and sublayering as we shall see. Figure 8.2 illustrates the mapping of the B-ISDN layers to the OSI layers and the sublayers of the PHY, ATM, and ATM Adaptation layers that we describe in detail later.

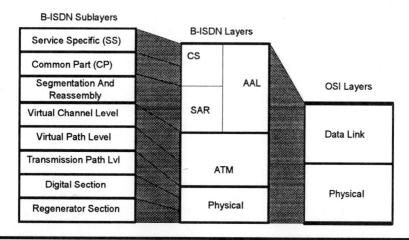

Figure 8.2 B-ISDN Layers and Sublayers and OSI Layers

It is interesting to look at the number of instances of defined standardized protocols or interfaces that exist for each layer, and whether their target implementation is in hardware or software. Figure 8.3 depicts the number of instances at each layer by boxes with the arrows on the right hand side showing how the layers are either more hardware- or software-intensive. The arrows illustrate the fact that ATM implementations move from being hardware-intensive at the lower layers (PHY and ATM layer) to software-intensive at the higher layers (AALs and higher layers). This shows how ATM is the pivotal protocol, for which there is only one instance, for a potentially large number of physical media, several AALs, and an ever-expanding set of higher layer functions. The inverted pyramid on the left-hand side of Figure 8.3 illustrates this concept. In other words, ATM allows machines with different physical interfaces to transport data independently of the higher layer protocols using a common, well-defined protocol amenable to high performance and cost-effective hardware implementation.

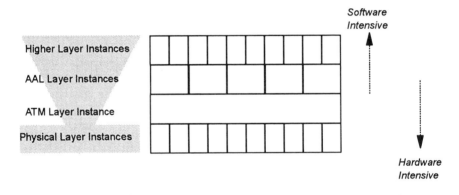

Figure 8.3 ATM Protocol Model Hardware to Software Progression

Now the journey begins up through the layers of the B-ISDN/ATM protocol model, starting with the physical layer.

8.2   PHYSICAL (PHY) LAYER

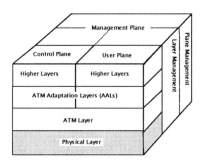

This section covers the key aspects of the PHYsical (PHY) Layer as they relate to the remainder of the book. The PHY Layer provides for transmission of ATM cells over a physical medium that connects two ATM devices. The PHY Layer is divided into two sublayers: the Physical Medium Dependent (PMD) sublayer and the Transmission Convergence (TC) sublayer. The TC sublayer transforms the flow of cells into a steady flow of bits and bytes for transmission over the physical medium. The PMD sublayer provides for the actual transmission of the bits in the ATM cells.

8.2.1   Physical Medium Dependent (PMD) Sublayer

The PMD sublayer provides for the actual clocking of bit transmission over the physical medium. There are three standards bodies that have defined the physical layer in support of ATM: ANSI, CCITT/ITU-T, and the ATM Forum. We summarize each of the standardized interfaces in terms of the interface clocking speed and physical medium below.

8.2.1.1   ANSI Standards

ANSI standard T1.624 currently defines three single-mode optical ATM SONET-based interfaces for the ATM UNI:

- ✎ STS-1 at 51.84 Mbps
- ✎ STS-3c at 155.52 Mbps
- ✎ STS-12c at 622.08 Mbps

ANSI T1.624 also defines operation at the DS3 rate of 44.736 Mbps using the Physical Layer Convergence Protocol (PLCP) from the 802.6 Distributed Queue Dual Bus (DQDB) standard.

8.2.1.2 CCITT/ITU-T SDH Recommendations

CCITT/ITU-T recommendation I.432 defines two optical Synchronous Digital Hierarchy (SDH)-based physical interfaces for ATM which correspond to the ANSI rates mentioned in the last section. These are:

- ❖ STM-1 at 155.520 Mbps
- ❖ STM-4 at 622.08 Mbps

Since the transport rates (and the payload rates) of the SDH STM-1 and STM-4 correspond exactly to the SONET STS-3c and STS-12c rates, interworking should be simplified. ITU-T standardizes additional electrical, physical interface rates of the following type and speeds:

- ▷ DS1 at 1.544 Mbps
- ▷ E1 at 2.048 Mbps
- ▷ DS2 at 6.312 Mbps
- ▷ E3 at 34.368 Mbps
- ▷ DS3 at 44.736 Mbps using PLCP
- ▷ E4 at 139.264 Mbps

8.2.1.3 ATM Forum Interfaces

The ATM Forum has defined four physical layer interface rates. Two of these are interface rates intended for public networks and are the DS3 and STS-3c standardized by ANSI and the ITU-T. The SONET STS-3c interface may be supported on an OC-3, either single-mode or multimode fiber. The following three interface rates and media are for private network application:

- ⌘ FDDI-based at 100 Mbps
- ⌘ Fiber Channel-based at 155.52 Mbps
- ⌘ Shielded Twisted Pair (STP) at 155.52 Mbps

The FDDI-based PMD and fiber channel interfaces both use multimode fiber, while the STP interface uses type 1 and 2 cable as specified by EIA/TIA 568. The ATM Forum is specifying ATM cell transmission over common building wiring, called Unshielded Twisted Pair (UTP) types 3 and 5.

8.2.2 Transmission Convergence (TC) Sublayer

The TC sublayer converts between the bit stream clocked to the physical medium and ATM cells. On transmit, TC basically maps the cells into the Time Division Multiplexing (TDM) frame format. On reception, it must delineate the individual cells in the received bit stream, either from the TDM frame directly, or via the Header Error Check (HEC) in the ATM cell header. Generating the HEC on transmit and using it to correct and detect errors on receive are also important TC functions. Another important function that TC performs is cell rate decoupling by sending idle cells when the ATM layer has not provided a cell. This is a critical function that allows the ATM layer to operate with a wide range of different speed physical interfaces.

We cover two examples of TC mapping of ATM cells: direct mapping to a SONET payload, and the PLCP mapping to a DS3. We then cover the use of the Header Error Check (HEC) and why it is so important. We then complete our description of the TC sublayer with an illustration of cell rate decoupling using unassigned cells.

8.2.3 Examples of TC Mapping

In this section we give an example of direct and Physical Layer Convergence Protocol (PLCP) mapping by the Transmission Convergence (TC) sublayer of the physical layer.

8.2.3.1 SONET STS-3c Direct Mapping

The SONET mapping is performed directly into the SONET STS-3c (155.52 Mbps) Synchronous Payload Envelope (SPE) as defined in Chapter 3 and as shown in Figure 8.4. ATM cells fill in the STS-3c payload continuously since an integer number of 53-octet cells do not fit in an STS-3c frame. This results in better efficiency than carriage of M13-mapped DS3s, or even VT1.5 multiplexing over SONET. The Data Communications Channel (DCC) overhead is not used on the User-Network Interface (UNI). The ATM layer uses the HEC field to delineate cells from within the SONET payload. The cell transfer rate is 149.760 Mbps. The mapping over STS-12c is very similar in nature. The difference between SONET and SDH is in the TDM overhead bytes.

8.2.3.2 DS3 PLCP Mapping

Figure 8.5 illustrates the DS3 mapping using the Physical Layer Convergence Protocol (PLCP) defined in IEEE 802.6. The ATM cells are enclosed in a 125 μs frame defined by the PLCP, which is defined inside the standard DS3 M-frame. The PLCP mapping transfers 8 KHz timing across the DS3 interface which is somewhat inefficient in that the cell transfer rate is only 40.704 Mbps, which utilizes only about 90% of the DS3's approximately 44.21-Mbps payload rate. Note that the BIP-8 indicator is computed over the POH and associated ATM cells of the previous PLCP frame.

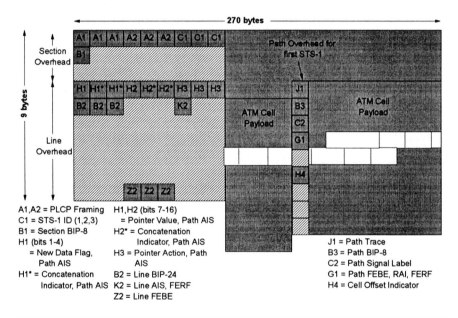

Figure 8.4 B-ISDN UNI Physical Layer — STS-3c

8.2.4 TC Header Error Check (HEC) Functions

The Header Error Check (HEC) is a 1 byte code applied to the 5 byte ATM cell header. The HEC code is capable of correcting any single-bit error in the header. It is also capable of detecting many patterns of multiple-bit errors. The TC sublayer generates HEC on transmit and uses it to determine if the received header has any errors. If errors are detected in the header, then the received cell is discarded. Since the header tells the ATM layer what to do with the cell, it is very important that it not have errors; otherwise it might be delivered to the wrong user, or an undesired function in the ATM layer may be inadvertently invoked.

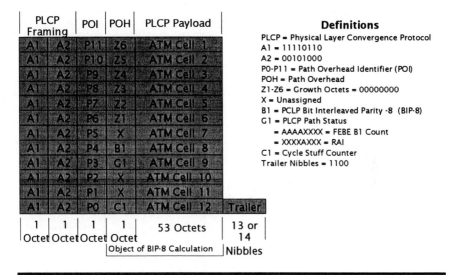

PLCP Framing			POI	POH	PLCP Payload	
A1	A2	P11	Z6	ATM Cell 1	
A1	A2	P10	Z5	ATM Cell 2	
A1	A2	P9	Z4	ATM Cell 3	
A1	A2	P8	Z3	ATM Cell 4	
A1	A2	P7	Z2	ATM Cell 5	
A1	A2	P6	Z1	ATM Cell 6	
A1	A2	P5	X	ATM Cell 7	
A1	A2	P4	B1	ATM Cell 8	
A1	A2	P3	G1	ATM Cell 9	
A1	A2	P2	X	ATM Cell 10	
A1	A2	P1	X	ATM Cell 11	
A1	A2	P0	C1	ATM Cell 12	Trailer

1 Octet | 1 Octet | 1 Octet | 1 Octet | 53 Octets | 13 or 14 Nibbles

Object of BIP-8 Calculation

Definitions

PLCP = Physical Layer Convergence Protocol
A1 = 11110110
A2 = 00101000
P0-P11 = Path Overhead Identifier (POI)
POH = Path Overhead
Z1-Z6 = Growth Octets = 00000000
X = Unassigned
B1 = PCLP Bit Interleaved Parity -8 (BIP-8)
C1 = PLCP Path Status
   = AAAAXXXX = FEBE B1 Count
   = XXXXAXXX = RAI
C1 = Cycle Stuff Counter
Trailer Nibbles = 1100

Figure 8.5 B-ISDN UNI Physical Layer — DS3

The TC also uses HEC to locate cells when they are directly mapped into a TDM payload. The HEC will not match random data in the cell payloads when the 5 bytes that are being checked are not part of the header. Thus, it can be used to find cells in a received bit stream. Once several cell headers have been located through the use of HEC, then TC knows to expect the next cell 53 bytes later. This process is called *HEC-based cell delineation* in standards.

8.2.5 TC Cell Rate Decoupling

The TC sublayer performs a cell rate decoupling, or speed matching function, as well. Physical media that have synchronous cell time slots (e.g., DS3, SONET, SDH, STP, and the Fiber Channel-based method) require this function, while asynchronous media such as the FDDI PMD do not. As we shall see in the next section, there are special codings of the ATM cell header that indicate that a cell is either *unassigned* or *idle*. All other cells are *assigned* which correspond to the cells generated by the ATM layer. Figure 8.6 illustrates this operation between a transmitting device and a receiving ATM device. The transmitter multiplexes multiple VPI/VCI cell streams, queueing them if an ATM slot is not immediately available. If the queue is empty when the time arrives to fill the next synchronous cell time slot, then the TC sublayer inserts an unassigned or idle cell. The receiver extracts unassigned or idle cells and distributes the other, assigned cells to the destinations.

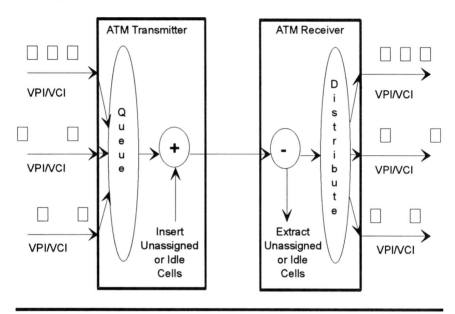

Figure 8.6 Cell Rate Decoupling Using Unassigned Cells

ITU-T Recommendation I.321 places this function in the TC sublayer of the PHY layer and uses idle cells, while the ATM Forum places it in the ATM layer and uses unassigned cells. This presents a potential low-level incompatibility if different systems use different cell types for cell rate decoupling. Look for ATM systems that support both methods to ensure maximum interoperability. The ITU-T model views the ATM layer as independent of whether or not the physical medium has synchronous time slots.

8.3   ATM LAYER — PROTOCOL MODEL

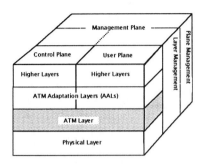

This section moves to the focal point of B-ISDN, the Asynchronous Transfer Mode (ATM) Layer. First the relationship of the ATM layer to the physical layer and its division into a Virtual Path (VP) and Virtual Channel (VC) level are covered in detail. This is a key concept, which is the reason Chapter 7 presented several analogies. Several examples are provided in this section portraying the role of end and intermediate systems in a real-world setting rather than just a formal model. This is accomplished by explaining how the ATM layer VP and VC functions are used in intermediate and end systems in terms of the layered protocol model. An example is then provided showing how intermediate systems perform ATM VP or VC switching or cross-connection, and how end systems pass cells to the ATM Adaptation Layer (AAL).

8.3.1   Physical Links and ATM Virtual Paths and Channels

A key concept is the construction of ATM Virtual Paths (VPs) and Virtual Channels (VCs). Figure 8.7 illustrates this derivation based on ITU-T Recommendation I.311. The physical layer is composed of three levels: regenerator section, digital section, and transmission path as shown in the figure. At the ATM layer we are only concerned about the transmission path because this is essentially the TDM payload that connects ATM devices. Generically, an ATM device may be either an endpoint or a connecting point for a VP or VC. A Virtual Path Connection (VPC) or a Virtual Channel Connection (VCC) exists only between endpoints as shown in the figure. A VP link or a VC link can exist between an endpoint and a connecting point or between connecting points. A VPC or VCC is an ordered list of VP or VC links, respectively.

8.3.1.1   VC level

The *Virtual Channel Identifier (VCI)* in the cell header identifies a single VC on a particular Virtual Path (VP). Switching at a VC connecting point is done based upon the combination of virtual path and VCI. A *VC link* is defined as a unidirectional flow of ATM cells with the same VCI between a VC connecting point and either a VC endpoint or another VC connecting point. *A Virtual Channel Connection (VCC)* is defined as a concatenated list of VC links. A VCC defines a unidirectional flow of ATM cells from one user to one or more other users.

A network must preserve cell sequence integrity for a VCC; that is, the cells must be delivered in the same order in which they were sent. A Quality of Service (QoS) is associated with a VPC. Chapter 12 defines QoS in detail.

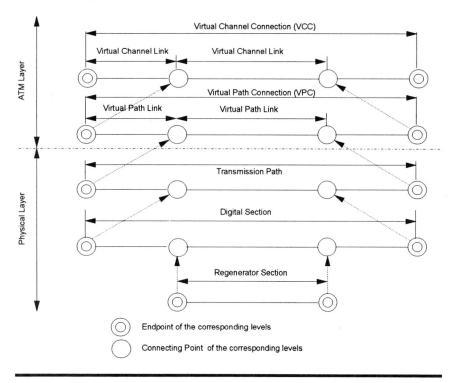

Figure 8.7 Physical Layer, Virtual Paths, and Virtual Channels

8.3.1.2 VP Level

Virtual Paths (VPs) define an aggregate bundle of VCs between VP endpoints. A *Virtual Path Identifier (VPI)* in the cell header identifies a bundle of one or more VCs. A VP link provides unidirectional transfer of cells with the same VPI between VP endpoints or connecting points. Switching at a VP connecting point is done based upon the VPI — the VCI is ignored. A VP link is defined as a VP between a VP connecting point and either a VP endpoint or another VP connecting point. A Virtual Path Connection (VPC) is defined as a concatenated list of VP links. A VPC defines a unidirectional flow of ATM cells from one user to one or more other users.

Standards do not require a network to preserve cell sequence integrity for a VPC; however, the cell sequence integrity requirement of a VCC still applies. A Quality of Service (QoS) is associated with a VPC. If a VPC contains VCCs in different QoS classes, then the VPC assumes the QoS of the VCC with the highest quality.

8.3.2 Intermediate Systems (IS) and End Systems (ES)

As a more concrete illustration of connecting and endpoint functions, Figure 8.8 looks inside an Intermediate System (IS) and End System (ES) with reference to the ANSI T1.627-defined protocol model. The OSI terms Intermediate System (IS) and End System (ES) and this interpretation were adapted from ANSI T1.627. This example depicts an intermediate network node (IS) connecting two ATM CPE devices (ESs). The ATM Adaptation Layer in End System 1 (ES1) generates cell payloads for a number of connection endpoints at the boundary between the AAL and the ATM layers, which is called the ATM Service Access Point (SAP). The ATM entity in ES1 multiplexes these virtual connections and passes ATM cells to the PHYsical (PHY) layer at the interface labeled PHY-SAP. The bit stream is passed between the PHY layer in ES1 and the Intermediate System (IS) over the physical interface. The IS demultiplexes the ATM connections and applies each to a connecting point in the IS ATM entity. This action is not specified in the protocol model, which is the reason we show it by dashed lines. The connecting point in the ATM entity translates the VPI and/or the VCI depending on whether it is a VP or VC connecting point, determines the outgoing physical interface, and performs other ATM layer functions that we will define later. The IS ATM entity then multiplexes these onto the outgoing PHY-SAP for transfer to the destination — End System 2. The PHY layer in ES2 delivers these to its ATM entity via the PHY-SAP. The ES2 ATM entity demultiplexes the cells and delivers these to the endpoint of the corresponding VP or VC via the ATM-SAP.

The ATM layer requires that cell sequence integrity be preserved. This means that cells are delivered to intermediate connecting points and the destination endpoint in the same order in which they were transmitted.

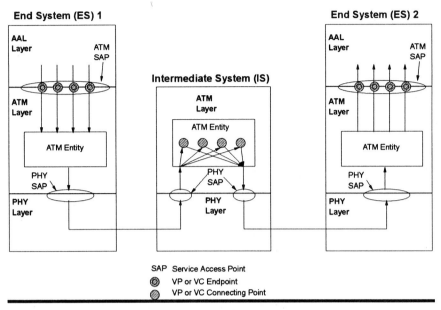

Figure 8.8 Intermediate System (IS)/End System (ES) VP/VC Functions

8.3.3   VP and VC Switching and Cross-Connection

This section provides a specific example of VP and VC endpoints and connecting points in intermediate and end systems for VP links, VPCs, VC links, and VCCs. Figure 8.9 depicts two end systems (or CPE) and an intermediate system (or switch). The endpoint and connecting points are shown using the terminology from Figure 8.7. The transmission path, virtual path, and virtual channel are shown as a nested set of *pipes* using the convention introduced in Chapter 7 from ITU-T Recommendation I.311. The transmission path PHY layer carries Virtual Paths (VPs) and Virtual Channels (VCs). These may be uni-directional, or bi-directional. Our example shows end systems (or CPE) that have both VP and VC endpoints. The left-hand-side end system, or CPE, originates a VP with VPI x and two VCs with VCI a and c.

The intermediate system contains VP and VC switching functions as shown in Figure 8.9. The intermediate system VP switching function translates the VPI from x to y since VPI x is already in use on the physical interface to the destination end system. All of the VCIs within VPI x are automatically connected to VPI y. This simultaneous switching of a large number of VCs within a VP is the principal reason for the standardization of VPs. If only a single level of addressing were used this function would not be possible. VC switching operates within VPs as illustrated by the other VC connection The VC switching function translates the received VCI a to an outgoing VCI b on VPI x for delivery to the destination. VCI c from VPI y is switched to some other destination. Similarly VCI a is extracted from another physical interface and/or VPI on the switch and placed in VPI x for delivery to the end system.

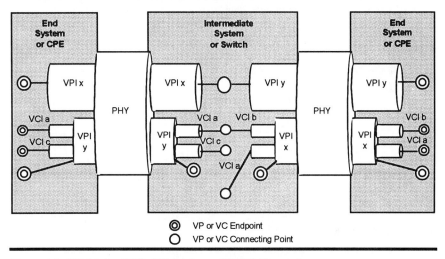

Figure 8.9 VP Link, VPC, VC Link, and VCC Example

8.4 ATM LAYER AND CELL — DEFINITION

Now for a detailed look inside the ATM cell header and the meaning of each field. The User-Network and Network-Network Interfaces (UNI and NNI) are defined first, followed by a summary of the ITU-T Recommendation I.361, ANSI, and ATM Forum definition of the cell structure at the ATM UNI and NNI. The basic functions of the ATM layer are then introduced, and each function is described in detail.

8.4.1 ATM UNI and NNI Defined

Figure 8.10 defines the ATM cell structure at the User-Network Interface (UNI) and the Network Node Interface (NNI). The ATM UNI occurs between the user equipment or End System (ES), or Broadband Terminal Equipment (B-TE), and either the Terminal Adapter (TA) or Network Termination (NT), or Intermediate System (IS).

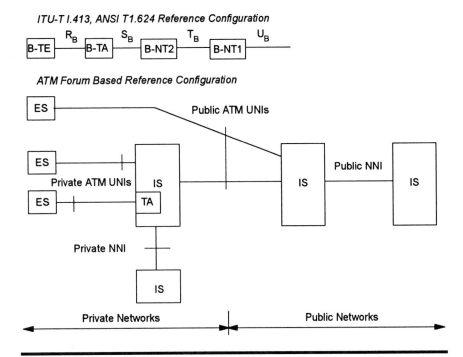

Figure 8.10 ATM UNI and NNI Reference Configuration

The ATM Forum terminology of private and public UNIs is mapped to the ITU-T reference point terminology in this figure. The ATM UNI may be a private ATM UNI, which would occur at the R or S reference points in ITU-T Recommendation I.413 and ANSI T1.624, or a public ATM UNI, which would occur at reference points T or U as shown in the figure. The Network Node Interface (NNI) defined in ITU-T Recommendation I.113 is normally thought of as the standard interface between networks, which will most likely also be the interface used between nodes within a network. The ATM

Forum distinguishes between an NNI used for private networks and public networks as shown in the figure,

8.4.2 ATM Cell Structure at the UNI and NNI

Two standardized coding schemes exist for cell structure: the User-to-Network Interface (UNI) and the Network-Node, or Network-to-Network Interface (NNI). The UNI is the interface between the user [or customer premises equipment (CPE)] and the network switch. The NNI is the interface between switches or between networks. UNI and NNI coding schemes are introduced and each field is defined in this section. ITU-T Recommendation I.361 is the basis of these definitions, with further clarifications given in ANSI T1.627 and the ATM Forum UNI and Broadband Inter-Carrier Interface (B-ICI) specifications.

8.4.2.1 ATM UNI Cell Structure

Figure 8.11 illustrates the format of the 53-byte ATM cell at the User-Network Interface (UNI). The cell header contains a logical address in two parts: an 8 bit Virtual Path Identifier (VPI) and a 16-bit Virtual Channel Identifier (VCI). The cell header also contains a 4-bit Generic Flow Control (GFC), 3-bit Payload Type (PT), and a 1-bit Cell Loss Priority (CLP) indicator. The entire header is error-protected by a 1-byte Header Error Check (HEC) field. This section details the meaning of each header field. A fundamental concept of ATM is that switching occurs based upon the VPI/VCI fields of *each* cell. Switching done on the VPI only is called a Virtual Path Connection (VPC), while switching done on both the VPI/VCI values is called a Virtual Channel Connection (VCC). VPCs/VCCs may be either provisioned as Permanent Virtual Circuits (PVCs), or established via signaling protocols as Switched Virtual Circuits (SVCs). SVCs involve the control plane for the UNI, as covered in Chapter 9.

8.4.2.2 ATM NNI Cell Structure

Figure 8.12 illustrates the format of the 53-byte ATM cell at the Network Node Interface (NNI). The format is identical to the UNI format with two exceptions. First, there is no Generic Flow Control (GFC) field. Secondly, the NNI uses the 4 bits used for the GFC at the UNI to increase the VPI field to 12 bits at the NNI as compared to 8 bits at the UNI. SVCs involve the control plane for the NNI, as covered in Chapter 9.

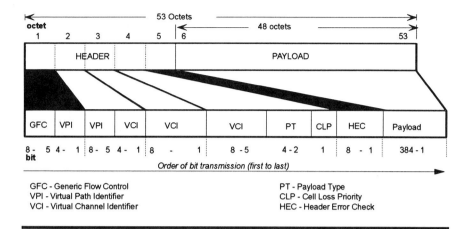

Figure 8.11 ATM User-Network Interface (UNI) Cell Structure

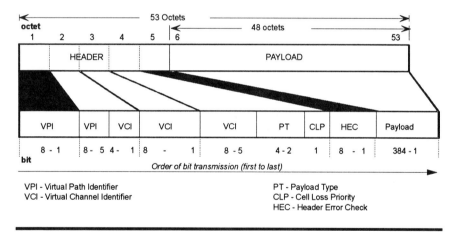

Figure 8.12 ATM Network Node Interface (NNI) Cell Structure

8.4.2.3 Definition of ATM Cell Header Fields

This section provides a description of each header field.

Generic Flow Control (GFC) is a 4-bit field intended to support simple implementations of multiplexing. In the early 1990s, GFC was being specified to implement a DQDB-like, multiple-access-type protocol. However, it appears unlikely that this type of GFC will be standardized. The current standards define the *uncontrolled* mode, where the 4-bit GFC field is always coded as zeroes. If too many non-zero GFC values are received, layer management should be notified. Chapter 12 covers the status of GFC standardization.

Cell Loss Priority (CLP) is a 1-bit field that indicates the loss priority of an individual cell.

Payload Type (PT) is a 3-bit field that discriminates between a cell payload carrying user information or one carrying management information. A later section details the coding of the payload type field.

The Header Error Control (HEC) field provides error checking of the header for use by the Transmission Convergence (TC) sublayer of the PHYsical layer as defined earlier in this chapter.

8.4.3 ATM Layer Functions

This section details the key functions of the ATM layer. The ATM Layer provides many functions, including:

- Cell Construction
- Cell Reception and Header Validation
- Cell Relaying, Forwarding, and Copying Using the VPI/VCI
- Cell Multiplexing and Demultiplexing Using the VPI/VCI
- Cell Payload Type Discrimination
- Interpretation of Pre-defined Reserved Header Values
- Cell Loss Priority Processing
- Support for Multiple QoS Classes
- Usage Parameter Control (UPC)
- Explicit Forward Congestion Indication (EFCI)
- Generic Flow Control
- Connection Assignment and Removal

Cell construction, reception and header validation, and several examples of relaying, forwarding, and copying were already covered in earlier examples. Subsequent sections cover descriptions of payload type discrimination, interpretation of predefined header values, and cell loss priority processing. Part 5 covers the key topics of support for multiple QoS classes, UPC, EFCI, GFC, and connection assignment and removal.

8.4.4 Relaying and Multiplexing Using the VPI/VCI

As shown through several earlier examples, the heart of ATM is in the use of the VPI and VCI for relaying or switching. ATM also effectively performs multiplexing and demultiplexing of multiple logical connections with different quality requirements using the fixed-length ATM cell.

The number of bits allocated in the ATM cell header limit each physical UNI to support of no more than $2^8 = 256$ virtual paths and each physical NNI to support of no more than $2^{12} = 4096$ virtual paths. Each virtual path can support no more than $2^{16} = 65,536$ virtual channels on the UNI or the NNI.

Although the UNI and NNI cell formats specify 8 and 12 bits for the VPI, respectively, and 16 bits for the VCI on both interfaces, an implementation need only support a smaller number of the lower order bits in the VPI and VCI. Thus, a real ATM application may differ markedly from the above maximums. This means that the number of virtual paths and virtual channels actually supported in a live ATM network may be far less than the maximum numbers defined above. This has important implications in interoperability if one ATM device expects the next ATM device to operate on VPI/VCI bits, but that device ignores these bits. One way to handle this is to allow each system to query the other about the number of bits that are supported. This function is supported in the ATM Forum Interim Local Management Interface (ILMI) and presented in further detail in Chapter 19.

8.4.5 Meaning of Preassigned Reserved Header Values

A key function of the ATM Layer is the identification of preassigned, reserved header values. Figure 8.13 shows the preassigned (also called pre-defined) header field values for the UNI. The 4-bit GFC field can be used with all of these values. The ITU-T has reserved the first 16 VCIs for future assignment as preassigned, reserved header value functions. Other portions of the book cover the use of these specific header values as indicated below.

Usage	VPI*	VCI	PT	CLP
Unassigned Cell	00000000	00000000 00000000	XXX	0
Idle Cell *	00000000	00000000 00000000	000	1
Reserved for Physical layer *	00000000	00000000 00000000	PPP	1
Meta-signalling (I.311)	XXXXXXXX	00000000 00000001	0A0	C
General broadcast signaling	XXXXXXXX	00000000 00000010	0AA	C
Point-point signaling	XXXXXXXX	00000000 00000101	0AA	C
Segment OAM F4 Flow Cell	YYYYYYYY	00000000 00000011	0A0	A
End-to-end OAM F4 Flow Cell	YYYYYYYY	00000000 00000100	0A0	A
Segment OAM F5 Flow Cell	YYYYYYYY	ZZZZZZZZ ZZZZZZZZ	100	A
End-to-End OAM F5 Flow Cell	YYYYYYYY	ZZZZZZZZ ZZZZZZZZ	101	A
Resource Management Cell	YYYYYYYY	ZZZZZZZZ ZZZZZZZZ	110	A

* Defined as invalid	X = "Don't Care"	A = Use by appropriate function
pattern by ATM Forum	Y = Any VPI value	C = Originator set CLP
	Z = Any non-zero VCI	P = Reserved for PHY Layer

Figure 8.13 Preassigned, Reserved Header Values

The physical layer usage of ATM cells is still in the process of standardization. Metasignaling, general broadcast, and point-to-point signaling are defined in Chapter 9. The OAM cell flows are detailed in Chapters 20 and 21. The current proposals for use of the resource management cell are described in Chapter 14. The use of the unassigned and idle cell types was described earlier in this chapter. The NNI has an additional 4 bits in the VPI field. The NNI preassigned, reserved header fields have not been completely standardized. The current version of the ATM Forum B-ICI specification only requires that the F4 OAM flows, point-to-point signaling, invalid patterns, and unassigned cells be supported.

8.4.6   Meaning of the Payload Type (PT) Field

Figure 8.14 depicts Payload Type (PT) encoding. We see the first bit is an AAL indication bit (currently used by AAL5 to identify the last cell), the second bit indicates upstream congestion, and the third bit discriminates between data and operations cells. Payload types carrying user information may also indicate whether congestion was experienced by Explicit Forward Congestion Indication (EFCI) or whether the cell contains an indication to the AAL protocol. The management information payload type indicates whether the cell is either a segment or end-to-end Operations Administration and

Maintenance (OAM) cell for a VCC or a Fast Resource Management (FRM) cell.

Use of EFCI and resource management cells is covered in Chapter 14. Use of the AAL_indicate PT is covered in the next section of this chapter. Chapters 20 and 21 detail OAM cell usage.

PT Coding	PT Coding
000	User Data Cell, EFCI = 0, AAL_indicate = 0
001	User Data Cell, EFCI = 0, AAL_indicate = 1
010	User Data Cell, EFCI = 1, AAL_indicate = 0
011	User Data Cell, EFCI = 1, AAL_indicate = 1
100	OAM F5 segment associated cell
101	OAM F5 end-to-end associated cell
110	Resource management cell
111	Reserved for future functions

EFCN = Explicit Forward Congestion Notification
AAL_indicate = ATM-layer-user-to-ATM-layer-user indication

Figure 8.14 Payload Type (PT) Encoding

8.4.7 Meaning of the Cell Loss Priority (CLP) Field

A value of 0 in the Call Loss Priority (CLP) field means that the cell is of the highest priority — or in other words, it is the least likely to be discarded. A value of 1 in the CLP field means that this cell has low priority — or in other words, it may be selectively discarded during congested intervals in order to maintain a low loss rate for the high-priority CLP=0 cells. The value of CLP may be set by the user or by the network as a result of a policing action. In Part 5, the CLP bit, its uses, and its implications are covered in great detail.

8.5   ATM ADAPTATION LAYER (AAL) — PROTOCOL MODEL

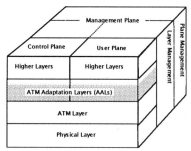

CCITT Recommendations I.362 and I.363 define the next layer of the ATM/B-ISDN protocol stack, the ATM Adaptation Layer (AAL). AAL service class attributes and example applications will be covered first, followed by the generic AAL protocol model. The Common Part (CP) AALs format and protocol are desccribed in detail with an example of each.

8.5.1   The AAL Protocol Structure Defined

The B-ISDN protocol model adapts the services provided by the ATM Layer to those required by the higher layers through the ATM Adaptation Layer (AAL). Figure 8.15 depicts the structure and logical interfaces of the AAL. Services provided to higher layers by an AAL Service Access Point (SAP) are shown at the top of the figure, across which primitives regarding the AAL Protocol Data Units (AAL-PDUs) are passed. The AAL is divided into the Convergence Sublayer (CS) and the Segmentation And Reassembly (SAR) sublayer. The CS sublayer is further subdivided into Service Specific (SS) and Common Part (CP) components. The SSCS sublayer may be null, which means that it need not be implemented. The CPCS must always be implemented along with the Segmentation And Reassembly (SAR) sublayer. These layers pass primitives regarding their respective PDUs between them as labeled in the figure, resulting in the passing of SAR-PDU primitives (which is the ATM cell payload) to and from the ATM layer via the ATM-SAP.

This protocol model may seem somewhat abstract now; however, specific examples clarifying these concepts will soon be provided. This chapter provides an explanation of the Common Part CS (CPCS) and SAR model, leaving the next chapter to give specific examples of the SSCS for the control plane that map this model to specific message formats. It is nearly impossible to generalize the CS and SAR functions since, as we shall see, there are significant differences between every AAL.

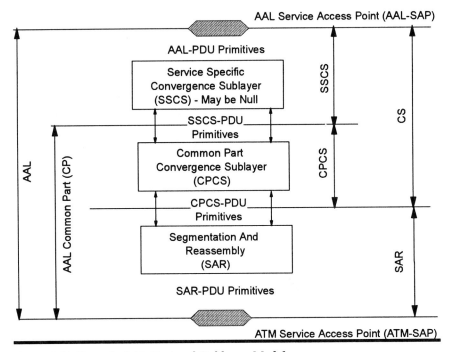

Figure 8.15 Generic AAL Protocol Sublayer Model

The protocol primitives will not be covered in detail in this book. Instead, their actions resulting in the transfer of PDUs either between sublayers or across a SAP will be viewed. Standards use the primitives: request, indicate, response, and confirm. Refer to the definition in Chapter 3 of the OSI layering principles for a further explanation and references on this subject.

8.5.2   AAL Service Attributes Classified

ITU-T Recommendation I.362 defines the basic principles and classification of AAL functions. The attributes of the service class are the timing relationship required between the source and destination, whether the bit rate is constant or variable, and the connection mode is connection-oriented or connectionless. Figure 8.16 depicts the four currently defined AAL service classes, labeled A through D, summarized as follows:

📖 Class A — constant bit-rate (CBR) service with end-to-end timing, connection-oriented

📖 Class B — variable bit-rate (VBR) service with end-to-end timing, connection-oriented

📖 Class C — variable bit-rate (VBR) service with no timing required, connection-oriented

📖 Class D — variable bit-rate (VBR) service with no timing required, connectionless

Attribute	Service Class			
	Class A	Class B	Class C	Class D
Timing relation between source and destination	Required		Not Required	
Bit Rate	Constant	Variable		
Connection Mode	Connection-Oriented			Connection-less
AAL(s)	AAL1	AAL2	AAL3/4 or AAL5	AAL3/4 or AAL5
Example(s)	DS1, E1, nx64 kbps emulation	Packet Video, Audio	Frame Relay, X.25	IP, SMDS

Figure 8.16 ATM ITU ATM/B-ISDN Service Classes

The mapping of service classes to AALs is only partially complete in the standards. The next section indicates the AAL(s) that can support the attributes of the defined AAL service class and also gives several application examples for each service class and AAL.

8.6   ATM ADAPTATION LAYER (AAL) — DEFINITION

AAL1 through AAL4 were initially defined by the CCITT to directly map to the AAL service classes A through D.   ITU-T Recommendation I.363 states the standards for the AALs. AAL1 has been defined by the ITU-T and further clarified in the ANSI T1.630 standard for Continuous Bit Rate (CBR) applications. The history of AAL development for Variable Bit Rate (VBR) services is interesting. Initially, AAL3 was being developed for connection-oriented services and AAL4 for connectionless services. As the details were being defined, it was realized that AAL3 and AAL4 were common enough in structure and function that they were combined into a single class called AAL3/4. A newcomer, AAL5, was conceived by the computer industry in response to perceived complexity and implementation difficulties in the AAL3/4 AAL, which had become aligned with the IEEE 802.6 Layer 2 PDU (L2_PDU). Initially, AAL5 was named the Simple Efficient Adaptation Layer (SEAL) for this reason. AAL5 was adopted by the ATM Forum, ANSI, and the CCITT in a relatively short time frame compared to the usual standards process and has become the predominant AAL of choice in a great deal of data communications equipment. AAL5 is currently standardized for the transport of signaling messages and frame relay and is described in greater detail in Chapters 9 and 17. AAL3/4 will likely be chosen for the support of SMDS since it is essentially identical to the IEEE 802.6 L2_PDU.

We describe the Common Part Convergence Sublayer (CPCS) and Segmentation And Reassembly (SAR) sublayer for each of the currently standardized Common Part (CP) AALs:

* AAL1 — Constant Bit-Rate (CBR) traffic
* AAL3/4 — Variable Bit-Rate (VBR) traffic
* AAL5 — Lightweight Variable Bit-Rate (VBR) traffic

8.6.1 AAL1

AAL1 specifies how TDM-type circuits can be emulated over an ATM network. Circuit emulation is specified in detail for DS1, DS3, and nxDS0 support in ANSI T1.630. AAL1 supports circuit emulation in one of two modes: the Synchronous Residual Time Stamp (SRTS) or Structured Data Transfer (SDT) method. The SRTS method supports transfer of a DS1 or DS3 digital stream, including timing. SDT supports an octet-structured nxDS0 service.

Figure 8.17 depicts the CPCS for AAL1 in support of SRTS and SDT.

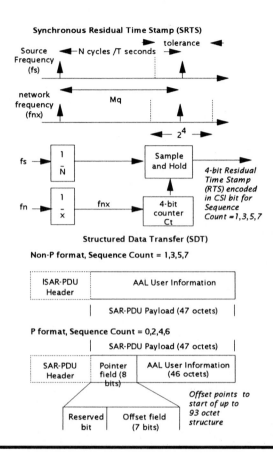

Figure 8.17 AAL1 Common Part Convergence Sublayer (CPCS)

A key concept in SRTS is that both the origin and destination have a very accurate frequency clock of frequency *fn*. The signal (e.g., DS1) has a service clock frequency *fs* with the objective being to pass sufficient information via the AAL so that the destination can reproduce this clock frequency with a high degree of accuracy. The method standardized for doing this is illustrated in the top part of Figure 8.17. The network reference clock *fn* is divided by *x* such that $1 \leq fnx/fs \leq 2$. The source clock is divided by *N* as shown in the figure to sample the 4-bit counter *Ct* driven by the network clock *fnx* once every N=3008=47*8*8 bits generated by the source. This sampled counter output is transmitted as the Residual Time Stamp (RTS) in the SAR PDU. ITU-T Recommendation I.363 and ANSI T1.630 show how this method can accept a frequency tolerance for the source frequency of 200 parts per million (ppm).

The Structured Data Transfer (SDT) method is more straightforward, as shown in the bottom part of Figure 8.17. SDT has two modes depending upon whether the sequence number is odd or even. The SDT CPCS uses a 1-octet pointer field in even sequence numbers of the 47-octet SAR-PDU payload to indicate the offset into the current payload of the first octet of an nxDS0 payload. The value of n may be as large as 92 in the P-format since the pointer is repeated every other cell when supporting AAL1.

Figure 8.18 depicts the SAR for AAL1. The 1 octet of overhead is broken down into four fields as shown in the figure. Since 1 octet is used by the AAL 1 SAR, this leaves 47 octets for user data. There are two major fields: the Sequence Number (SN) and the Sequence Number Protection (SNP) field. The 3-bit sequence count is incremented sequentially by the origin. The receiver checks for missing or out-of-sequence SAR-PDUs and generates a signal alarm when this occurs. The Convergence Sublayer Indication (CSI) bit in the SN field is used differently in the SRTS and SDT modes. In SRTS mode, the 4-bit RTS is sent in odd-sequence-numbered PDUs. In SDT mode, the CSI bit is used to indicate if the pointer field is present in even-sequence-numbered SAR-PDUs. The 3-bit CRC field computes a checksum across the SN field. The parity bit represents even parity across the first 7 bits in the 1-octet SAR-PDU overhead.

The sequence number is critical to proper operation of AAL1 since an out-of-sequence or missing SAR-PDU will disrupt at least 47 octets of the emulated circuit bit stream. A well-defined procedure is standardized to correct many problems due to bit errors in the sequence number field, or to accurately detect errors that are not corrected. The operation at the receiver is illustrated in the state machine at the bottom of Figure 8.18. While in the correction mode the receiver can correct single-bit errors using the CRC, but, if after CRC correction the parity check fails, then either a single, or multiple, bit error has been detected and the receiver switches to detection mode. The receiver stays in detection mode until no error is detected and the sequence number is sequential (i.e., valid).

8.6.2   Example of DS1 Circuit Emulation Using AAL1

Figure 8.19 shows an example of a transmitter using AAL1 operating in SRTS mode to emulate a DS1 digital bit stream created by a video codec. Recall from Chapter 4 that a DS1 frame has a 193-bit frame structure that repeats 8000 times per second. The DS1 frame has 1 framing bit and 192 user data bits as shown coming out of the codec on the left-hand side of the figure with time running from top to bottom. The Convergence Sublayer (CS) computes the Residual Time Stamp (RTS) once every 8 cell times and provides this to the SAR sublayer for insertion in the SAR header. The 193-bit frames are packed into the 47-octet (376 bits) SAR Protocol Data Units (PDUs) by the Segmentation And Reassembly (SAR) sublayer. The SAR sublayer then adds the sequence number, inserts the data from the CS and computes the CRC and parity over the SAR header, and passes the 48-octet SAR-PDU to the ATM layer. The ATM layer adds the 5-byte ATM header and outputs the sequence 53-byte cells shown in the figure. The process at the receiver is analogous to that shown here, except the steps are reversed.

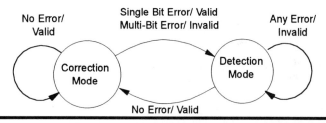

	SN Field even parity checked CRC corrected		SNP Field		
Cell Header	CSI bit	Sequence Count	CRC Field	Parity bit	SAR-PDU Payload
5 octets	1 bit	3 bits	3 bits	1 bit	47 octets

Figure 8.18 AAL1 Segmentation And Reassembly (SAR) Sublayer

8.6.3 AAL2

AAL2 specifies ATM transport of connection-oriented circuit and VBR high bit-rate packetized audio and video. The current standard at time of publication was not well defined. AAL2 may become a key protocol in future ATM implementations requiring support for variable bit-rate audio and video. The second Motion Photographic Experts (MPEG) video encoding standard, called MPEG2, can be operated at a variable bit rate. The standardization and specification of interoperable video and audio encoding using ATM is currently an active area of work. Some approaches are investigating the use of either AAL5 or AAL1 to provide this function.

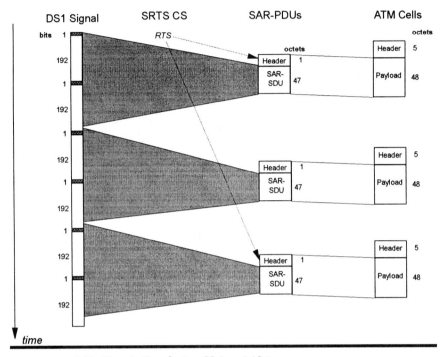

Figure 8.19 DS1 Circuit Emulation Using AAL1

8.6.4 AAL3/4

AAL3 and AAL4 are combined into a single Common Part (CP) AAL3/4 in support of Variable Bit Rate (VBR) traffic, both connection-oriented or connectionless. Support for connectionless service is provided at the Service Specific Convergence Sublayer (SSCS) level. For example, Chapter 17 contains an example of how the IEE80.26 L3_PDU could be carried over AAL3/4 in a manner that would interwork with 802.6 L2_PDUs.

Figure 8.20 depicts the CPCS-PDU for AAL3/4. The header has three components as indicated in the figure. The 1-octet Common Part Indicator (CPI) indicates the number of counting units (bits or octets) for the Buffer Allocation Size (BASize) field. The sender inserts the same value for the 2-octet Beginning Tag (BTag) and the Ending Tag (ETag) so that the receiver can match them as an additional error check. The 2-octet BASize indicates to the receiver how much buffer space should be reserved to reassemble the CPCS-PDU. A variable-length PAD field of between 0 and 3 octets is inserted in order to make the CPCS-PDU an integral multiple of 32 bits to make end system processing simpler. The trailer also has three fields as shown in the figure. The 1-octet ALignment field (AL) simply makes the trailer a full 32 bits to simply the receiver design. The 1-octet ETag must have the same value as the BTag at the receiver for the CPCS-PDU to be considered valid. The length field encodes the length of the CPCS-PDU field so that the pad portion may be taken out before delivering the payload to the CPCS user.

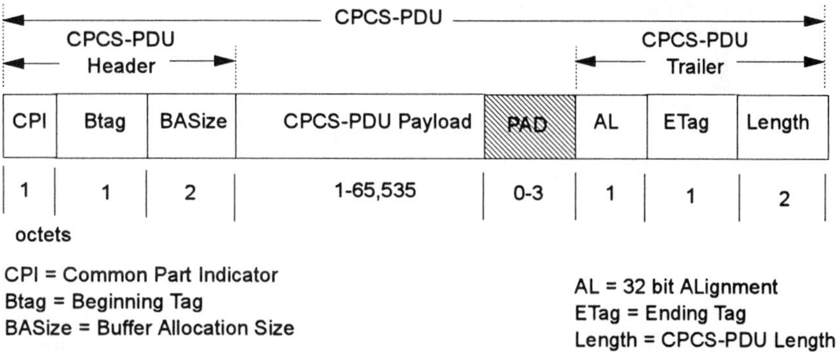

Figure 8.20 AAL3/4 CPCS Sublayer

Figure 8.21 depicts the SAR for AAL3/4. The SAR-PDU encoding and protocol function and format are nearly identical to the L2_PDU from IEEE 802.6. The SAR-PDU has a 2-octet header and trailer. The header contains three fields as shown in the figure. The 2-bit Segment Type (ST) field indicates whether the SAR-PDU is a Beginning Of Message (BOM), a Continuation Of Message (COM), an End Of Message (EOM), or a Single Segment Message (SSM). The 2-bit Sequence Number (SN) is incremented by the sender and checked by the receiver. The numbering and checking begins when an ST of BOM is received. The 10-bit Multiplex IDentification (MID) field allows up to 1024 different CPCS-PDUs to be multiplexed over a single ATM VCC. This is a key function of AAL3/4 since it allows multiple logical connections to be multiplexed over a single VCC. This function is essentially the same one used in the 802.6 L2 protocol where there was effectively no addressing in the cell header. The MID is assigned for a BOM or SSM segment type. The trailer has two fields. The 6-bit Length Indicator (LI) specifies how many of the octets in the SAR-PDU contain CPCS-PDU data. LI has a value of 44 in BOM and COM segments, and may take on a value less than this in EOM and SSM segments.

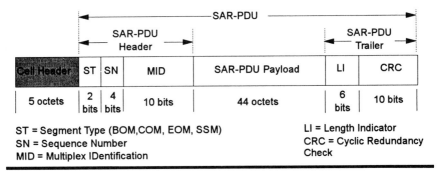

Figure 8.21 AAL3/4 Segmentation And Reassembly (SAR) Sublayer

8.6.5 AAL3/4 Multiplexing Example

Figure 8.22 depicts a data communications terminal that has two inputs with two 98-byte (or octet) packets arriving simultaneously destined for a single ATM output port using the AAL3/4 protocol. On the left-hand side of the figure two 98-byte packets are shown arriving simultaneously. Two parallel instances of the CPCS sublayer encapsulate the packets with a header and trailer. These are then passed to two parallel Segmentation And Reassembly (SAR) processes that segment the CPCS-PDU on two different MIDs, resulting in a BOM, COM, and EOM segment for each input packet. Because all of this occurred in parallel, the ATM cells resulting from this process are interleaved on output. This is the major additional function of AAL3/4 over AAL5, as we shall see by comparison with the AAL5 example later in this chapter. This also introduces complexity in AAL3/4 that is not present in AAL5.

8.6.6 AAL5

The Common Part (CP) AAL5 supports Variable Bit Rate (VBR) traffic, both connection-oriented or connectionless. Support for connectionless or connection-oriented service is provided at the Service Specific Convergence Sublayer (SSCS) level. For example, Chapter 9 describes the operation of signaling SSCS protocols over

AAL5, while Chapter 17 provides examples of frame relay and multiprotocol (including IP) operation over AAL5.

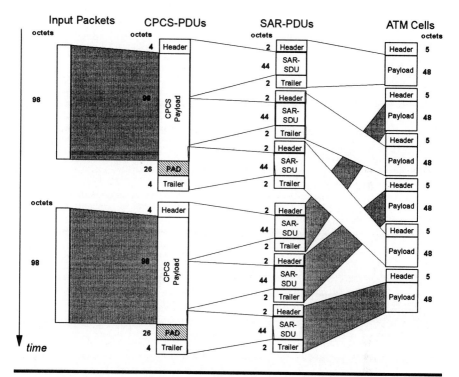

Figure 8.22 Multiplexing Example Using AAL3/4

Figure 8.23 depicts the CPCS for AAL5. The payload may be any integer number of octets in the range of 1 to 2^{16}-1 (65,535). The Padding field is of a variable length chosen such that the entire CPCS-PDU is an exact multiple of 48 so that it can be directly segmented into cell payloads. The User-to-User (UU) information is conveyed between AAL users transparently. The only current function of the Common Part Indicator (CPI) is to align the trailer to a 64-bit boundary, with other functions for further study. The length field identifies the length of the CPCS-PDU payload so that the PAD can be removed. Since 16 bits are allocated to the length field, the maximum payload length is 2^{16}-1 = 65,535 octets. The CRC-32 detects errors in the CPCS-PDU. The CRC-32 is the same one used in IEEE 802.3, IEEE 802.5, FDDI, and Fiber Channel.

Figure 8.24 depicts the SAR for AAL5. The SAR-PDU is simply 48 octets from the CPCS-PDU. The only overhead the SAR sublayer makes use of is the Payload Type code points for *AAL_indicate*.

AAL_indicate is zero for all but the last cell in a PDU. A nonzero value of AAL_indicate identifies the last cell of the sequence of cells indicating that reassembly should begin. This was intended to make the reassembly design simpler and make more efficient use of ATM bandwidth, which was the root of the name for the original AAL5 proposal [1], called the Simple Efficient Adaptation Layer (SEAL).

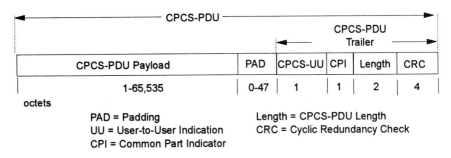

Figure 8.23 AAL5 Common Part Convergence Sublayer (CPCS)

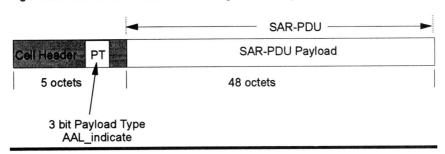

Figure 8.24 AAL5 Segmentation and Reassembly (SAR) Sublayer

8.6.7 AAL5 Multiplexing Example

Figure 8.25 depicts the same example previously used for the AAL3/4 example to illustrate a major difference in AAL5. The figure depicts a data communications terminal that has two 98-byte packets arriving simultaneously, destined for a single ATM output port, this time using the AAL5 protocol.

On the left-hand side of the figure the two 98-byte packets are shown arriving simultaneously. Two parallel instances of the CPCS sublayer add a trailer to each packet. Note that the entire packet does not have to be received before it can begin the SAR function as would be required in AAL3/4 to insert the correct Buffer Allocation

Size (BASize) field. The packets are acted on by two parallel Segmentation And Reassembly (SAR) processes which segment the CPCS-PDU into ATM cells. In our example these are destined for the same VPI/VCI, and hence only one can be sent at a time. This implementation is simpler than AAL3/4, but is unable to keep the link as fully occupied as the additional multiplexing of AAL3/4 could if the packets arrive much faster than the rate at which SAR and ATM cell transmission occur.

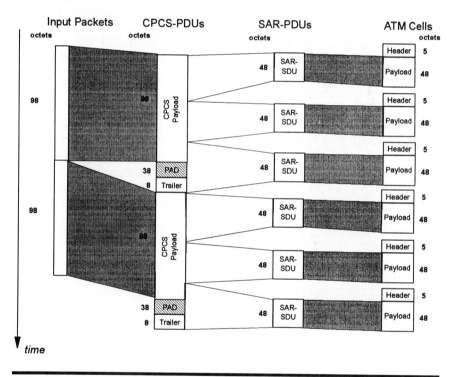

Figure 8.25 Multiplexing Example Using AAL5

8.7 REVIEW

This chapter covered the foundations of B-ISDN: the PHYsical (PHY) layer, the ATM layer, and the Common Part (CP) ATM Adaptation Layers (AALs). The chapter started with an overview of how these layers fit into the overall B-ISDN protocol model, and went on to investigate the sublayer structure of the PHY and ATM Adaptation layers. The PHY layer broke down into Physical Medium Dependent

(PMD) and Transmission Convergence (TC) sublayers. Examples of how the PMD supports different physical media and interface rates and how the TC sublayer effectively makes the PHY layer appear as a pipe that can transfer cells at a maximum rate to the ATM layer were covered. The ATM layer protocol model was covered, including an explanation of Virtual Path (VP) and Virtual Channel (VC) links, connections, and concepts complete with network examples from several points of view. The ATM cell was dissected in detail, clearly defining every field in the header, reviewing some of the basic functions, and identifying where to reference detailed treatment of particular aspects of ATM in other areas of this book. The AAL protocol model was then introduced in terms of the Convergence Sublayer (CS) and the Segmentation And Reassembly (SAR) sublayer, and the CS sublayer was further divided into a Service Specific (SS) and Common Part (CP). The CPCS and SAR were explained in detail focusing on the three AALs that are currently standardized: AAL1, AAL3/4, and AAL5. Definitions of the format and operation, as well as an example, were provided. An example of emulating the DS1 bit stream from a video codec using AAL1 was provided, and the multiplexing that can be performed by AAL3/4 and AAL5 was compared. Finally, two examples illustrated how AAL3/4 can make more efficient use of a VCC in some cases than AAL5 through increased overhead and complexity.

8.8 REFERENCE

[1]   T. Lyon, "Simple and Efficient Adaptation Layer (SEAL)," ANSI T1S1.5/91-292, August 1991.

9

User, Control, and Management Planes

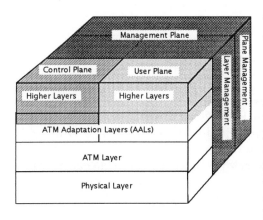

This chapter provides an overview of the higher layers and the Service Specific Convergence Sublayer (SSCS) portion of the ATM Adaptation Layer (AAL). It includes the user and control planes, along with the management plane, as illustrated in the B-ISDN cube on the left. A summary of higher-layer user plane function and purpose sets the stage for a summary of the user plane protocols covered in later chapters. Of course; a key point is that the principal purpose of the control and management planes is to support the services provided by the user plane. Next, the chapter moves on to the very important control plane, which is central in performing the functions needed in a Switched Virtual Connection (SVC) service. The B-ISDN signaling protocol, Service Specific Coordination Function (SSCF), and Service Specific Connection-Oriented Protocol (SSCOP) SSCS protocols are then covered in detail. Finally, the management plane is covered as an introduction to the detailed coverage provided in Part 7. The management plane is composed of overall plane management and management of each of the user and control plane layer components.

9.1 USER PLANE OVERVIEW

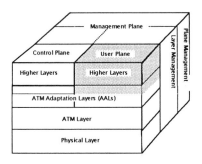

As shown in the shaded portion of the figure to the left this chapter first covers the general purpose and function of the user plane from a high level. The state of standardization in the Service Specific Convergence Sublayer (SSCS) and higher layers of the user plane are summarized as an introduction to a more detailed treatment later in the book. In fact, a great deal of standardization work still remains to be done in the area of higher-layer user plane functions.

9.1.1 User Plane — Purpose and Function

The protocol cube figure at the introduction of this chapter shows that the user plane spans the PHYsical Layer (PHY), ATM Layer, ATM Adaptation Layer (AAL), and higher layers. The AAL and higher layers provide meaningful interfaces and services to end user applications such as; frame relay, Switched Multimegabit Data Service (SMDS), Internet Protocol (IP), other protocols, and Application Programming Interfaces (APIs) which are detailed in Chapters 11 and 17.

It is also important to note that the control and management planes exist in support of the user plane, in a manner similar to that developed for ISDN, as described in Chapter 4. The control plane provides the means to support the following types of connections on behalf of the user plane:

 * Switched Virtual Connections (SVCs)
 * Permanent Virtual Connections (PVCs)

SVCs and PVCs can be either point-to-point, point-to-multipoint, multipoint-to-point, or multipoint-to-multipoint Virtual Path Connections (VPCs) or Virtual Circuit Connections (VCCs), as defined in Chapter 4. A VPC or VCC provides a specified Quality of Service (QoS) with a certain bandwidth defined by traffic parameters in an ATM layer traffic contract, as defined in Chapter 12.

The entire B-ISDN architecture must support the user's application needs to be successful.

9.1.2   User Plane — SSCS Protocols

To date, two Service Specific Convergence Sublayer (SSCS) protocols have been developed specifically for the user plane:

* Frame Relay SSCS
* SMDS SSCS

Chapter 17 details both of these. There is no SSCS required for support of IP or circuit emulation over ATM since the common part AAL directly supports them. There is some discussion and the likely possibility that the Service Specific Connection-Oriented Protocol (SSCOP) defined for signaling could be used to provide an assured data transfer service in the user plane. It is anticipated that SSCS protocols will be developed for the following user-driven applications:

☺   Desktop-quality video
☺   Entertainment-quality video
☺   Multicast LAN support
☺   LAN emulation
☺   Reliable data delivery (like the X.25 capability)
☺   Interactive, cooperative computing support
☺   Database Concurrency, Commitment, and Recovery (CCR) function support

A large amount of additional standardization work is required to support the applications listed above and put a real smile on the user's face. It is likely that many more SSCS protocols will be required to support these applications before the B-ISDN protocol suite is mature.

9.1.3   User Plane — Higher Layers

The area of higher layer protocol support for ATM has only one standard to date, which underscores the need for further standards work:

☺ IETF RFC 1483 — Multiprotocol Encapsulation Over ATM

Chapter 17 presents IETF RFC 1483 in detail, complete with examples of how it enables current applications to use ATM. That the higher-layer standardization is not as far along as the lower layers is to be expected. The foundation must first be built before adding the walls and finally the roof. Of course, work continues on the foundational layers by defining new physical layers, new ATM capabilities, and possibly even new AALs in the ever expanding B-ISDN/ATM mansion. There are some key activities in progress that Chapters 6 and 11 summarize:

☺ ATM Forum LAN Emulation working group
☺ ATM Forum System Aspects and Applications (SAA) working group
☺ ATM Forum Private Network-Network Interface (P-NNI) working group
☹ IETF work group supporting IP over ATM
☹ IETF work group for routing over ATM networks

It is suggested that users follow the progress of these activities and provide inputs to these groups on requirements. Users need these functions, but they are not defined yet; that is why they don't have smiles on their faces in the above list. You have a voice in the standards process — use it!

9.2 CONTROL PLANE AAL OVERVIEW

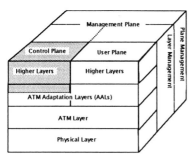

The control plane handles all virtual connection-related functions, most importantly the Switched Virtual Circuit (SVC) capability. The control plane also performs the critical functions of addressing and routing. The higher-layer and service-specific AAL portions of the signaling protocol have recently reached an initial level of standardization. This chapter covers these functions as indicated by the shaded portions of the B-ISDN cube in the figure on the left.

The signaling protocol architecture of the control plane is very similar to that of Narrowband Integrated Services Digital Network (N-ISDN) as depicted in Figure 9.1.

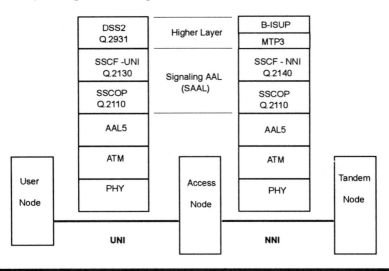

Figure 9.1 Overview of Control Plane Architecture

The specifications for the Signaling AAL (SAAL) are being developed in the ITU-T and are being adopted by ANSI and the ATM Forum. ITU-T Recommendation Q.2931 (previously called Q.93B) specifies the B-ISDN signaling on the ATM UNI. Q.2931 was derived from both the Q.931 UNI signaling protocol specified for N-ISDN, and the Q.933 UNI signaling protocol for frame relay. The formal name for the ATM UNI signaling protocol is the Digital subscriber Signaling System 2 (DSS2), while the name for ISDN UNI signaling was DSSS1. ITU-T Recommendation Q.2130 (previously called Q.SAAL.2) specifies the Service Specific Coordination Function (SSCF) for the UNI. ITU-T Recommendation Q.2110 (previously called Q.SAAL.1) specifies Service Specific Connection-Oriented Protocol (SSCOP). The ISDN User Part (ISUP) is being adapted in a similar way to broadband as the UNI protocol was in defining the broadband NNI signaling which is called B-ISUP. The B-ISUP protocol operates over the Message Transfer Protocol 3 (MTP3), identical to that used in Signaling System 7 (SS7) for out-of-band N-ISDN and voice signaling. This will allow B-ISDN network signaling the flexibility to operate over existing signaling networks or directly over new ATM networks. The series of ITU-T Recommendations Q.2761 through Q.2764 specify the B-ISUP protocol. ITU-T

Recommendation Q.2140 specifies the SSCF at the NNI. The NNI signaling uses the same SSCOP protocol as the UNI.

9.3 CONTROL PLANE ADDRESSING AND ROUTING

There are two capabilities that are critical to a switched network: addressing and routing. *Addressing* occurs at the ATM VPI/VCI level and at the logical network level. Since the VPI/VCI is unique only to a physical transmission path, there is a need to have a higher level address that is unique across at least each network. Ideally, the address should be unique across all networks in order to provide universal connectivity. Once each entity involved in switching virtual connections has a unique address, there is another even more onerous problem of finding a route from the calling party to the called party. This problem is solved by using *routing*.

9.3.1   ATM Layer VPI/VCI Level Addressing

The signaling protocol automatically assigns the VPI/VCI values to ATM addresses and physical ATM UNI ports based upon the type of SVC requested according to the following set of rules: either point-to-point or point-to-multipoint. A physical ATM UNI port must have at least one unique ATM address. An ATM UNI port may also have more than one ATM address.

Recall that a VCC or VPC is defined in only one direction; that is, it is simplex. A point-to-point SVC (or a PVC) is a pair of simplex VCCs or VPCs: a forward connection from the calling party to the called party, and a backward connection from the called party as illustrated in Figure 9.2. The forward and backward VCC or VPC can have different traffic parameters. A point-to-point SVC is defined by the forward and backward VPI (and VCI for a VCC) as well as the ATM address associated with the physical ATM UNI ports on each end of the connection. The VPI(/VCI) assignment can be different for the forward and backward directions of a VPC or VCC at the same end of the connection as well as being different from the other end of the connection. A convention where the VPI (and VCI for a VCC) is identical at the same end of a connection may be used.

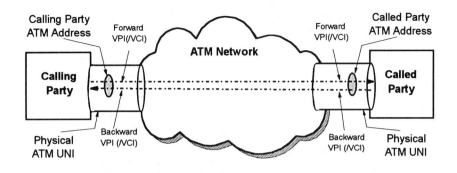

Figure 9.2 Point-to-Point Switched Virtual Connection (SVC)

A point-to-multipoint SVC (or PVC) is defined by the VPI and the ATM address associated with the physical ATM UNI port of the root node, and the ATM address and VPI(/VCI) for each leaf node of the connection, as shown in Figure 9.3.

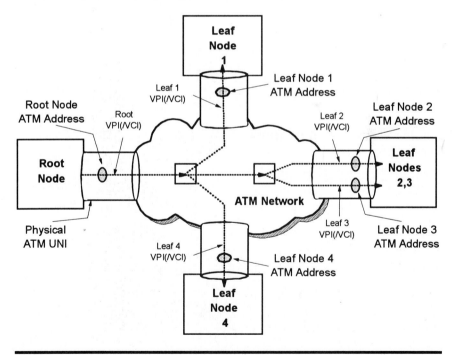

Figure 9.3 Point-to-Multipoint Switched Virtual Connection (SVC)

There is essentially only a forward direction because the backward direction is allocated zero bandwidth. Note that more than one

VPI/VCI value and ATM address can be assigned to a physical interface as part of the point-to-multipoint connection. This means that the number of physical ATM UNI ports is always less than or equal to the number of logical leaf endpoints of the point-to-multipoint connection. The implementation of a point-to-multipoint connection should efficiently replicate cells within the network. A minimum spanning tree is an efficient method of constructing a point-to-multipoint connection, as illustrated in Figure 9.3.

9.3.2   Desirable Attributes of ATM Layer Addressing

A set of desirable attributes should be followed when designing an ATM Layer addressing scheme. Of course; each address must be unique. The desirable attributes of the addressing scheme include at least the following:

☞ Simplicity
☞ Automatic assignment
☞ Efficient usage of the address space
☞ Ease of managing changes in addresses
☞ Extensibility of the addressing scheme

9.3.3   ATM Control Plane (SVC) Addressing

Currently two types of ATM Control Plane (SVC) addressing plans are being considered in the standards bodies to identify an ATM UNI address: the Network Service Access Point (NSAP) format defined in ISO 8348, CCITT X.213, and CCITT E.164 standards.

Figure 9.4 summarizes the current version of the NSAP addressing plans from the ATM Forum UNI version 3.0 specification. The Authority and Format Identifier (AFI) identifies which of the formats is being used and designates the authority that allocates the Data Country Code (DCC). Each address is composed of an Initial Domain Identifier (IDI) and a Domain Specific Part (DSP). These may be subject to change as a result of further study due to the strong relationship between addressing and routing.

Three IDI formats are currently specified by the ATM Forum: DCC, ICD, and E.164. The Data Country Code (DCC) specifies the country in which an address is registered as defined in ISO 3166. The International Code Designator (ICD) identifies an international organization as administered by the British Standards Institute. E.164 specifies ISDN and telephone numbers which will be defined later in this section.

Within each of these domains lies the Domain Specific Part (DSP). The DSP is identical for the DCC and ICD formats. The DSP Format Identifier (DFI) specifies the meaning of the remainder of the address. The Administrative Authority (AA) field identifies the organization that is responsible for administering the remainder of the address, for example a carrier, private network, or manufacturer. The remainder of the DSP is identical for all domains. The Routing Domain (RD) identifies a unique domain within the IDI format — further subdivided using the 2-byte AREA field. The End System Identifier (ESI) and SELector (SEL) portions of the DSP are identical for all IDI formats as specified in ISO 10589. The ESI can be globally unique, for example, a 48-bit IEEE MAC address. The SELector (SEL) field is not used by routing but may be used by End Systems (ES).

a) Data Country Code (DCC) Format

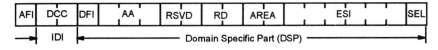

b) International Code Designator (ICD) Format

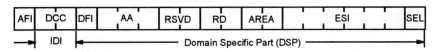

c) E.164 ATM Address Format

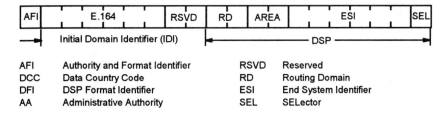

AFI	Authority and Format Identifier	RSVD	Reserved
DCC	Data Country Code	RD	Routing Domain
DFI	DSP Format Identifier	ESI	End System Identifier
AA	Administrative Authority	SEL	SELector

Figure 9.4 ATM Forum Addressing Plans

Figure 9.5 summarizes the CCITT E.164 numbering plan format. The international number is coded in Binary Coded Decimal (BCD) and is padded with zeroes on the left-hand side to result in a constant length of 15 digits. There is a Country Code (CC) of one to three digits as standardized in CCITT Recommendation E.163. The remainder of the address is a Nationally Significant Number (NSN). The NSN can be further broken down as a National Destination Code (NDC) and Subscriber Number (SN). The North American Numbering Plan (NANP) is a subset of E.164. The NDC currently corresponds to an area code and switch NXX identifier for voice applications. Further standardization of the E.164 number for data and other applications, such as SMDS, is in progress.

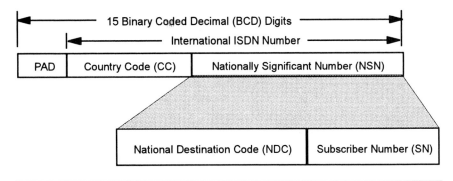

Figure 9.5 CCITT E.164 Numbering Plan Format

9.3.4 Basic Routing Requirements

Cells from the same VPC or VCC must follow the same route, defined as the ordered sequence of physical switch ports which the cells traverse from source to destination. A route is established in response to the following events:

* A PVC is newly provisioned
* An SVC connection request is made
* A failed PVC is being automatically reestablished

A route is cleared in response to the following events:

* A PVC disconnect order is processed
* A failure is detected on a restorable PVC
* An SVC disconnection request is made
* Call clearing in response to a failure

The route traversed should minimize a cost function including, but not limited to, the following factors:

+ Delay
+ Economic expense
+ Balance utilization (when multiple links are present between a node-pair)

9.3.5   Desirable Routing Attributes

There are desirable attributes to follow when designing an ATM layer routing scheme. Attributes of the routing scheme include at least the following:

* Simplicity
* Automatic determination of least-cost route(s)
* Ease of managing changes in the network in terms of new links and nodes
* Extensibility of the routing scheme to a large network

9.3.6   A Simple ATM Layer VCC Routing Design

A simple routing design for VCCs utilizes routing based upon the VPI value only. Each physical node is assigned a VPI value, which means that it is a VPC endpoint as illustrated in Figure 9.6.

Every node can route traffic to a destination node using a VPC connecting point with the VPI corresponding to the destination node number. This routing is accomplished by each node — knowing that the tandem nodes will connect this VPC through to the destination node.

The principal advantage of this method is that it is very simple — no VPI or VCI translation is required. This method has several disadvantages; it is inefficient since VPIs are allocated on routes that are not used, and it limits the number of VPCs that can be assigned to user applications.

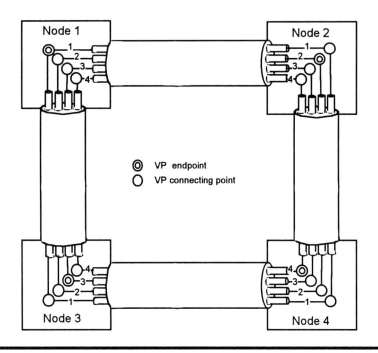

Figure 9.6 Illustration of Simple VPC-Based Routing

9.4 CONTROL PLANE — PROTOCOL MODEL

This section introduces the protocol model for the signaling Service Specific Convergence Sublayer (SSCS), and also summarizes the B-ISDN UNI protocol for signaling.

9.4.1 Layered SSCS Model

Figure 9.7 illustrates the protocol model for the Signaling AAL (SAAL). The Common Part AAL (CP-AAL) is AAL5 as defined in Chapter 8. The SSCS portion of the SAAL is composed of the following two protocols:

&ain Service Specific Coordination Function (SSCF)
&ain Service Specific Connection-Oriented Protocol (SSCOP)

The SAAL primitives are provided at the SAAL Service Access Point (SAP). The CP AAL5 interfaces with the ATM layer at the ATM SAP. There is a one-to-one correspondence between an SAAL SAP and an ATM SAP. The signaling SSCF and SSCOP protocols and the CP-AAL are all managed as separate layers, by corresponding layer management functions as indicated on the left-hand side of Figure 9.7. Layer management is responsible for setting parameters in the individual layer protocols and monitoring their state and performance. Plane management coordinates across the layer management functions so that the overall end-to-end signaling capability is provided.

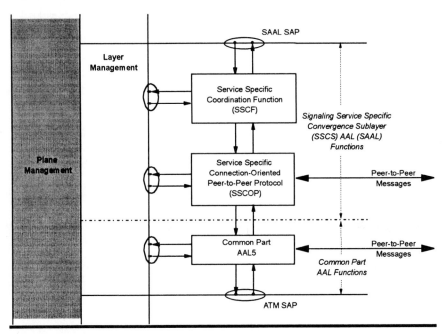

Figure 9.7 Signaling AAL (SAAL) Layered Protocol Model

9.4.2 Service Specific Coordination Function (SSCF)

The Service Specific Coordination Function (SSCF) provides the following services to the Signaling AAL (SAAL) user:

- ❖ Independence from the underlying layers
- ❖ Unacknowledged data transfer mode
- ❖ Assured data transfer mode

❖ Transparent relay of information
❖ Establishment of connections for assured data transfer mode

The SSCF provides these capabilities primarily by mapping between a simple state machine for the user and the more complex state machine employed by the SSCOP protocol.

9.4.3 Service Specific Connection-Oriented Protocol (SSCOP)

The Service Specific Connection-Oriented Protocol (SSCOP) is a peer-to-peer protocol which performs the following functions:

❧ Guaranteed sequence integrity, or ordered delivery
❧ Error correction via error detection and retransmission
❧ Receiver-based flow control of the transmitter
❧ Error reporting to layer management
❧ Keep alive messaging when other data is not being transferred
❧ Local retrieval of unacknowledged or enqueued messages
❧ Capability to establish, disconnect, and synchronize an SSCOP connection
❧ Transfer of user data in either unassured or assured mode
❧ Protocol level error detection
❧ Status reporting between peer entities

SSCOP is a fairly complicated protocol, but is specified in the same level of detail as a successful protocol like HDLC. The unassured mode is a simple unacknowledged datagram protocol, similar to the User Datagram Protocol (UDP).

Much of the richness of SSCOP is provided in the assured data transfer mode. A connection must be established before any data can be transferred. Figure 9.8 illustrates an example of the SSCOP retransmission strategy. The error detection capability of AAL5 determines whether a frame is successfully received. SSCOP requires that the transmitter periodically poll the receiver as a keep-alive action and to determine if there is a gap in the sequence of successfully-received frames. The receiver must respond to the poll, and if more than a few poll responses are missed, the transmitter will take down the connection. A key feature is where the receiver identifies that one or more frames are missing in its sequence, as illustrated in the figure. The transmitter then only resends the missing frames. Chapter 16 shows how this selective reject type of retransmission protocol significantly reduces unproductive retransmissions for very high-speed links, such as those in ATM networks, as compared to a "Go-Back N" retransmission strategy employed in X.25, which is described in Chapter 5.

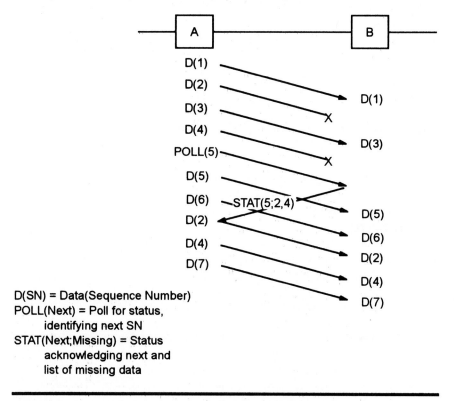

Figure 9.8 Example of SSCOP Retransmission Strategy

SSCOP PDUs also employ a 24-bit sequence number that allows very high sustained rates to be achieved in a window flow controlled protocol such as that described in Chapter 14, for example, the Transmission Control Protocol (TCP).

9.5   CONTROL PLANE — SIGNALING FUNCTIONS

The control plane signaling functions currently defined by the ATM Forum and the ITU-T are first described, identifying the common functions and differences. Next, the plans for the next phase of signaling functions identified by the ITU-T are summarized.

9.5.1 Signaling Functions —Current

The current version of the ATM Forum UNI version 3.0 signaling specification and the ITU-T Q.2931 standard are closely aligned. First the functions of Q.2931 are summarized, followed by a description of how the ATM Forum UNI 3.0 both subsets Q.2931, and defines additional functions.

The major functions defined in ITU-T Recommendation Q.2931 are:

① Point-to-point connection setup and release
① VPI/VCI selection and assignment
① Quality of Service (QoS) class request
① Identification of calling party
① Basic error handling
① Communication of specific information in setup request
① Subaddress support
① Specification of Peak Cell Rate (PCR) traffic parameters
① Transit network selection

The ATM Forum UNI version 3.0 specification does not require the following capabilities from Q.2931:

✕ No alerting message sent to called party
✕ No VPI/VCI selection or negotiation
✕ No overlap sending
✕ No interworking with N-ISDN
✕ No subaddress support
✕ Only a single transit network may be selected

The ATM Forum UNI version 3.0 specification defines the following capabilities in addition to Q.2931:

🖳 Support for a call originator setup of a point-to-multipoint call
🖳 Extensions to support symmetric operation
🖳 Addition of sustainable cell rate and maximum burst size traffic parameters

⌨ Additional information elements for point-to-multipoint endpoints
⌨ Additional NSAP address structures

There is a general intention stated by both the ITU-T and the ATM Forum to align the specifications of future releases.

9.5.2 Signaling Functions — Next Phase

The ITU-T has specified a capability set 2 that defines a set of capabilities that will be standardized next. These capabilities must first be described in more detail before protocol standardization can begin. The following general functions are part of capability set 2:

▶ Specification of a call model where each call may have multiple connections, for example, in multimedia
▶ Support for a distributed point-to-multipoint call setup protocol
▶ Renegotiation of traffic parameters during the course of a connection
▶ Support for multipoint and multipoint-to-point calls
▶ Specification of metasignaling which establishes additional connections for signaling

9.6 CONTROL PLANE — SIGNALING PROTOCOL

This section provides an overview of the signaling messages and their key parameters. The basics of the signaling protocol are then presented. Finally, the signaling messages and protocol in action are illustrated by an example of a point-to-point call setup and release and the establishment of a point-to-multipoint call.

9.6.1 The Signaling Messages

The following sections summarize key parameters for the following major signaling messages from the Q.2931 protocol as defined by the ATM Forum UNI specification version 3.0:

Point-to-Point Connection Control:

☎ Call Establishment Messages
     CALL PROCEEDING
     CONNECT ACKNOWLEDGE
     SETUP
☎ Call Clearing Messages
     RELEASE
     RELEASE COMPLETE
☎ Status Messages
     STATUS ENQUIRY
     STATUS (Response)

☎ Global Call Reference Related Messages:
     RESTART (All)
     RESTART ACKNOWLEDGE
     STATUS

☎ Point-to-Multipoint Connection Control:
     ADD PARTY
     ADD PARTY ACKNOWLEDGE
     ADD PARTY REJECT
     DROP PARTY
     DROP PARTY ACKNOWLEDGE

Each signaling message has a number of Information Elements (IE), some of which are Mandatory (M) and others of which are Optional (O). The key mandatory information elements used in the protocol are:

♪ ATM user cell rate requested
♪ Called party number
♪ Connection identifier (assigned VPI/VCI value)
♪ QoS Class requested

The messages related to a particular call attempt each contain a common mandatory information element, the *call reference,* that is unique on a signaling interface. All messages must also contain an information element for their type, length, and protocol discriminator (i.e., the set from which these messages are taken). There are an even larger number of optional parameters, such as:

◐ Broadband bearer capability requested
◐ Broadband lower and higher-layer information

- ❤ AAL parameters
- ❤ Called party subaddress
- ❤ Calling party number and subaddress
- ❤ Transit network selection
- ❤ Cause Code
- ❤ Endpoint reference identifier and endpoint state number

9.6.2 The Signaling Protocol

The signaling protocol specifies the sequence of messages that must be exchanged, the rules for verifying consistency of the parameters, and the actions to be taken in order to establish and release ATM layer connections. A significant portion of the specification is involved with handling error cases, invalid messages, inconsistent parameters, and a number of other unlikely situations. These are all important functions since the signaling protocol must be highly reliable in order for users to accept it.

Signaling protocols may be specified in several ways: via narrative text, via state machines, or via a semigraphical Specification Definition Language (SDL). The ATM Forum UNI specification version 3.0 uses the narrative method. For complicated protocols, such as Q.2931, a very large sheet of paper would be needed to draw the resulting state machine in a manner such that a magnifying glass is not required to read it. The SDL allows a complicated state machine to be formally documented on multiple sheets of paper in a tractable manner. For example, Q.921 and SSCOP are specified using SDL.

The Q.2931 protocol is based upon the ISDN Q.931 and Frame Relay Q.933 protocols. Since a large amount of expertise has been built up over the years on these subjects, the prospects for implementing the rather complex Q.2931 protocol in an interoperable manner are encouraging.

9.6.3 Point-to-Point Call Setup and Release Examples

Two simple, but relevant, examples for a point-to-point call, call setup and call release, illustrate the basic aspects of the ATM Forum- and Q.2931-based signaling protocol for the ATM UNI. Figure 9.9 illustrates the point-to-point call setup example. The examples employ: a calling party with ATM address A on the left, a network shown as a cloud in the middle, and the called party with ATM address B on the right. Time runs from top to bottom in all of the examples. The calling party initiates the call attempt using a SETUP message indicating B as the called party number. The network routes the call to the physical interface on which B is connected and outputs a SETUP message indicating that the VPI/VCI that should be used if the call is accepted. Optionally, the SETUP message may communicate the identity of the calling party A. The called party accepts the call attempt by returning the CONNECT message, which is propagated back to the originator by the network as rapidly as possible in order to keep the call setup time low. The CONNECT ACKNOWLEDGE message is used from the network to the called party and from the calling party to the network as the final stage of the three-way handshake to ensure that the connection is indeed active.

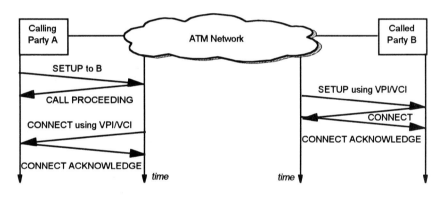

Figure 9.9 Point-to-Point Call Setup Example

Figure 9.10 illustrates the point-to-point call release example. The reference configuration and conventions are the same as in the call setup example above. Either party may initiate the release process. This example illustrates the calling party as the one that initiates the disconnect process by sending the RELEASE message. The network then propagates the RELEASE message across the network to the other party B. The other party acknowledges the RELEASE request by returning a RELEASE COMPLETE message, which is then propagated back across the network to the calling-party RELEASE originator. This two-way handshake completes the call release process.

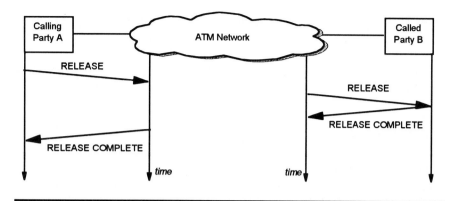

Figure 9.10 Point-to-Point Call Release Example

9.6.4   Point-to-Multipoint Call Setup Example

Figure 9.11 now illustrates an example of setting up a point-to-multipoint call from an originator (root) node A to two leaves, specifically two leaf nodes B and C connected to a local ATM switch on a single ATM UNI, and a third leaf node D connected to a separate ATM UNI. Node A begins the point-to-multipoint call by sending a SETUP message to the network requesting setup of a point-to-multipoint call identifying leaf node B's ATM address. In the example, node A requests a call SETUP to node B, and the network responds with a CALL PROCEEDING message in much the same way as a point-to-point call. The network switches the call attempt to the intended destination and issues a SETUP message to node B identifying the assigned VPI/VCI. The first leaf node then indicates its intention to join the call by returning a CONNECT message that the network in turn acknowledges with a CONNECT ACKNOWLEDGE message. The network informs the calling root node A of a successful addition of party B through a CONNECT and CONNECT ACKNOWLEDGE handshake as shown in the figure.

Continuing on with the same example, the root node requests that party C be added through the ADD PARTY message, which the network relays to the same ATM UNI as party B through the ADD PARTY message to inform the local switch of the requested addition. Party C responds with an ADD PARTY ACKNOWLEDGE message that is propagated by the network back to the root node A. The root node A requests that the final leaf party D be added through an ADD PARTY message. The network routes this to the UNI connected to party D, and issues a SETUP message since this is the first party on this ATM UNI. Node D responds with a CONNECT message to which the network responds with a CONNECT ACKNOWLEDGE message. The fact that leaf party D has joined the point-to-multipoint call is communicated to the root node A through the ADD PARTY ACKNOWLEDGE message.

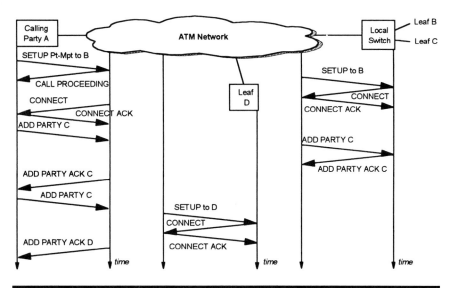

Figure 9.11 Point-to-Multipoint Call Setup Example

The leaves of the point-to-multipoint call may be removed from the call by the DROP PARTY message if one or more parties would remain on the call on the same UNI, or by the RELEASE message if the party is the last leaf present on the same UNI. The root node should drop each leaf in turn and then release the entire connection.

9.7 MANAGEMENT PLANE

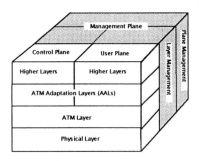

The management plane covers the layer management and plane management functions as shown in the B-ISDN cube on the left. layer management interfaces with the PHYsical, ATM, ATM Adaptation Layer (AAL), and higher layers. Plane management is responsible for coordination across layers and planes in support of the user and control planes through layer management facilities. This makes sure that everything works properly. Layer management will be discussed first, followed by plane management. This, however, is only an introduction to the more detailed treatment of these subjects in Part 7.

9.7.1 Layer Management

Layer management has a management interface to the PHYsical, ATM, AAL, and higher layer protocol entity in both the control and user planes as depicted in Figure 9.12. This two-dimensional view is constructed by cutting the B-ISDN cube open from the back and then folding it out flat. This view illustrates the oversight role of plane management as well. Plane management only interfaces with layer management, which provides interfaces to the user and control plane layers. Standards for these management interfaces are being defined by the ITU-T and ANSI for telecommunications equipment using the Common Management Information Protocol (CMIP), and by the IETF for data communications equipment using the Simple Network Management Protocol (SNMP).

Layer management has the responsibility for monitoring the user and control plane for faults, generating alarms, and taking corrective actions, as well as monitoring for compliance to the performance stated in the traffic contract. The operation and maintenance information functions found within specific layers are handled by layer management. These functions include fault management, performance management, and configuration management as covered in Part 7. The standards for PHY layer management are very mature. The standards for ATM layer fault and performance management are nearing the first stage of usability as presented in Chapters 19 and 20. Standardization for management for the AAL and higher layers is just beginning.

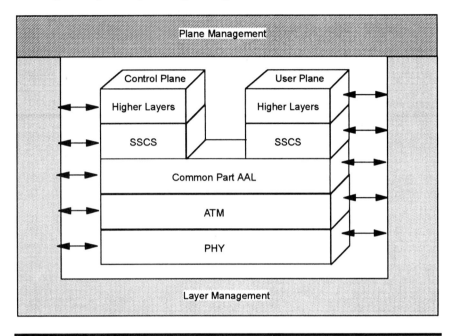

Figure 9.12 Layer Management Relation to User and Control Planes

9.7.2 Plane Management

Plane management has no defined structure, but instead performs management functions and coordination across all layers and planes in the entire system. The Telecommunication Management Network (TMN) architecture developed by the ITU-T for managing all types of telecommunications networks is being extended to perform the B-ISDN plane management role as described in Chapter 19.

9.8 REVIEW

This chapter defined B-ISDN from the top-down perspective in terms of the user, control, and management planes. Topics to be introduced in detail in Chapter 17 regarding Service Specific Convergence Sublayer (SSCS) protocols and in Chapter 11 regarding higher layers were summarized. Focus then turned to the higher layers and SSCS of the control plane. Key concepts of addressing and routing in the control plane were defined and then explained through the control plane protocol model. An overview of the suite of B-ISDN signaling protocols was then covered. This included the Service Specific Coordination Function (SSCF) and Service Specific Connection-Oriented Protocol (SSCOP) SSCS protocols that operate over the Common Part AAL5. Finally, the management plane was introduced in terms of its plane management and layer management components as an introduction to the detailed coverage provided in Part 7.

9.9 REFERENCE

[1]   ITU-T, "Signaling Capability Set 2 Document," 1993.

4

ATM Hardware and Software

This section provides the reader with an overview of ATM hardware and software. It begins by detailing the three predominant classes of ATM hardware: Central Office (CO)-based switches; Campus ATM switches; and local devices including routers, switches, hubs, and end systems. A comparison of switch vendors is also provided. As ATM moves to the desktop, there is a need to add ATM interface cards to powerful workstations. A look at why it will be quite some time before ATM moves into your personal computer is next. The role of ATM cross-connects and PBXes is then covered. For ATM to integrate with the existing voice structure at the premises, ATM interfaces for the Private Branch Exchange (PBX) are required. A discussion is included on the development of ATM chip sets. The chapter ends with a discussion on the direction IBM is taking with ATM and its next generation of ATM hardware. Chapter 11 covers the implications of ATM networking and applications. Two industry implementations of IP over ATM are explored, along with their implications on end system hardware and software. Key ATM applications are covered, followed by a summary of industry forum and standards activity progress to resolve key issues of ATM implementation.

10

ATM Hardware and Switching

This chapter reviews several variations of ATM switching, concentration, hubbing, and end system interfaces. The myriad of ATM hardware is divided into three major categories: Central Office (CO)-based switches, Campus (Ca) switches, and local ATM switches. This chapter then presents the results of an intense analysis of vendors in the ATM equipment marketplace at the time of publication. The coverage then moves to the class of local or ATM premises equipment: routers, switches, hubs, and end systems, offering anything from an ATM interface card to a full ATM switching architecture. A review is then provided of workstation ATM interface cards are emerging on the market which provide speeds of 100 Mbps and 140 Mbps to the desktop! It may be some time before ATM interface cards replace the Ethernet card commonly seen in PCs today, but it is on the radar screen. Proprietary and standard ATM chip sets appearing on the scence are reviewed. Finally, there is no doubt that IBM has played an important part in network-wide computing over the last three decades, and their new version of ATM networking products will most likely continue this tradition.

10.1   SWITCH MODELS

ATM switches have key characteristics that distinguish how they will operate. First, a switch architecture is either blocking, virtually non-blocking, or non-blocking. Next, the switch fabric at the heart of the machine is covered. Finally, the buffering method and its implications on performance are discussed. These three elements comprise the core of switching architecture.

10.1.1 ATM Switch Blocking Performance

A figure of merit, called *blocking,* attributed to ATM switches has been adapted from circuit switches. In circuit switches, if an inlet channel can be connected to any unoccupied output channel, up to the point where all inlets are occupied, then the switch is said to be strictly non-blocking. Another typical assumption is that the distribution of the inlet channels needing connection to specific outlet channels is uniformly distributed and random. Circuit switches are often then specified as *virtually non-blocking,* meaning that a small blocking probability occurs as long as no more than a certain fraction of inlet channels are in use.

This blocking concept has been extended to ATM switches. Keep in mind that an ATM switch utilizes a different paradigm than the circuit switch. When an inlet is connected to an outlet in a circuit switch, bandwidth is reserved and completely isolated from other connections. This is generally not true in an ATM switch — there are virtual connections (VPCs or VCCs) which arrive on input ports destined for potentially different output ports. Cell loss can occur, depending upon the statistical nature of this virtual connection traffic as it is handled by the cell switching method and buffering strategy. Cell loss in an ATM switch is the analog to call blocking in a circuit switch. The assumption is usually made that arriving virtual connection traffic is uniformly and randomly distributed across the outputs. The switch performance is then normally cited as virtually non-blocking (sometimes called non-blocking), meaning that up to a certain input load a very low cell loss ratio occurs.

The ATM switch blocking performance is sensitive to the switch architecture and the source traffic assumptions. This is an important practical consideration since, depending upon your traffic characteristics, one switch type may be better and less expensive than another.

10.1.2 Switch Architectures

Figure 10.1 illustrates several of the more common published switch architectures, or *switch fabrics,* implemented in the current and foreseen ATM switches. Of course, some switches are hybrids of these designs, and often one larger switch fabric is used to connect another, smaller switch fabric to yield one larger overall switch fabric. Each architecture is described in terms of its complexity, maximum overall speed, scalability, ease of support for multicast, blocking level, and other unique attributes.

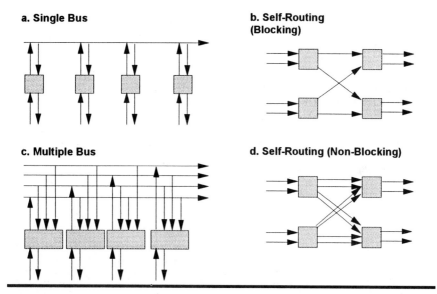

Figure 10.1 Example Switch Architectures

The single bus shown in Figure 10.1a is the simplest switch type. Basically ports are connected to a single bus, which can be implemented by a large number of parallel circuit board traces. The total speed of such a bus is usually in the range of 1 to 10 Gbps. There is some complexity introduced by the need for bus arbitration, which in combination with the buffering strategy controls the blocking level. Multicast is relatively easy to do since all output ports are "listening" to a common bus.

The multiple bus switch fabric shown in Figure 10.1 b extends the single bus concept by providing a broadcast bus for each input port. This eliminates the need for bus arbitration, but shifts additional requirements on controlling blocking to the outputs. In general, each bus runs at slightly greater than the port card speed (say 100 to 600 Mbps), and uses multiple circuit traces on a shared backplane that all the other cards plug into. Therefore the switch bandwidth is comparable to that of the single bus switch. An early version of this switch architecture was called the "knockout" switch because the outputs could receive cells from only a limited number of inputs at the same time in an attempt to make the architecture more scalable. For smaller switches every output may be able to receive from every input simultaneously. Another method is to employ arbitration to ensure that the output port does not receive too many cells simultaneously; however, this creates a need for input buffering and adds complexity. Multicast is natural in this switch type as it was for the single bus since each input is broadcasting to every output.

The self-routing switch fabrics shown in Figure 10.1c and 10.1d, such as the Batcher Banyan or cross point networks (see Chapter 4), have more complicated internal elements; however, these can be scaled to larger sizes due to the regular nature of the individual switching elements in Very Large Scale Integration (VLSI) implementations. These types of networks have been the subject of a great deal of research and investigation. See [10] for an in-depth review of these switch architectures. These switches generally do not support multicast well, and require either a separate copy network or special processing in the internal switching elements. If the self-routing network runs at the same speed as the input ports, then blocking can be quite high. Self-routing networks generally have some buffering within the switching elements. An augmented self routing fabric basically runs the internal matrix at a faster speed, or has multiple connections between switching elements.

Table 10.1 summarizes a comparison of the characteristics for these switch fabric architectures.

Table 10.1 ATM Switch Characteristics

Characteristic	Single Bus	Multiple Bus	Self-Routing	Augmented Self-Routing
Complexity	Low	Medium	Higher	Higher
Maximum Speed	1-10 Gbps	1-10 Gbps	1-200 Gbps	1-200 Gbps
Scalability	Poor	Better	Good	Best
Pt-to-MultiPt Support	Good	Good	Poor	Poor
Blocking Level	Low	Low-Medium	Medium	Low
Unique Attributes	Inexpensive	Inexpensive	Amenable to VLSI	Amenable to VLSI

Space division crosspoint switches introduced in Chapter 4 can also be used as an ATM switch fabric. The crosspoints are rearranged every cell time in a crosspoint switch.

10.1.3 Switch Buffering Methods

The buffering strategy employed in the switch also plays a key role in the switch blocking (i.e., cell loss) performance. Figure 10.2 illustrates various ATM switch buffering strategies analyzed and compared in the following sections.

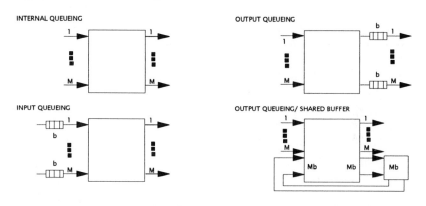

Figure 10.2 Switch Buffering Methods

The following parameters define the notation in the figure:

M = number of ATM switch ports
b = number of effective buffer positions per port

Chapter 4 gave an example of internal queueing in the section on address switching. Switch fabrics built with internal queueing have the potential to scale to large sizes. On the other hand, it is difficult to provide other functions — such as priority queueing, large buffers, and multicast — with internally queued switching fabrics.

Input queueing was simple to implement; however, it can suffer from a fatal flaw. With input queueing, when the cell at the Head Of Line (HOL) cannot be switched through the fabric, all the cells behind it are delayed. Chapter 15 presents an analysis that shows input queueing limits switch throughput to only 50 to 60% of the port speed. Therefore, input queueing alone is not adequate for many applications, and is usually employed in conjunction with other queueing methods.

As shown in Chapter 15 output queueing is theoretically optimal, and shared output queueing is the best in terms of achieving the maximum throughput with the fewest cell buffer positions. Most

ATM switches employ at least some output queueing for this reason. Real ATM switches may have a combination of input, output, and internal queueing.

10.1.4 Other Aspects of ATM Switches

Other factors which are important in comparing switch architectures are listed below.

- **Modularity**, which is defined as the incremental number of ports that can be added to a switch.
- **Maintainability**, which measures the isolation of a disruption on the remainder of the switch.
- **Availability**, which means that the operation continues in the presence of single or multiple faults.
- **Complexity**, often measured by logic gate counts, chip pin-out, and card pin-out in comparing switch implementations.
- **Flexibility**, which covers the capability to implement further packet processing functions easily.
- **Expandability**, which considers the maximum number of switch ports supportable by the architecture.

10.2 HARDWARE CATEGORIES

ATM technology has appeared in many of the major internetworking devices used today: switches, routers, bridges, hubs, multiplexers, and ATM interface cards for high-end workstations. Thus, ATM is disintegrating the fine line between local, campus and wide-area networking using a broadband, cell-based architecture that scales well in distance, speed, and network size..

ATM has found its way into three primary hardware environments: the Central Office (CO), the Customer Premises Environment (CPE) or Campus (Ca), along with other devices such as routers, switches, and hubs. CO-based switches often tout throughput in excess of 5 Gbps, while campus switches may claim throughput of less than 5 Gbps. However, throughput alone does not distinguish a CO switch from a Campus switch.

Central Office (CO) switches are generally larger and more industrial strength than their CPE counterparts. The CO environment often dictates DC power and the capability to scale to a large number of ports. In contrast, the customer premises environment often connotes a smaller capacity, reduced scalability, AC power, and less processing power then the CO switch. Premises ATM equipment, such as switches, routers, hubs, and bridges, also play an important role in ATM networking.

Figure 10.3 illustrates the different roles of a CO ATM switch and a CPE, or campus ATM switch. Starting with the customer premises ATM-ready router or hub, which allows Clients (C) and Servers (S) to communicate in a virtual network. A local ATM switch interconnects these ATM routers in what is called a collapsed backbone. Workstations and servers may be directly connected to the local ATM switch forming a high-performance work group. Local ATM capable PBXes may also be connected for access to public circuit data and voice services. Chapter 11 covers these local applications in more detail. The local ATM switch accesses the Wide Area Network (WAN) via a DS3 ATM User-to-Network Interface (UNI), which is connected to the ATM CO switch supporting a public ATM service. The CO switch can also switch the video and voice traffic to the voice switches and other networks, as shown in the figure.

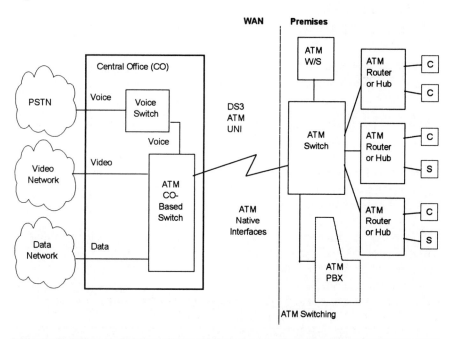

Figure 10.3 Campus and Central Office (CO) Switching Roles

10.2.1 Central Office (CO) ATM Switches

Central Office (CO) ATM switches are the backbone of an ATM network, often requiring throughput in excess of 5 Gbps. Typically they contain entirely ATM native (UNI) interfaces. The CO ATM switches set up calls for the CPE switches, much like CO voice switches set up calls for CPE PBXs. The CO environment dictates DC power and "rack and stack" physical , or more appropriately, the ability to expand both processing and port capacity.

Figure 10.4 illustrates a typical CO switch, manufactured by Digital Switch Corporation. The DSC MegaHub iBSS has single shelf 4 Gbps Broadband Switching Modules (BSMs). Up to ten BSMs can be interconnected to increase overall switching capacity. This switch supports DS1, DS3, OC-3, and OC-12 interfaces for ATM, Frame Relay and SMDS services.

Figure 10.4 DSC MegaHub iBSS Switch

10.2.2 Campus ATM Switches

Campus ATM switches are typically smaller than their CO counterparts, typically less than 5 Gbps, but providing many interfaces not found in the CO version, such as native LAN (i.e., Ethernet and Token Ring), MAN (i.e., FDDI and DQDB), SNA, X.25, and voice. Some provide protocol conversion, LAN emulation, and virtual networking. Campus switches and CPE typically run on AC (i.e., wall outlet) power, have smaller port capacity and less processing power than the CO switch.

Figure 10.5 illustrates a typical campus ATM switch manufactured by Synoptics. The Synoptics Lattiscell ATM switch has up to 5 Gbps of virtually non-blocking switching capacity. The Lattiscell switch supports DS3, OC3, 100 Mbps 4B/5B optical fiber, and 155 Mbps 8B/10B opritcal fiber interfaces carrying ATM cells. A network of Lattiscell switches is controlled by a proprietary Synoptics Connection Management System (CMS).

Figure 10.5 Synoptics Lattiscell ATM Switch

10.3 ATM SWITCH COMPARISON

This section defines major attributes of CO, CPE, and Campus ATM switches and summarizes how industry leaders compare. Many vendors responded to a survey for this book providing information on products commercially available as of 2Q94. A few vendors either chose not to respond or did not have products available by the 2Q94 timeframe. Current publications were used for details on vendor announcements for those vendors who did not respond. At the time of publication, IBM provided a summary of their ATM direction.

10.3.1 Manufacturers and Product Names

The ATM CO switch vendors who responded to our survey included: the ADC Telecommunications, Inc. A^{2}DS (Loral Data Systems Cell Packet Switch CPS-100), General Datacomm (GDC) Apex DV2, Siemens EWSXpress 3100 ATM Core Switch, Stratacom BPX, and the TRW BAS 2010C.

ATM campus switch vendors who responded to our suvey included: the Cascade B-STDX 9000, Cisco ATM switch, DSC MegaHub iBSS, LightStream 2010™, NEC Model-5 and Model-10, Newbridge Network Inc. 36150 ATMnet, and the Northern Telecom (NT) Magellan Gateway Switch. Several of these vendors provided graphics or photographs of their switches, presented throughout this chapter. Table 10.2 lists the vendors covered in this section, their product names, their switch sizes and types. The type is indicated as CO for Central Office and Ca for Campus. Some switches are categorized as either the CO or campus type.

There were a few vendors who did not have public information available on their products. These include the Alcatel Data Network 1100 HSS, Ascom Timeplex, AT&T GCNS2000, Motorola 6950 SoftCell, NET ATMX, and Telematics AToM1E6, to name a few.

10.3.2 Switch Architecture

The key attribute for switch architecture is the degree to which it is virtually non-blocking. ADC Telecommunications, Inc. A^{2}DS, Cascade B-STDX 9000, Cisco ATM switch, DSC MegaHub iBSS, General Datacomm (GDC) Apex DV2, LightStream 2010™, NEC

Model-5 and Model-10, Newbridge Network Inc. 36150 ATMnet, Northern Telecom (NT) Magellan Gateway Switch, Siemens EWSXpress 3100 ATM Core Switch, and the TRW BAS 2010C all responded as non-blocking, although some were more specific as to the probability of blocking, such as $1*10^{-10}$ — virtually non-blocking. The Stratacom switch offers non-blocking performance when operating at 4.43 Gbps, but can also offer virtually nonblocking 9.6-Gbps full duplex throughput if required. All other vendors did not state the degree of blocking that their switches achieved.

10.3.3 Buffer Capacity

Buffer capacity defines the total number of input, internal and output storage positions for cells. More buffer capacity can mean better throughput for protocols such as Transmission Control Protocol (TCP), however large buffers can create more latency and delay during peak traffic conditions. As a basic rule, the larger the buffers, the less traffic will be dropped, but the greater the delay. Buffer capacity can be measured in cells per port, cells per switch, or in the delay for a certain line rate. To compare the two, DSC offers 4096 cells corresponding to a 1.25 seconds delay at a DS1 rate, 16,384 cells equating to about 170 ms worth of delay at the DS3 rate, and 65,535 cells corresponding to about 186 ms worth of delay at an OC-3 rate.

TRW reported a switch-wide total of 846,000 cells. Stratacom measures buffer capacity as 24,000 cells per DS3 port. Cascade defines a maximum capacity of 32,000 cells. Cisco reported 16,000 cells. DSC specified 65,535 cells per OC-3c, 16,384 cells per DS3/E3, and 4096 cells per DS1/E1. Siemens specified 7680 cells per 64x64 switching network (shared). NEC offers 128 cells per output buffer on the Model 10, and 1000 cells per line on the Model 5. ADC measures buffer capacity is 1.3 seconds for an ATM DS3 port and 0.3 seconds for an ATM OC-3 port. GDC buffers vary depending on the series card (C or H) used. Both cards can support two 31-cell deep FIFO buffers per input port and an additional 31-cell deep buffer on the output side. The H series cards support an additional 1100-cell deep buffer for all remaining traffic. All other vendor buffering schemes were not provided or not available.

Table 10.2 ATM Switch Vendors

Switch Vendor	Switch Name	Size (Gbps)	Type
ADC	A^{2}DS	5	CO
Alcatel Data Networks	1100 HSS	10	CO
AT&T	GCNS2000	*	CO
Cascade	B-STDX 9000	1.2	Ca
Cisco	ATM Switch	2.5	Ca
DSC	MegaHub iBSS	4+	CO/Ca
FORE	ASX-100	*	*
Fujitsu	FETEX-150	*	CO
GDC	APEX DV2	6.4	CO
GTE	SPANet	1.2	CO
LightStream	2010	2.0	Ca/CO
Motorola	6950 SoftCell	1	Ca
NEC	Model-5 and Model 10	2.4	Ca/CO
NET Adaptive	ATMX	1.2	Ca
Newbridge	36150 ATMnet	2.5	Ca/CO
NCR	Univercell	1.6	Ca
NT	Magellan	1.2	CO
Siemens	EWSXpress 3100	10	CO
Stratacom	BPX	9.6	CO
Synoptics	LattisCell	2.5	Ca
Telematics	AToM 1E6	2.5	Ca
TRW	BAS 2010C	20	CO

10.3.4 Buffering Method

The buffering method can be input, fabric, or output. Synoptics and TRW use input, output, and fabric buffering. Cisco, GDC, LightStream, NEC (Model 5), and Stratacom use input and output queueing. ADC, Cascade, DSC, NEC (Model 10), and Newbridge use output buffering. Siemens and NT use fabric buffering. All other vendor buffering methods were not provided or not available.

10.3.5 Switching Delay

Delay is measured (in μs) as the total one-way delay through the switch. Large amounts of delay may effect some forms of traffic adversely, such as video and voice- or session-oriented traffic, such as SNA traffic.

The delay varies widely based on switch fabric. Stratacom quotes 1.2 μs, NEC (both models) <5μs, Cascade <10μs, DSC <15 μs (without internal switch congestion), Synoptics 20 μs, TRW 26 μs, GDC <40 to 50 μs, Cisco and NT offer 40 μs average delay, and Siemens 250 μs (average). ADC quotes a maximum 16-cell delay. All other vendor delay figures were not provided or not available.

10.3.6 ATM Interfaces

The typical ATM UNI physical interfaces found in these switches include DS1/E1, DS3/E3, OC-3/STM-1, OC-12/STM-4, 4B/5B optical fiber, and 8B/10B fiber as defined in Chapter 8. Synoptics supports DS3/E3, OC-3/STM-1, 4B/5B, and 8B/10B. Siemens supports DS1/E1, DS3/E3, OC-3/STM-1, and OC-12/STM-4. Stratacom supports DS1/E1 and DS3/E3, with near-future plans for both OC-3/STM-1 and OC-12/STM-4. TRW supports DS3/E3 and OC-3/STM-1, with plans to support OC-12/STM-4 in the near future. DSC supports DS3/E3, OC-3/STM-1, and OC-12/STM-4. Cascade supports both DS1/E1 and DS3/E3 interfaces. Newbridge supports DS1/E1, DS3/E3, OC-3/STM-1, and 4B/5B interfaces. Cisco supports DS3/E3, OC-3/STM-1, and 4B/5B, with OC-12/STM-4 in development and DS1/E1 supported through an external ATM DSU with plans for future integration. GDC supports DS1/E1, DS3/E3, OC-3/STM-1, E2, X.21, X.27/RS449, and HSSI, with future plans to support OC12C/STM-4. LightStream and NEC (Model 10) support DS1/E1, DS3/E3, and OC-3/STM-1 interfaces. The NEC Model 5 supports OC-3/STM-1 and 4B/5B interfaces only. NT supports DS1 Isochronous, DS3, OC-3, and 4B/5B UNI interfaces. Motorola Codex supports nx64, DS1/E1, and DS3 interfaces. All other vendor interfaces supported were not provided or not available.

10.3.7 Maximum Number of Ports

It is also important to note the maximum number of ports, by speed, supported by the switch. Most CO switch providers offer only Wide Area Network (WAN) ports, leaving the Local Area Network (LAN) support to the CPE and premises switch vendors. Most Campus switches provide both Wide Area Network (WAN) and Local Area Network (LAN) ports.

Stratacom offers 36 DS3 ports in their broadband shelf, while supporting 360 ports in their narrowband shelf. DSC supports up to 150 OC-3, 360 DS3, or 1024 DS1 ports. Motorola Codex offers up to 1000 LAN and WAN ports, of which six are DS3s and 64 are DS1s. NEC (Model 10) can support up to 96 DS1, 48 DS3, or 60 OC-3 WAN ports. The NEC Model 5 can support up to 17 ATM TAXI or OC-3 ports. Siemens supports up to 62 WAN ports, expandable to 256 ports. TRW supports 256 DS3, 64 OC-3, or 16 OC-12 ports (future). LightStream offers up to 72 DS1s and 15 DS1s. ADC offers up to 64 WAN ports (from one to four per module, with a maximum of 16 modules). Cascade offers up to 140 DS1 WAN ports. Cisco offers up to 30 LAN ports and up to 40 WAN ports (with up to 300 channels). GDC offers up to 64 LAN or WAN ports (including Ethernet, Token Ring, and TAXI). Alcatel offers up to 29 ports at speeds from 64 kbps to 155 Mbps. Synoptics offers up to 2 DS3 WAN and up to 16 LAN ports, including OC-3, TAXI, and 8B/10B. NT offers up to eight 155-Mbps WAN ports. Newbridge supports up to 16 ports. All other vendor port maximums were not provided or not available.

10.3.8 AAL Support

What ATM Adaptation Layer (AAL) types are supported (i.e., AAL1, AAL3/4, and AAL5)? ADC, Cascade, DSC, GDC, Siemens, Stratacom, and Synoptics all support AAL1, 3/4, and 5. Cisco and NEC (both models) support AAL 3/4 and AAL 5. LightStream and Newbridge support AAL 1 and AAL 5. NT supports AAL 1 for its DS1 interface. Only the TRW switch did not support standard AAL classes of service. AAL support information from all other vendors was not provided or not available.

10.3.9 Interworking Interfaces

ATM switches also support interworking for a variety of LAN, MAN, and WAN interfaces, such as: Ethernet, Fast Ethernet, FDDI, Frame Relay (FR), SMDS, and circuit emulation. The ADC and Stratacom switches support SMDS, circuit emulation, and frame relay. Cascade supports SMDS, circuit emulation, and FR, as well as a FRAD functionality. DSC supports SMDS, circuit emulation, and frame relay. Cisco supports Ethernet, SMDS, Fast Ethernet (planned), frame relay, Token Ring, FDDI, X.25, SDLC, and IBM Channel. LightStream supports Ethernet, circuit emulation, FDDI, frame relay, and frame forwarding. Siemens supports SMDS, circuit emulation, frame relay, and DXI. Newbridge supports Ethernet, circuit emulation, Local ATM interfaces, and JPEG video. Motorola Codex supports circuit emulation and frame relay. Alcatel supports frame relay, SMDS, and X.25. NT supports TAXI and circuit emulation. TRW did not support any of the interfaces mentioned above. All other vendor support for interworking interfaces was not provided or not available.

Figure 10.6 illustrates the Nippon Electric Company (NEC) NEAX61 Model 10 switch and local control terminal. This switch has 2.4 Gbps of non-blocking switch capacity, and supports DS1, DS3, and OC-3 interfaces. The NEC Model 10 also supports Ethernet, Frame Relay, and SMDS.

Figure 10.7 illustrates the GDC APEX DV2 switch, offering the greatest number and variation of interfaces of all switches surveyed.

10.3.10   Number of Quality of Service (QoS) Classes (Priorities)

The number of QoS Classes supported by an ATM switch varies. Multiple classes of service allow the assignment of lower priority or higher priority to different types of traffic. Delay-sensitive traffic can be given a higher priority over delay-insensitive traffic. Quality of Service classes are in addition to, and should not be confused with, the use of the Cell Loss Priority (CLP) bit which supports two levels of loss prioritization, as will be detailed in Part 5.

ADC supports up to 64 QoS classes, Stratacom 32 QoS classes, and TRW and NT both support eight. Cascade, Cisco, DSC, GDC, and LightStream all support four QoS classes. Siemens supports three QoS classes. The NEC Models 10 and 5 switches support two and four QoS classes, respectively, and Synoptics supports two. All other vendor QoS class support information was not provided or was not available at the time of publication.

Figure 10.6 NEC NEAX61- Model 10 Switch

Figure 10.7 GDC APEX-DV2 Switch

10.3.11   PVC and SVC Support

ADC supports both ATM PVCs and SVCs compliant with the ATM Forum and Q.93B specifications. Cascade supports ATM PVCs and FR SVCs. TRW plans to support SVCs when the standards become available. Cisco, DSC, GDC, LightStream, Siemens, and Synoptics support both ATM PVCs and SVCs. Newbridge and both NEC models support PVCs and plan to support SVCs. NT, Stratacom, and

'TRW support ATM PVCs. All other vendor support for PVCs and SVCs was not provided or not available.

10.3.12   Point-to-Multipoint Support

Does the switch support point-to-multipoint connections? This is sometimes referred to as multicast. ADC, Cascade, Cisco, DSC, GDC, LightStream, Newbridge, both NEC models, NT, Siemens, Synoptics, and TRW all support point-to-multipoint connections. Often, this feature is supported in software at lower speed switch fabrics and in hardware at higher speeds. Stratacom has included point-to-multipoint in its future plans. Information on support for point-to-multipoint from all other vendors was not provided or not available.

10.3.13   Traffic Control

It is also important to know the number of traffic parameters that are monitored per connection. TRW supports 8, while Stratacom supports nearly 20 per connection, most for network management purposes. Siemens supports four per connection. The NEC Model 10 supports several, and Cascade supports one traffic parameter per ATM connection — at the peak cell rate. DSC also supports four traffic control parameters: PCR, SCR, CDV, and MBS as defined in Chapter 12. GDC supports these four and more. Cisco monitors three traffic parameters per connection, while NT monitors two. Information on traffic control from all other vendors was not provided or not available.

10.3.14   Congestion Control

Does the switch support congestion control for CPE? ADC, Cascade, Cisco, DSC, GDC, LightStream, NT, Siemens, Stratacom, and TRW all support selective cell discard for CPE. Both NEC models support selective discard based on connection priority, not CLP bit. Support for congestion control from all other vendors was not provided or not available. No vendor quoted support for EFCI.

Figure 10.8 illustrates the Stratacom Broadband Packet eXchange (PBX) switch. The BPX has over 9.6 Gbps of switching capacity with a number of interfaces. Stratacom provides a proprietary, closed loop congestion control scheme called "Foresight" on both its Frame Relay product (the IPX) and the BPX.

Figure 10.8 Stratacom BPX Switch

10.3.15   VPI/VCI Bits

ATM switches vary in the number of VPI and VCI address bits they support. TRW supports 8 VPI bits and 12 VCI bits on the NNI, and 16 VCI bits on the UNI and NNI. Stratacom uses all VPI and VCI addressing bits, but uses 4 of the VPI address bits to pass along control and connection management information. Both NEC models also support all VPI/VCI bits. Cascade supports 20 bits across the VPI and VCI, while GDC supports all bits. ADC supports up to 15 total VPI bits and VCI bits. DSC supports 16 VCI and 8 VPI bits on the UNI, and 16 VCI and 12 VPI bits on the B-ICI. Cisco supports 24 bits, 8 VPI and 16 VCI. NT supports 24 bits on the UNI and 28 bits on NNI. LightStream supports 0 VPI bits and 15 VCI bits. Siemens supports 4000 VCCs per port. Information on support for VCI and VPI bits from all other vendors was not provided or not available.

10.3.16   Bridging

CPE or campus switches usually handle classic LAN traffic and encapsulation. ATM Campus switches currently support proprietary encapsulated bridging of LAN protocols over ATM. This method is commonly used for protocols which cannot be routed, such as NetBios, DEC LAT, and SNA LLC2. Cisco, GDC, Newbridge, and Synoptics were the only Campus switch vendors reviewed that supported LAN bridging. GDC supports AAL5 for bridging or routing through their Quad AUI Ethernet card. LightStream also supports AAL5 for MAC level "self-learning" bridging and IP routing. Information on support for bridging encapsulation from all other vendors was not provided or not available.

10.3.17   Virtual LAN Capability

Some Campus ATM switches support a virtual LAN capability. "Virtual LANs" use software to define "virtual workgroups" which logically band together multiple users on different LANs in the same virtual LAN segment. Thus, each workstation can belong to multiple virtual LANs. Cisco, GDC, Newbridge, Siemens, and Synoptics were the only switch vendors to support virtual workgroups. This capability can be related to the bridging function. Information on support for virtual LAN capability from all other vendors was not provided or not available.

10.3.18   Network Management

Common network management protocols supported include the Simple Network Management Protocol (SNMP), Common Management Information Protocol (CMIP), or proprietary protocols.

LightStream, NT, Stratacom, and TRW all support SNMP. Cascade, DSC, and Siemens support both SNMP and CMIP. ADC and GDC support SNMP, CMIP, and their proprietary Network Management System (NMS). Synoptics supports SNMP and their proprietary NMS. Newbridge supports CMIP and CPSS (a proprietary NMS) and plans to support SNMP. The NEC Model 10 supports CMIP, while the NEC Model 5 supports SNMP. Almost all vendors offer some form of proprietary NMS as an option, such as Stratacom's StrataView Plus workstations employing an Informix (SQL) database. GDC's NMS 3000 offers one of the only object-oriented, hierarchical database constructed systems available today. Cisco supported the greatest suite of network management protocols, including SNMP, CMIP (via a translator), IBM NetView, LAN Net Manager, DEC MOP, and Windows PC Configuration Builder. Network management information from all other vendors was not provided or not available.

10.3.19   Redundancy

ATM hardware may offer some level of redundancy at the node, card, and module level. M for N (abbreviated M:N) redundancy is defined as M spares for a set of N active components, such as CPUs, port cards, power supplies, or switching matrix. ADC, Cascade, DSC, GDC, Newbridge, NT, and Stratacom all offer 1:1 redundancy on shared elements, while TRW offers both 1:1 and 1:N redundancy, LightStream offers 1:1 redundancy for their switch matrix and network processor boards, and the NEC Model 10 offers full redundancy. The Siemens' switch is fully redundant. Information on redundancy from all other vendors was not provided or not available.

10.3.20   Availability

All information on switches detailed above was publicly available at the time of publication. Note that other capabilities include automatic network configuration after the addition or removal of hosts, switches, trunks, links, or any other piece of common equipment; automated ATM network routing, and ATM address resolution; detection and propagation of topology changes; and link load balancing.

As a side note, out of all the vendor information submitted for this text, GDC, NEC, and Stratacom provided the most detailed and thorough explanation of their products and features.

10.3.21   Additional Information

There were a few other vendors who, while not having detailed information available, are worth mentioning. DEC has announced planned support for ATM switches, hubs, and workstation adapters. The DEC FDDI-based GIGASwitch has the capability to support DS3/E3 and OC-3 ATM interface modules in up to 22 ports. The DEChub 900 also supports ATM interfaces and operates at up to 2 Gbps. Workstation TURBOchannel adaptors allow DEC 3000 AXP workstations to communicate via an ATM over SONET interface supporting PVCs and SVCs. DEC also has PCI adapter cards planned for 1995, and a switch called AN2 which may support up to 12.5 Gbps in the works.

NCR plans to support its Univercell family of workgroup (4-port, 1.5-Mbps port speed, 0.5 Gbps switching speed), departmental (8-port, 51- to 622-Mbps port speed, 1.6-Gbps switching speed), and backbone (12-port, 51- to 622-Mbps port speed and 3.2-Gbps switching speed) ATM switches, to be released in that order, or 3.08-Mbps, 1.6-Gbps, and unknown in that order. NCR SmartHUB XE supports an ATM module which provides 100-Mbps port speed and 0.8-Gbps switching speed.

Telematics International Inc. will be introducing their AToM 1E6 backbone ATM switch and Neutron local ATM access device in late 1994. These devices plan to support up to 2.5-Gbps throughput and 16 slots to support DS1, DS3, or OC-3 interfaces. See References 4, 5, 6, 7, 8, and 9 for more details on these ATM switches.

10.4   LOCAL ATM — ROUTERS, SWITCHES, HUBS, AND END SYSTEMS

The various types of ATM-based equipment that are either available now or under development for use primarily in the local area are reviewed in this section.   In some cases this also spans both the metropolitan and wide areas.   The principal categories are those in place today: switches, routers, and hubs.   There are also a number of other, usually lower level, functions that can be used in a building block manner, such as multiplexers, concentrators, CSU/DSUs, and bridging   devices,   are   also   summarized.   Representative manufacturers and products are mentioned as examples, but this is by no means an exhaustive list.   See References 1 and 2 for more information,   contacts,   and   a   review   of   specific   ATM   CPE manufacturers.

10.4.1 Evolution of Routers, Bridges, and Hubs

Perhaps the greatest driving factor for routers, bridges, and hubs was the LAN.   Due to diverse markets, technologies, and protocol suites, a need evolved to make diverse LANs speak one language (or at least provide a translation between similar languages on similar types of LANs — e.g., Ethernet to Ethernet, Token Ring to Token Ring).   When bridges and routers first came along, they were designed to deal with local area networking.   Now the functions of both begin to merge, and with the use of increased processor speeds and technologies (i.e., Reduced Instruction Set (RISC) processors), with reduced costs, they begin to support the local and wide area networking from low access speeds of DS0 up to DS1 and even DS3 speeds and beyond through ATM and SONET interfaces.

Routers, bridges, and hubs each provide functions which can either be provided separately or together in one piece of equipment.   Each provides protocol support for certain levels of the OSI reference model.   Bridges provide level 1 and 2 support; routers and hubs, 1, 2, and 3; and a device called a gateway, 1 through 7.   Figure 10.9 relates the OSI Reference Model (OSIRM) layers to bridges, routers, and hubs.

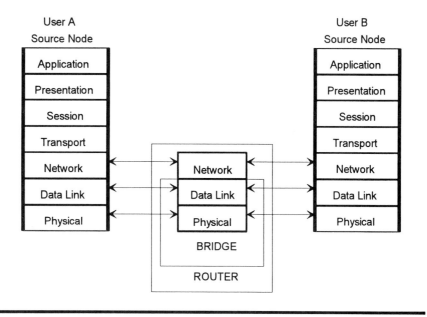

Figure 10.9 Router Communications via the OSIRM

10.4.2 Local ATM Switches

Local ATM switches are connected together in a network, and may also interconnect with other networks as shown in Figure 10.10. Usually a private switched network is connected to one or more public switched networks. Even the name *network* used in switching as opposed to *subnetwork* points to the fact that interworking is more difficult in switched networks than in routed networks. Switches are connection-oriented devices. Users interface to switches and communicate the connection request information via a user-network signaling protocol. Between switches an interswitch protocol may be used. Networks are interconnected via a more complex network-network signaling protocol. Signaling functions may be emulated by network management protocols where individual ATM cross-connects are made.

Examples of local ATM switch products were reviewed in detail in the last section.

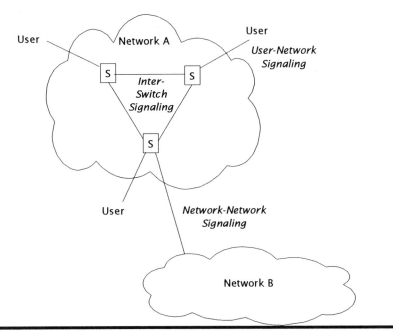

Figure 10.10 Switching Interfaces, Functions, and Architecture

10.4.3 ATM Routers

A router handles multiple protocols and therefore may need to look inside a sequence of ATM cells in order to perform this function. Figure 10.11 illustrates the range of interfaces and scope of routing. Routers are often connected by real or virtual circuits. Routers have very sophisticated software, and now are being delivered with special-purpose hardware to increase packet routing throughput. Current routers can forward on the order of 100,000 IP packets per second.

A key function of routers is automatically discovering the addresses of devices connected to a network of routers using an interior routing protocol, or even to a network of networks using an exterior routing protocol. Indeed even the naming of networks as subnetworks as a part of a larger network has proven to be a very scalable concept. Packets are routed based upon the destination address, sometimes using the source address as well, or even an end-to-end route specification. Routers connect dissimilar protocols by way of routing and data protocol conversion. Routers handle both connection-oriented and connectionless services. Routers also interconnect dissimilar media via media conversion. Routers also continually monitor the state of the links that interconnect routers in a network, or links with other networks through a variety of routing protocols, such as the Routing Information Protocol (RIP), the Border Gateway Protocol (BGP), or the Open Shortest Path First (OSPF) routing protocol. Through these protocols, routers can discover network topology changes and provide dynamic rerouting.

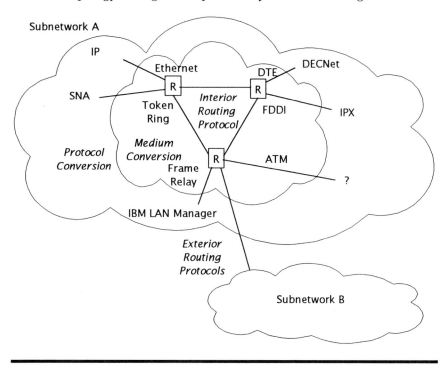

Figure 10.11 Routing Interfaces, Functions, and Architecture

Modern routers will have support for ATM LAN emulation, ATM interface and trunk cards, and/or ATM switching capabilities. ATM-based routers can function as both access devices and as full-blown ATM switches. As access devices they accept multiple protocols and either route them to another port on the router, or convert them to ATM cells for transport over an ATM network. They can also route LAN packets to ATM switches via the Data Exchange Interface (DXI), where an external ATM CSU/DSU then converts the DXI packet into a stream of ATM cells as described in Chapter 17. As switches, ATM routers also have the capability to switch ATM cells between multiple ATM interface cards. Two prime examples of routers with ATM capabilities include the Cisco 7000 and Wellfleet Backbone nodes.

The Cisco Systems 7000 offers DS3/E3, OC-3/STM-1, and 4B/5B ATM interfaces. Up to 30 LAN or 40 WAN (up to 300 channels) can be supported, including Ethernet, Token Ring, FDDI, frame relay, SMDS, ATM, X.25, and ISDN. It supports AAL 3/4 and AAL5, and up to four QoS classes. PVCs, SVCs, and point-to-multipoint are supported, along with three traffic parameters per connection for traffic control. It also supports selective cell discard and 24 total VPI/VCI bits. Routing and bridging are done in compliance with RFC 1483 and LAN emulation standards, and virtual LAN workgroups are supported. An entire suite of network management protocols are supported, including SNMP, CMIP, NetView, LAN Net Manager, DEC MOP, and Windows PC Configuration Builder.

The Wellfleet Backbone Node offers DS1/E1 and DS3/E3 DXI and standard OC-3/STM-1 ATM UNI interfaces. Up to 52 LAN or WAN ports can be supported, including Ethernet, Token Ring, FDDI, frame relay, SMDS DXI, ATM, X.25, and the Point-to-Point Protocol (PPP). It supports AAL 3/4 and AAL5, and up to 3 QoS classes. PVCs and multicast capability are included, along with bridging encapsulation. Eight VPI and 15 VCI address bits are supported. SNMP and a proprietary network management system are supported, with 1:3 power supply redundancy.

10.4.4 ATM Hubs

Figure 10.12 illustrates the wiring collection, segmentation, and network management function a hub performs. Usually many Ethernet or Token Ring twisted pair lines, in some cases FDDI, and in the future ATM over twisted pair to individual workstations, are run to a hub usually located in the wiring closet. Hubs allow administrators to assign individual users to a resource (e.g., an Ethernet segment), shown as an ellipse in the figure, via network management commands. Lower level hubs are often connected in a hierachy to higher level hubs, sometimes via higher speed protocols such as FDDI, and in the future ATM over optical fiber and high-grade twisted pairs. Hubs are often employed in a hierarchical manner to concentrate access for many individual users to a shared resource, such as a server or router as shown in Figure 10.12. The highest level hubs are candidates for a collapsed backbone architecture based on ATM, also supporting high-speed access to shared resources such as routers and switches. These subjects will be covered in detail later in this chapter. If a user is added, changed, or moved from one office to another, the administrator performs the corresponding action to add, change, or move the user.

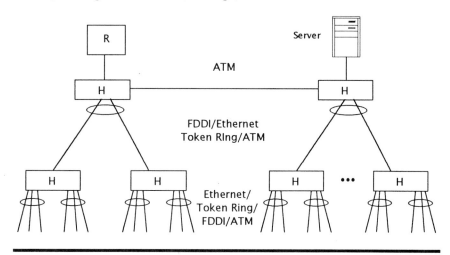

Figure 10.12 Hub Interfaces, Functions, and Architecture

Router and hub ATM interface cards support a wide range of industry standard interfaces, such as UTP, DS1, DS3, 100 Mbps (4B/5B), 140 Mbps (4B/5B0, 155 Mbps Fiberchannel, OC-3, or even OC-12. They typically come equipped with their own microprocessors. Hub cards have a broad range of functionality, including PVCs and SVCs, multicast (point-to-multipoint) and broadcast, AAL 3/4 and AAL 5 processing, and guaranteed QoS; support TCP/IP and ATM APIs, and are SNMP MIB and CMIP compliant. ATM adapters are designed to support a variety of system buses, such as EISA, ISA, VME, SBUS, TURBOChannel, NeXTbus, NuBus, GIO32 and GIO64, Microchannel, PCI, and FutureBus. Also look for hooks into the HP700, Novell, and Microsoft APIs. The leading vendors with ATM interface cards include Cabletron, cisco Systems, Synoptics, and Wellfleet. Another example is the 3COM hub and router ATM support through a CellBuilder module for LAN frame to ATM cell conversion.

10.4.5 End System Interfaces

What makes a user require an ATM interface on their end system? Users typically first try to segment their existing Ethernet LAN into smaller and smaller segments until they reach the threshold of Ethernet switching which provides 10 Mbps to each user. Desktop videoconferencing — the integration of the phone, computer, and video — is becoming a reality for many corporations, which will require the capabilities of ATM.

Providing ATM to the desktop is a controversial issue for some. Many hardware vendors such as SUN, HP, SGI, DEC, and NeXT have purchased or developed ATM interface adapter cards and are also pursuing the application interfaces to their existing software operating systems and network management packages.

For some users, ATM to the desktop is too costly at this point. However, the price for an ATM interface card is dropping rapidly, and in a few years may equal the cost of today's LAN adapter cards. Although ATM speeds are greater than many current applications, the rapid growth in computer performance and increased need for communication will change this situation. When the workstation becomes faster than the current LAN, then an upgrade is in order. An upgrade to ATM will be the choice of many.

In the Campus environment, bandwidth is close to free, especially once fiber is run. Therefore, the cost of the local networking equipment and workstation interface cards is important.

Adapter cards are also available that can encapsulate IPX and DECnet. ATM network interface cards (NICs) can be purchased for little more than $1000, and prices are decreasing. In fact, the Yankee Group predicts sales of ATM LAN equipment will be $38.5 million in 1993 and $619 million in 1996.

There are a limited number of native ATM interfaces that can be installed in current end systems. This type of interface is necessary for the high-performance ATM work groups where the high bandwidth, multiple classes of service, and flexibility of ATM require extension to the desktop.

★ Fore Systems adapter cards and device driver software are available for Sun, IBM, SGI, VME, and DEC computers.

★ N.E.T. compatible adapter cards and driver software are available for Sun Sparcstations by Interphase, N.E.T. and EISA bus PCs adapter cards by Network Systems Corporation and Transwitch.

★ Synoptics compatible adapter cards and driver software for Sun Sparcstations are made available by Network Peripherals and EISA bus PCs adapter cards by Transwitch.

This is currently a relatively short list, but it is expanding. Almost every adapter card works optimally with only one manufacturer's switch, but there is a degree of interchangability enabled by standards.

10.4.6 Other Local ATM Devices

This section covers the following other categories of local ATM devices: ATM multiplexers/concentrators, ATM bridging devices, and ATM CSU/DSUs. Figure 10.13 presents a pictorial representation of the interfaces and function of these devices. An ATM multiplexer takes multiple, often lower-speed ATM interfaces and concentrates them into a smaller side of often higher speed ATM trunk interfaces. A bridging device takes a bridgeable protocol, such as Ethernet or Token Ring, and connects it over an ATM network. This makes it appear to the user devices as if they were on the same shared medium, or segment, as shown by the ellipse in the figure. A Channel Service Unit/Digital Service Unit (CSU/DSU) takes the frame-based ATM DXI interface over a High-Speed Serial Interface (HSSI) and converts it into a stream of ATM cells. Chapter 17 defines the operation of the ATM DXI protocol in detail.

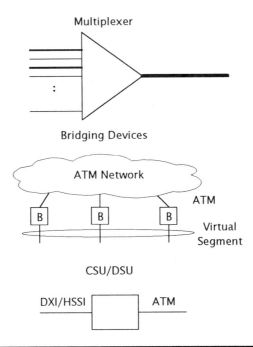

Figure 10.13 ATM Multiplexers, Bridges, and CSU/DSUs

10.4.6.1 ATM Multiplexers / Concentrators

A multiplexer, or concentrator, takes multiple ATM interfaces as input and concentrates these into a single ATM interface. A switch can also perform multiplexing; however, multiplexers are usually less expensive than switches because that have fewer functions. Examples of ATM multiplexers or concentrators include:

★ ADC Kentrox ATM Access Concentrator with a DS1 ATM UNI interface (AAC-1) and a DS3 interface (AAC-3), FR DXI, ATM DXI, and SMDS DXI

★ W/ATM announced by Digital Link

★ Fujitsu SMX-6000 ATM Service Multiplexer with DS3 and OC-3 WAN cell conversion to a public ATM network switch

★ Hitachi AMS 5000 ATM access node system for multiplexing Time Warner ATM home sets to AT&T ATM network switches

★ CRAY announced joint venture with FORE Systems to develop an HIPPI-to-ATM interface.

10.4.6.2 ATM Bridging Devices

A bridging device encapsulates a bridged protocol, such as Ethernet, and emulates the encapsulated protocol's bridging functions. These include self-learning and self-healing capabilities. Examples of ATM bridging devices are the Ethernet bridging port cards in the Newbridge 36150 and the Lightstream 2010.

10.4.6.3 ATM CSU/DSUs

A Channel Service Unit/ Data Service Unit (CSU/DSU) performs the conversion from a HSSI DTE/DCE interface operating at up to 50 Mbps utilizing the frame-based ATM DXI protocol to an ATM UNI interface as defined in Chapter 17. Examples of ATM CSU/DSU products include:

★ ADC Kentrox ADSU

★ Digital Link DL3200.

10.4.7 The RUSH to ATM

The almost Universal rush of Router, Switch, and Hub (resulting in the dyslexic acronym RUSH) manufacturers to embrace ATM in the industry trade press is dizzying. Simultaneously, it appears that everyone is moving into the others area of functionality. Indeed, many products have a split personality: hubs and switches with router cards, hubs doing switching, and routers doing switching. It also seems like the new wave of ATM hardware will contain all of these features and functions. This is evidenced by strategic partnerships like the Cisco Systems and Cascade joint development efforts. Where will it all lead? Chapter 23 addresses this topic in a discussion on the future of ATM.

10.5   ATM IN THE PC

ATM is not likely to make its way into the mid-1990s generation of PC. First, the PCs of the mid-1990s are internal bandwidth constrained to an average of 33-50 MHz buses, where the ATM LAN offers speeds of 100 Mbps and 140 Mbps. And even if the bus speed could be matched, peripheral speed could not. ATM-ready PCs will eventually show up, and this will make multimedia to the desktop a reality.

10.6   ATM IN THE PBX

Users, vendors, and carriers are also adopting the ATM switching approach to PBXes. PBXes provide automatic setup of circuits between telephone sets today, which most current ATM switches do not plan to do. Traditional PBX call processing and call control are slowly being built into ATM switching architectures. The PBX vendors do not see this happening, in fact the opposite — they see ATM-ready PBX products replacing ATM LAN switches. The most likely scenario will be a coexistence of ATM-ready PBXs and ATM switches with call processing and control capability in both the campus and wide area. ATM interfaces are beginning to appear in PBXs. One example is the Siemens and ROLM product.

10.7 ATM IN CROSS-CONNECTS

ATM is also moving into the cross-connect arena. Witness the plans for Tellabs to integrate ATM switching into the Tellabs Titan 5500 SONET cross-connect system. Also witness DSC's announced intent to develop ATM technology in its cross connect family. Whether the distinction between a switch and a cross-connect that developed in the era of TDM will apply to ATM remains to be seen. Usually a switch is more intelligent (and expensive) than a cross-connect for a comparable port speed.

10.8 ATM CHIP SETS

This section cites a few examples of ATM chip set development. The basic 802.6 Physical Layer Convergence Protocol (PLCP) chip sets closely map to the DS3 ATM UNI. Many of the hardware vendors have joined the semiconductor firms in producing both ATM and AAL layer chipsets. SONET chips are also available to work together with ATM chips. Let's take a look at a few current ATM chip set examples.

Integrated Device Technology Inc. (IDT) has announced a 51.84 ATM transceiver using the Partial Response Class IV (PRIV) IBM technology. This will provide ATM to desktops, workstations, and servers over UTP-3 and UTP-5 cable. IBM and HP also support this proposal.

Fibermux and LSI Logic Corp have teamed to develop a chip set called ATMosphere, designed to provide LAN switching and virtual LAN management for migration to ATM. The chip set is composed of a 4-chip ASIC, 16-port input, 16-port output, crosspoint ATM switch. The chip supports 16 full-duplex 200-Mbps non-blocking channels for a total capacity of 1.2-Gbps.

LSI Logic ATMizer Megacore provides a completely reprogrammable ATM termination solution in one chip. This chip contains a RISC-based ATM Processing Unit (APU) which allows customers to update their ATM products as the standards evolve [3].

Early leaders in ATM chip set development include NET Adaptive, FORE Systems, and MPR Teltech. National Semiconductor has also entered the race and is now a primary manufacturer of ATM chip sets, adapter boards, and motherboards.

Pacific Microelectronics Centre and Sierra Semiconductor Corp. (PMC-Sierra) led the effort in the development of a PHY layer LAN chip set based on SONET/SDH and ATM in the SONET-ATM UseR Network (SATURN). DS3 ATM, SMDS, and STS-3 interface chip sets soon followed.

In another industry-leading move, PMC-Sierra and Mitel Semiconductor have formed the ATM Alliance to develop ATM chip sets. TI also plans to offer an ATM chip set.

In the international arena, Fujitsu Microelectronics has developed ATM chip sets for its current switch and future workstation ATM products. Oki, Mitsubishi, and Matsushita have also developed various ATM chip sets. The international member-formed Strategic Microelectronics Consortium (SMC) was conceived to develop ATM chip products, and now this consortium covers 90% of the Canadian microelectronics industry.

These are only a few examples as of time of publication. Pending the success of ATM, it is expected that in the next two years most major chip manufacturers will be developing some form of ATM chip set, for use in any of the ATM devices described in this chapter.

10.9 IBM'S ATM ENTRY

IBM announced in late 1993 a strategy for ATM capability to be implemented across most IBM networking products. IBM has defined a four-point ATM strategy:

1. Provide a range of products, such as: VLSI ATM chip sets, ATM interface adapters, ATM-capable workstations and servers, a Campus ATM switch, a WAN ATM switch, along with ATM system and network management platforms.

2. Offer IBM Broadband Network Services (BNS) consistent ATM technology based on control point technology architecture (submitted to forums for standardization).

3. Offer ATM technology that offers evolution to ATM while maintaining coexistence and interoperability with configurations of today.

4. Actively participate in the ATM Forum and standards bodies.

10.9.1 Broadband Network Services

The IBM Network Systems' Broadband Network Services (BNS) architecture, as shown in Figure 10.13, describes a set of control point services, access services, and transport services. Control point services include address resolution, bandwidth management, and reservation through QoS guarantees, route computation and load balancing, non-disruptive path switching and rerouting, and multicast group management. Access services provide the framework for the high-speed network to support a variety of standard open interfaces including ATM, frame relay, circuit emulation, and higher layer services. such as TCP/IP and APPN. Transport services provide the reliable, bandwidth managed pipes for user traffic. IBM will support the ATM Layer, and ATM Adaptation Layer (AAL), by a proprietary control point technology through the UNI. IBM plans to license BNS.

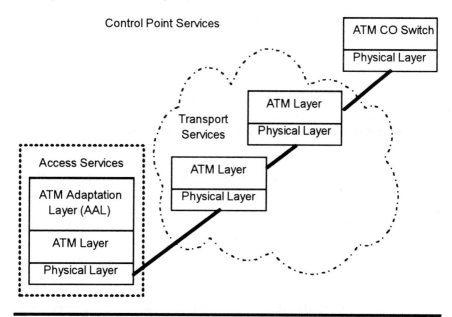

Figure 10.13 IBM Broadband Network Services

10.9.2 Upgrade with Minimum Pain

IBM plans to provide end-to-end solutions for the desktop, Campus, and public and private WAN environments, while preserving the customers' existing investments in wiring, network topologies and protocols, and applications. This is witnessed in IBM's proposal for customers to use their existing UTP, STP, and fiber through 25 Mbps carriage of ATM cells over voice-grade (category 3) Unshielded Twisted Pair (UTP).

10.9.3 Products

While the full impact of the IBM ATM strategy has not been reflected across the entire product line, announcements have been made for the following four products:

- 8250 Intelligent Hub with ATM support providing workstation access to mainframes via a 3172 FEP

- Campus ATM switch

- Private or Public WAN ATM Switch labeled the Transport Network Node (TNN) and running at 8 Gbps

- Network management support for the above three products

The TNN is planned to support HDLC, ISDN, Fiber Channel, CBR voice, and video. All ATM implementations will allow bandwidth to be dedicated between two ATM devices. Devices can use adapters of different speeds based upon application requirements. IBM also offers support for Packet Transfer Mode (PTM), the capability to support high speed-switching of variable-length packets.

Network management is performed through AIX NetView/6000 through an extended set of ATM applications using Management Information Base (MIB) management tools for SNMP and vendor-supplied MIBs. SMNP will be offered first, with later conformance to CMIP.

10.10 REVIEW

ATM hardware and switching systems take many forms. The chapter began with a discussion of switch network models — the building blocks of ATM hardware. ATM hardware was then broken down into large Central Office (CO) switches, followed by Campus switches which often form the core access for many private ATM networks and access to many public ATM networks. An exhaustive vendor analysis of most ATM switch products on the market by 3Q94 was presented. The categories of local ATM systems, such as routers, bridges, hubs, local ATM switches, multiplexers, concentrators, bridging devices, and CSU/DSUs, were then introduced. Some of the manufacturers and their products in each of these areas were listed. Many manufacturers in this industry segment are rushing towards ATM, but there is still a long road to achieving all of the functions that users demand. ATM in the PC or workstation, PBX, and cross-connect was discussed next. The last category of ATM devices discussed were the ATM chip sets, which are essential in ushering in an era of cost effective ATM. The chapter closed with a review of IBM's entry into the ATM market.

10.11 REFERENCES

[1] L. Gasman, "ATM CPE - Who is Providing What?," *Business Communications Review*, October 1993.
[2] J. Johnson, "ATM Networking Gear: Welcome to the Real World," *Data Communications*, October 1993.
[3] *Business Communications Review*, May 1993.
[4] *Telephony*, August 9, 1993.
[5] *Network World*, 1993.
[6] *Data Communications*, 1993, 1994.
[7] *Electronic Business Buyer*, Vol 19, Issue 9, September 1993.
[8] *Computer World*, 1993, 1994.
[9] *Broadband Networks & Applications*, 1994.
[10] P. Newman, "ATM Technology for Corporate Networks," *IEEE Communications Magazine*, April 1992
[11] A. Gurugé, "The Blue-Brick Road to ATM," *Business Communications Review*, October 1993

ATM End Systems:
Networking and Applications

This chapter covers the implications of ATM for networking and applications. This is probably one of the least understood, least standardized — yet most critical — areas to the success of ATM. The last chapter described local devices as either routers, switches, or hubs, and summarized other ATM-related devices such as CSU/DSUs and multiplexers. This chapter covers two specific examples developed for support of IP over ATM: one from Fore Systems, and the other from the IETF's classical IP over ATM work. ATM implications for end system hardware and system software are then explored. Some key ATM applications are then covered: multimedia and IP support for multiple QoS classes. Next, the foremost applications of ATM networking in the local environment are described, summarizing some key open issues. The chapter concludes with a summary of the industry forum and standards activities in progress to resolve these key issues and bring the realization of ATM closer to the user.

11.1   NETWORKING, ADDRESSING, AND ROUTING

This section begins by describing the general problems encountered in networking, addressing, and routing. Possible solutions to some of these issues are illustrated through the use of two specific examples for operating the connectionless Internet Protocol (IP), which was introduced in Chapter 5, over a connection-oriented ATM network. Examples covered include the Fore Systems product and the IETF RFC 1577 specified classical IP over ATM approach. Both examples use the Switched Virtual Circuit (SVC) capability as described in Chapter 9. The architecture and means by which addresses are administered, resolved, and utilized differ markedly between the examples. The Fore Systems approach uses a distributed, broadcast address resolution method, while the IETF approach utilizes a centralized route server approach.

11.1.1   Address Assignment and Resolution

A key requirement in any communication network is that unique addresses be assigned to each of the entities that want to communicate. This is the case in the telephone network where every phone in the world has a unique number, and in the 48-bit IEEE 802.3 Medium Access Control (MAC) assignments that are built into every Ethernet interface. Every host and user in the Internet is assigned a unique IP address.

Ensuring that the address assignments are unique and efficiently administering an address space presents some challenges. Not only must addresses be handed out, but a means for users to return addresses and request blocks of addresses is also required. Furthermore, if there is more than one administrative authority, then the scope of assignments allocated to each administration must be clearly defined. It is sometimes difficult to predict the demand for addresses. For example, area codes are realigned because the demand differs from what was forecast years ago. If an administrative authority hands out blocks of addresses too freely, then the network can run out of unique addresses well before the limit determined by the number of bits in the addresses, as has occurred with IP addresses.

Once you have your own address and the address of something that you wish to communicate with, how do you resolve the address of the desired destination into information about how to get it there?

The two examples of IP over ATM later in this chapter cover this subject in detail. First, consider the following simple analogy. Let's say that you have spoken to an individual on the telephone for the first time and have agreed to meet him or her at a party to which you both have been invited by the same host. Once you arrive at the party, you can find the individual (resolve the address) in one of two ways: you can jump up on stage, grab the microphone, and broadcast your presence, or you can locate the host and ask to be introduced to the individual. Broadcast is commonly used in shared-medium LANs to resolve addresses. A problem arises when the volume of broadcast traffic begins to approach the level of user traffic. The concept of having someone who has the information (the host) resolve the address for you is called a *route server* in data communications terminology. Note, however, that the route server's capacity is limited by its inherently centralized architecture.

11.1.2   Routing, Restoration, and Reconfiguration

Once the destination address is resolved, there is the issue of what is the best way to reach that address through a network of nodes. Each link in a network can be assigned a *cost*, with a routing objective to find a least-cost route. This cost may or may not be economic, but may reflect some performance information such as the delay or latency of the link. Routing algorithms generally exchange information about the topology, that is, the links that are connected and their associated costs, in one or two generic methods [6],[5]. The first is a *distance vector* algorithm where neighbor nodes periodically exchange vectors of the distance to every destination in the network. This process eventually converges on the optimal solution. The second is where each router learns the entire *link state* topology of the entire network. Currently this is done by flooding only changes to the link state topology through the network. Flooding involves copying the message from one node to other nodes in the network in a tree-like fashion such that only one copy of the message is received by every node. The link state approach is more complex, but converges much more rapidly. Convergence is the rate in which a network goes from an unstable state to a stable state. When the topology of the network changes due to a link or node failure, or the addition of a new node or link, this information must be updated at other nodes. The amount of time required to update all nodes in the network about the topology change is called the convergence time.

The distance vector method was used in the initial data communication networks such as the ARPANET and is used by the Internet's Routing Information Protocol (RIP). A key advantage of the distance vector is its simplicity. A key disadvantage of the distance vector protocol is that the topology information message grows larger with the network, and the time for it to propagate through the network increases as the network grows. Convergence times on the order of minutes are common.

The link state advertisement method is a more recent development and was designed to address the scalability issues of the distance vector method. When a link is added or deleted from the network, an advertisement of its cost is flooded through the network. Each node has complete knowledge of the network topology in time t (usually tens of milliseconds to seconds) after any change and computes the least-cost routes to every destination using an algorithm such as the Dijkstra algorithm [5]. Examples of link state advertisement routing protocols are the Internet's Open Shortest Path First (OSPF) [6] and the OSI IS-IS Routing Protocol [5], where IS stands for Intermediate System. Key advantages of these protocols are reduction of topology update information and decreased convergence times, saving sometimes on the order of seconds. A key disadvantage is the increased complexity of these methods and consequently increased difficulty in achieving interoperability.

Routing is a complicated problem, and the above descriptions serve as only a brief introduction sufficient for our purposes. Those readers interested in more detail should consult more detailed descriptions, such as Reference 5 for OSI-based routing and Reference 7 for IP-based routing.

11.1.3 LAN Emulation

The capability for ATM networks to emulate many desirable features of existing LANs is often a key requirement. Terminal Equipment (TE) should be able to interface to ATM LANs and be automatically recognized. Equipment configurations and connection should be capable of being moved, added, and/or changed by network management, or even automatically.

Many users view ATM as a unifying technology that will allow seamless interworking of the older, so-called legacy LAN systems (i.e., Ethernet, Token Ring, and even FDDI) with the latest, high-performance ATM technology. This will offer a competing technology with LAN emulation using ATM.

Many LAN users expect to have access to all of the available ATM bandwidth, if it is unused by other users, much as they do on LANs today. This expectation is in sharp contrast to the telephony-oriented notion of the traffic contract, as defined in Chapter 12, which reserves bandwidth in order to guarantee a performance level. ATM standards currently do not support this capability. Chapter 14 describes work in progress to specify means to meet the LAN users' desire to access the available bandwidth.

11.1.4   Fore Systems Approach

An example of how application software can utilize signaling is the method used by the Fore Systems customer premises ATM switch, called the ASX-100, as detailed in Ref. [1]. The following explanation is meant to be illustrative, and includes simplifications and extensions in an attempt to make the approach easier to understand. The Fore Systems ASX-100 Switched Virtual Circuit (SVC) capability is capable of tunneling over a full-mesh VPC Permanent Virtual Circuit (PVC) network, as shown in Figure 11.1. The ASX-100 essentially acts as a bandwidth manager, under control of the administrator and the end user applications. The ASX-100 initially supported IP-based services operating over an ATM SVC backbone using a prestandard protocol. The switch has since been upgraded to standard ATM SVC signaling.

The Fore Systems Simple Protocol for ATM Network Signaling (SPANS) SVC tunneling capability can be invoked from standard UNIX software socket calls for IP, or through a published Fore Systems Application Program Interface (API). This summary covers support for standard IP packets. The lower portion of Figure 11.1 depicts the physical network of three FORE Systems ASX-100 ATM switches interconnected via PVC VPCs using a SPANS NNI protocol. Each switch is assigned a unique switch number S. Each switch port is assigned a unique number P. The combination of switch and port number make up the ATM address for signaling. Each workstation or server is running UNIX operating system software much like that shown for the workstation with IP address A. Applications such as NFS, SNMP, FTP, TELNET, or SMTP as described in Chapter 5 make standard UNIX socket calls to the User Datagram Protocol (UDP) or the Transmission Control Protocol (TCP). The socket is identified by the combination of the port number for UDP or TCP and the IP address, and often an additional identifier called the protocol family (see Ref. 6). The application must first request that

TCP attempt to open a connection with the requested destination using a three-way handshake before application packets can be transferred.

Fore systems manufactures device driver hardware and software for workstations and servers that contain the functions shown in the white bubble in Figure 11.1. The IP traffic is passed to an IP handler, which has access to an ATM Adaptation Layer (AAL), ARP cache, and SPANS signaling processor. There is an ATM VCC (VPI/VCI) reserved for use by an Address Resolution Protocol (ARP) and the SPANS signaling protocol on each ATM UNI interface as shown in the figure. All other VCCs can be assigned to user connections in support of IP traffic. The operation of the device driver is now explained through a detailed example in the following narrative and illustrations for the network configuration and address assignments of Figure 11.1.

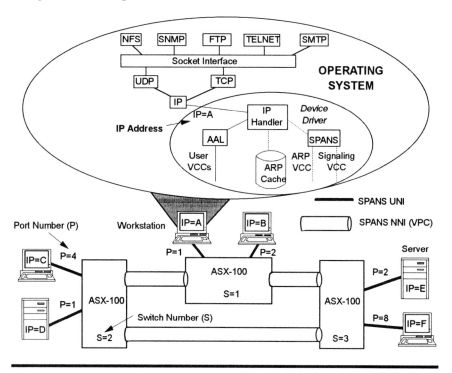

Figure 11.1 Fore Systems IP Support Configuration

When an application makes a standard socket call in an attempt to transfer data to the destination IP address, on a machine which has a Fore Systems adapter card and driver software, one or more IP packets are passed to the driver software as shown in step 1 of Figure 11.2. The Transmission Control Protocol (TCP) will send a connect packet and wait for a response before proceeding, while the Unacknowledged Datagram Protocol (UDP) just submits a packet. In this example, the IP packet from IP address *A* is destined for IP address *E*. The ARP cache defines what ATM VPI/VCI should be used for a particular destination IP address. The IP handler checks to see if the IP address is already in the ARP cache, and if so, the IP packet can be transferred using the VPI/VCI in ARP cache as shown in Figure 11.5 Otherwise the workstation generates an ARP request as shown in step 2 for the destination IP address (**E**), which in turn is broadcast by the ASX-100 to all other ASX-100s. Each port receives this ARP on the VPI/VCI reserved for ARPs and ARP responses as shown in the figure. The destination IP address *E* receives the ARP in step 2.

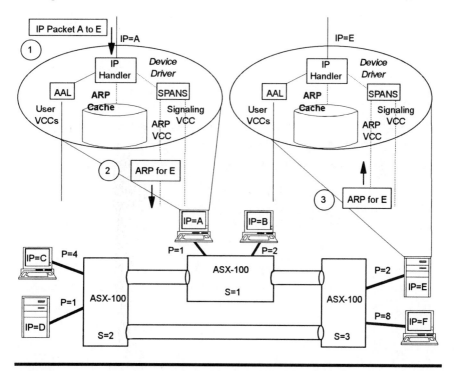

Figure 11.2 Fore Systems IP Support — IP Packet and ARP

After receiving the ARP, the destination IP address *E* sends an ARP response indicating that it is at the ATM address corresponding to switch S=3 and port P=2 as shown in step 4 of Figure 11.3. If no

ARP response is received, then TCP will try again to make the connection. UDP does not retry. Because the ARP request identifies the sender, the response is sent only to the source IP address A as shown in step 5. The source device driver can now set up the ATM connection from the origin port (S=1, P=1) to the destination port (S=3, P=2) using the SETUP message shown in step 6. The ASX-100 network routes this SETUP message through the network to switch S=3, port P=2, and outputs the message to the destination port as shown in step 7. The destination switch also assigns the VCC for this connection in the SETUP message. While the ARP and SVC process progresses, the adapter buffers any additional IP packets generated by the application until address resolution and signaling completes so that transmission over the ATM network can occur. Note that this may result in higher layer, such as TCP, time-outs. This type of architecture may also be well suited to future congestion control schemes where direct interaction with the end system application is highly desirable.

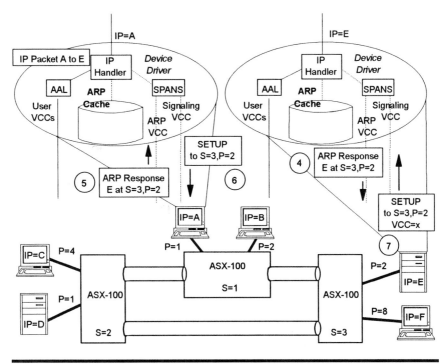

Figure 11.3 Fore Systems IP Support — ARP Response and SETUP

The destination device driver then responds with the CONNECT signaling message to communicate the fact that it has accepted the connection in step 8 as seen in Figure 11.4. Under abnormal circumstances, the destination would reply with a RELEASE which would cause the connection attempt to fail so that the source could inform the caller. Under normal circumstances, the ASX-100 message routes the VCC through the network and delivers a CONNECT message to the source device driver on the signaling VCC as shown in step 9. Switch 1 at this point uses the CONNECT message as an opportunity to assign the VCC numbered *y* to the connection setup for communications between IP addresses *A* and *E*. The source device driver updates its ARP cache to indicate that the way to reach the IP destination address from the IP source address *A* is through the use of VCC *y* in step 10.

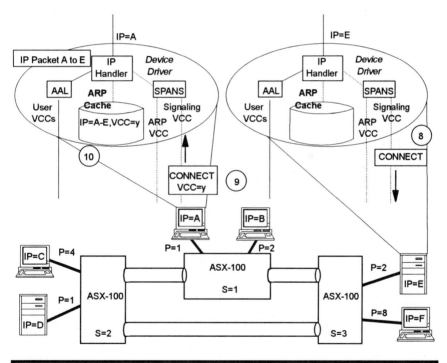

Figure 11.4 Fore Systems IP Support — CONNECT and ARP Cache

Upon successful completion of the preceding steps, IP packets can now be forwarded by the IP handler over the VCC associated with the IP source-destination address pair in ARP cache as shown in Figure 11.5. The forwarded IP packet is segmented by the ATM

Adaptation Layer (AAL) function, with each cell assigned the VPI/VCI value of y as taken from ARP cache for the source-destination pair A-E and as shown in step 11. By virtue of the ATM call setup process, the ASX-100 network switches the cells to the destination switch and port. It also outputs the cells using the translated VPI/VCI value of x as established at call setup time in step 12. The destination AAL function reassembles the IP packet, which the IP handler passes to the destination IP address operating system software in step 13. Currently, if there is inactivity for a long period of time (many minutes), then that ARP cache entry is deleted. If a failure occurs, or a device is disconnected from a switch, then the ASX-100 network signaling protocol (SPANS NNI) informs the connection endpoints to remove the entry from their ARP caches. The SPANS protocol establishes simplex, or one-way, ATM connections. A full duplex path is established as a pair of these processes, with the opposite direction being a mirror image of that described above.

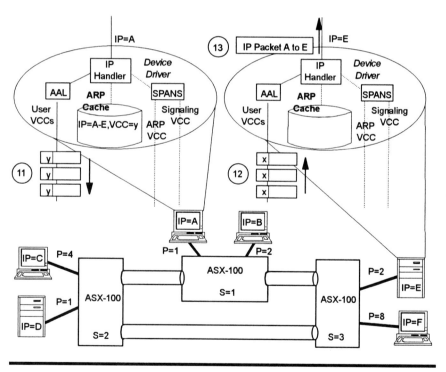

Figure 11.5 Fore Systems IP Support — Data Transfer Phase

The bandwidth allocated to VCCs for such IP connections is zero, which causes it to be served in the lower priority queue of the two 256-cell queues in the ASX-100. Therefore, IP connection requests are never refused; however, they may experience loss if the cell traffic being offered over VCCs with nonzero bandwidth leaves too little bandwidth for the IP/SVC traffic. This allows reserved traffic to be served at the highest priority, with the remaining bandwidth available for on-demand IP connections, such that the on-demand IP applications cannot interfere with reserved bandwidth applications.

11.1.5 IETF Classical IP over ATM Approach

Current ATM standards do not have a capability comparable to the group addressing feature of SMDS. The capability of allowing one address to be able to broadcast to all other addresses in the group effectively emulates a LAN, which is very useful in the exchange of LAN topology updates, Address Resolution Protocol (ARP) messages, and routing information. SMDS effectively does this through a Layer 3 function which is currently not specified for ATM.

IETF RFC 1577 specifies Classical IP over ATM for the use of ATM as a direct replacement for the "wires" interconnecting IP end stations, LAN segments, and routers in a Logical IP Subnetwork (LIS). These procedures are intended to make the operation of ARP and Inverse ARP interoperable over early ATM SVC and PVC networks, respectively. Additional work on addressing differences between the public and private ATM UNI, security and firewall procedures, and a scaleable routing protocol must be completed before progress can be made on other techniques.

A standard method for routers to automatically learn the IP address of the router on the far end of each ATM PVC is defined by draft RFC 1293 [4], called the *Inverse Address Resolution Protocol (Inverse ARP)*. Basically it involves a station sending an InARP message containing the sender's IP address over the ATM VCC PVC. This situation occurs in PVC networks when a PVC first initializes, or a router reloads its software because the VCC is known, but the IP address that can be reached on this VCC is not known. The station on the other end of the ATM VCC PVC then responds with its IP address, and an association between the IP addresses of the pair and the ATM VCC VPI/VCI on each ATM interface is established.

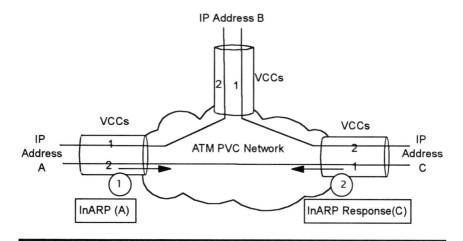

Figure 11.6 Example of Inverse ARP Procedure

Figure 11.6 illustrates the principle of Inverse ARP (InARP) over an ATM network comprised of VCC PVCs as shown. IP address A sends an inverse ARP over VCC 2 [InARP(A)] in the first step. This appears on VCC 1 to IP address C at the receive side. IP address C responds with its identity [InARP Response(C)] on VCC 1 in the second step. Now IP addresses A and C know that they can reach other by transmitting on ATM VCCs 2 and 1, respectively.

A route server is defined for each LIS in order to support ATM SVCs. Each station has a hardware address (that is specified in the signaling message) and the address of the route server located within the LIS. This server must have authoritative responsibility for resolving the ARP requests of all IP stations within the LIS. The IP stations act as clients to the route server, and are responsible for registering with the route server, aging their ARP table entries to remove old data, and making Address Resolution requests to the server. An IP stations registers by placing an SVC "call" to the ARP server. The route server then knows the IP and ATM address correspondence. The route server time stamps entries and may periodically test that the IP station is still there using a ping command. Old entries are "aged" using this time stamp such that unresponsive IP stations are removed. The key route server function is then to enable IP stations to resolve the association of an IP address with an ATM address. This differs markedly from the Fore Systems approach which was completely distributed and broadcast the ARP and ARP response packets, as compared with the centralized classical IP over ATM approach.

RFC 1577 specifies that implementations must support IEEE 802.2 Logical Link Control/SubNetwork Attachment Point (LLC/SNAP) encapsulation as described in RFC1483, which Chapter 17 covers. LLC/SNAP encapsulation is the default packet format for IP datagrams. The default Maximum Transfer Unit (MTU) size for IP stations operating over the ATM network is 9180 octets. Adding the LLC/SNAP header of 8 octets, this results in a default ATM AAL5 protocol data unit MTU size of 9188 octets.

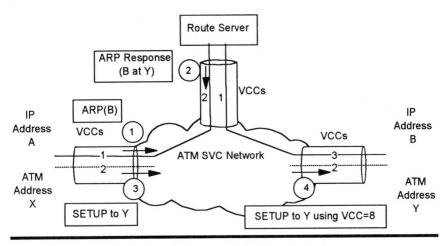

Figure 11.7 Classical IP over ATM — ARP and Setup Steps

The next example is classical IP over ATM as defined in IETF RFC 1577. IP addresses A and B have already made a Switched Virtual Connection (SVC) to the route server so that the route server knows the association of their IP and ATM addresses. Figure 11.7 illustrates the reference configuration and a simple example of the IETF's classical IP over ATM concept. Two interfaces are shown: one with IP address A and ATM address X and the other with IP address B and ATM address Y. There may be multiple IP addresses and ATM addresses assigned to an interface over the VCCs shown by solid lines in Figure 11.7. They have also registered the association of their IP and ATM addresses. IP addresses have a VCC over which ATM UNI signaling messages are sent as illustrated by the dashed line. When IP address A wishes to send data to IP address B, the first step is to send an Address Resolution Protocol (ARP) message to the route server. In the second step, the route server returns the ATM address Y of IP address B. In the third step, ATM address X originates a SETUP message to ATM address Y. The SETUP control message is switched through the ATM network and delivered at the destination ATM address Y on VCC 2. The switched ATM network makes a connection and requests a SETUP to ATM address Y using VCC 8.

Figure 11.8 illustrates the final steps in the classical IP over ATM scenario, which is nearly identical to the Fore Systems approach. In the fifth step, ATM address Y responds with a CONNECT message which is used by the switched ATM network to establish the VCC back to the originator, ATM address X. In the sixth step, a CONNECT message is sent by the ATM network to ATM address X indicating that VCC 4 should be used in its connection with ATM address Y and VCC 8. In the seventh and final step, communication between IP addresses A and B can occur over VCC 4 for IP address A and VCC 8 for IP address B as indicated in the figure. Either ATM address X or Y could release the ATM SVC by issuing a DISCONNECT message. There is a problem with IP here — the paradigm is that IP is connectionless and the signaling is connection-oriented, so normally a disconnect is not sent.

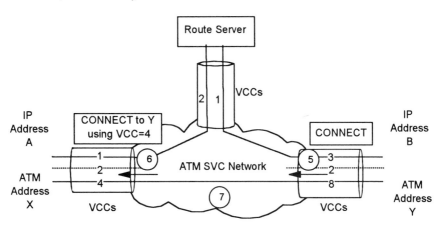

Figure 11.8 Classical IP over ATM — Connect and Data Transfer Steps

Initial ATM services could also support a point-to-multipoint configuration which could be used to emulate the above LAN broadcast functions for early application. This differs from a true multicast in that each endpoint would be the root of a point-multipoint tree. Therefore, cells on multiple VCCs would be received as part of the broadcast. Alternatively, the end application must replicate the broadcast packet on each point-to-point virtual channel, as described earlier.

11.1.6   Signaling Support Issues

The two previous examples of supporting IP over ATM rely heavily on ATM Switched Virtual Connection (SVC) capability. Several questions are left unanswered. How rapid must the ARP and call setup process operate for it to be useful to end applications? What should be the holding time duration before the switched connection is released to make efficient use of resources? What call attempt rate will these applications generate? How many active calls need to be supported? Will the Q.2931 protocol be able to support the required call setup time and attempt rate?

The answers to these questions will be answered through implementation experience and user feedback. Rapid call setup and the support for high call attempt rates will be critical factors in the success of the approach of supporting connectionless services by what is effectively rapid, dynamic connection switching.

11.2   END SYSTEM ATM HARDWARE AND SOFTWARE

This section covers ATM implications for End System (ES) hardware, systems software, and applications.

11.2.1   Hardware

Most applications inherently deal with variable-length data units, and do not make use of individual cells. Witness the fact that no current application can make optimal use of the ATM cell directly. One cell is usually too small to contain a data packet (even a single encapsulated TCP/IP acknowledgment!) but too large to transport voice without delay that may require echo cancellation. Some applications will undoubtedly be developed that make direct use of the ATM cell; however, most applications will require one or more layers and dimensions of adaptation.

The way in which a variable-length data unit is mapped to a fixed length cell is via an ATM Adaptation Layer (AAL) as described in Chapter 8. Not all AALs are created equal — some are easier and more efficient than others in support of application software and hardware. In particular, AAL5 began as the Simple Efficient Adaptation Layer (SEAL), originally proposed by computer manufacturers to the telephony-oriented standards organizations. In a relatively short period of time, AAL5 has achieved a wider degree of acceptance than the older AAL3/4 method, which is largely based upon 802.6. AAL5 allows the receiving hardware to determine when it is time to reassemble a packet and check the CRC by only looking at the cell header. In contrast, each cell using AAL3/4 must be checked for CRC and whether it is the last cell.

The manner in which the receiver link and memory access interact with a processor performing other operating system and application tasks can significantly affect the achievable communications throughput. Two extremes of an end system ATM communication adapter are summarized based on information derived from [8] and [9]. Figure 11.9 illustrates a simple and complex end system ATM/AAL adapter. Each adapter utilizes hardware for interface to the Physical (PHY) layer, ATM layer, and ATM Adaptation Layer (AAL), for example, using a Segmentation And Reassembly (SAR) integrated circuit to perform this function. The Direct Memory Access (DMA) system receives an entire packet from the SAR chip so that the processor is never involved in the PHY, ATM, or AAL functions.

The simple adapter shown in Figure 11.9 is typical of many Ethernet interfaces today. The processor is interrupted when a packet is received, causing a switch to the communications task, which then sets up the DMA transfer. The DMA system then transfers the packet data across the bus to the shared memory, during which time the processor must remain idle. The processor must also perform some memory management functions, such as allocating buffers for the packet data and processing the packet header so that it is placed in the corresponding logical queue. Finally, the processor can perform another task switch and return to operating system or application software execution until another packet arrives.

The additional or modified elements from the simple adapter that make up the complex adapter are shown as shaded boxes in Figure 11.9. An intelligent DMA system that does not require the processor to set up data transfers and a separate memory management unit to handle the housekeeping tasks will free up even more processor time. A key benefit of the intelligent DMA system is that the processor and memory management units have their own local memory. Thus, they can still function even when the shared memory is being accessed over the bus, for example, when DMA transfers occur.

The graph at the bottom of Figure 11.9 plots the effective throughput versus the packet arrival rate for 256-byte packets. The increase in effective throughput achieved by the more complex design is nearly 100 %! More complex adapters will be required to achieve close to maximum throughput.

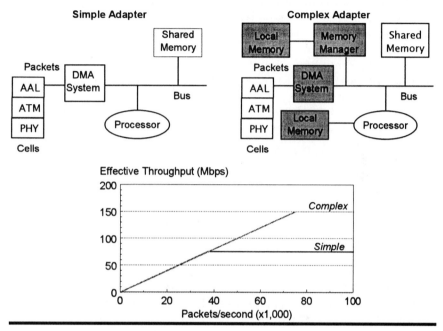

Figure 11.9 ATM Adapter Performance Comparison

11.2.2 Operating System Software

ATM software is defined as the higher layers of the B-ISDN protocol cube that are involved with the user, control, and management planes. The lower physical, ATM, and AAL common part layers will be implemented in Application Specific Integrated Circuit (ASIC) hardware over time, if ATM is successful, and will consequently reduce the cost of ATM implementation. In this scenario, the hardware becomes a commodity, and software becomes the value-added, differentiating feature of ATM equipment. Of course, this same effect will likely occur with the lower layers of ATM-based software in the future.

Focus now turns, in large part, to the software that exists in the end system equipment, and not on the system software that exists in the switches, bridges, routers, or hubs covered earlier in this and the previous chapter.

There is a critical need to align the connection-oriented ATM paradigm with the fundamentally connectionless nature of distributed data communications. As a rule, applications don't normally set up a connection before attempting to perform a function; they just request that the function be done and expect a response indicating that either it was done completely or not at all. The Fore Systems and classical IP over ATM approaches are similar in the sense that they attempt to (quickly) set up a connection after the first packet arrives. However, the industry needs to find a better method.

11.2.3   IP Support for Multiple QoS Classes

When multiple applications are multiplexed over a single IP address, which is carried by a single ATM VCC, then only the single QoS class supported for the ATM VCC is provided to all IP applications. One solution to this limitation would be to provide different IP addresses for different application QoS classes. However, this would require changes to many existing software systems. It may be possible to use an IP Type of Service field to indicate the QoS class. However, a large amount of software would require modification: additional protocol interfaces for TCP and UDP, Internet applications, and operating system software changes at the socket level.

Most LAN-based data communication protocols were developed under the assumption that the delivered quality needed only to be a best effort. Recent applications such as videoconferencing, multimedia, and interactive sessions require some guaranteed bandwidth and quality to be useful. Work has begun on a proposed ReSerVation Protocol (RSVP) to introduce these capabilities to the Internet Protocol (IP) [10]. RSVP defines a means for receivers to initiate reservations in multipoint topologies according to styles that support various audio and videoconferencing needs. Dynamic changes in the connection are supported by communicating reservation and actual usage state information throughout the network.

11.3 APPLICATION PROGRAMMING INTERFACES (APIs)

An Application Programming Interface (API) allows an application program to gain access to well-defined ATM-based capabilities. This is critical to close the gap between the ATM-based capabilities and the application in order for ATM to succeed. The objective of an API standard should be to define a framework for developing common APIs between applications and multiple operating systems platforms (OSs) running on diverse end systems and to provide effective access to the above capabilities.

11.3.1 Existing ATM APIs and Platforms

One example for support of an existing API, a UNIX socket call, was given for the Fore Systems switch earlier in this chapter. The dominant APIs today are UNIX named pipes, sockets, and streams, Microsoft's NDIS, NetBIOS, the Hayes AT command set, and SNA APPC. The current APIs run on UNIX, Windows/NT, Windows, and OS/2 end system platforms. These APIs work in network operating system environments such as Netware, Vines, Netbeui, AppleTalk, DECNET, and TCP/IP. In general, most APIs operate only in a very specific environment and support limited interoperability.

The device driver for an existing ATM API would ideally present the same software interface, therefore requiring no application change. If an application wanted to take advantage of additional capabilities, the API should be rich enough to support this. Many of the current APIs do not perform functions necessary in ATM, such as point-to-multipoint, specification of traffic parameters, support for connection-oriented signaling, and support for multiple Qualities of Service (QoS).

11.3.2 A Possible Set of ATM API Functions

The ATM API should provide access to capabilities such as the AAL for user data, the ATM layer, the ATM SVC protocol, the ILMI, or an ATM OAM flow. Figure 11.10 depicts these various interfaces. The potential functions for each capability are briefly described below.

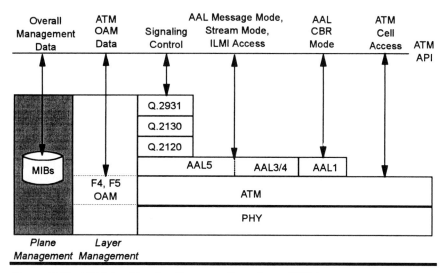

Figure 11.10 ATM Application Program Interfaces (APIs)

API access to the signaling layer should allow an application to set up and release a call in a very simple manner, hiding the complexities of the signaling protocol. The API should support point-to-point and multipoint connection types. The capability to add a release party in a point-to-multipoint connection should be provided. As other signaling capabilities are defined, these should also be extended to the application.

The API should provide access to the ILMI information. The ability to send messages to the management entity on the other end of the UNI should be provided. Also, the capability to query or set the local status should be provided, which is not supported via SNMP since all entries are read only in the ILMI MIB.

The API should provide direct access to the ATM Adaptation Layer (AAL), effectively emulating the AAL Service Access Point (SAP). Data transmission from the application to the AAL can either be in *message* or *stream* mode. Message mode takes an entire packet and completely fills cells prior to transmission, making it very efficient but possibly incurring delay in waiting for the next packet before sending the previous packet. Stream mode immediately sends the cells, minimizing delay while incurring a possible loss in efficiency. Thus, the mode used depends on the application's data transfer requirements. The API may also support a service where it periodically requests data from the application in support of a Continuous Bit-Rate (CBR) service. For example, this type of interface could effectively transfer clock to the application in support of video transfer.

API access to the F4 (for VPC) and F5 (for VCC) Operations, Administration, and Management (OAM) flows, which carry connectivity and testing information, should be supported on both a segment and end-to-end basis.

There should also be a method for providing direct access to the ATM cell flow, where cells are transmitted and received without processing or error checking. This could be used in a test and development environment, or could be made available as a high-performance interface to facilitate applications that can directly access the power of ATM.

11.3.3   API Standardization Efforts

The ATM Forum is working on an API specification. Study group 8 of the ITU-T is studying a Programming Communication Interface (PCI) for terminal systems that is closely related to this subject.

11.4   ATM-BASED APPLICATIONS

This section covers some end system features and business applications that are well suited to ATM: multimedia, collapsed backbones, client-server architectures, ATM workgroups, and seamless networking. These are commonly acknowledged as key ATM applications in the industry, technical, and trade press as shown by reference to recent articles.

11.4.1   Multimedia Applications

Much new computer equipment is capable of presenting and accepting multiple media, such as image, video, audio, text, tabular, and graphical elements. As end station computer equipment hardware and software for these capabilities becomes more widespread and cost-effective, it is anticipated that there will be an increasing demand to interconnect such multimedia workstations. Many business and commercial applications are envisioned for these types of capabilities, as described earlier in this book. See Refs. 11

and 12 for a further discussion of multimedia applications such as cooperative editing and spreadsheet work, desktop videoconferencing, interactive audiovisual computing, and entertainment.

The multiple QoS class support and the capability to reserve bandwidth and dynamically set up calls make ATM an ideal basis for the support of multimedia applications. One key additional function that will be needed in ATM is a means to associate multiple connections into a single multimedia call in a useful way. This is planned for standardization in the signaling protocol by the ITU in the 1995–1996 time frame. In the interim, proprietary implementations will yield insight into what actually meets end user requirements.

Multimedia applications interfaces and protocols are currently often proprietary since standards groups to address the issues that are raised have not even been formed in some cases. Therefore, early multimedia users will have to assemble networks, end systems, adapters, and software to meet their needs. As experience is gained and particular approaches are proven to be effective through the process of experience and competition, multimedia standards will continue to emerge through industry forums or de facto standardization.

11.4.2 Client-Server Architectures

Currently many servers are dedicated to a single Ethernet segment, or some more powerful servers may even have multiple Ethernet cards or even FDDI interfaces. Multiple interfaces at a lower speed will not achieve nearly the performance of a single, higher speed interface. This is true because there may be idle capacity on some interfaces in the set, while all bandwidth is accessible on a single, high-speed interface. Therefore, an ATM interface on the server in client-server-based applications is justified well before ATM interfaces on clients would be.

Servers need high bandwidths, not necessarily high throughput, to achieve low latency [15]. Clients may be able to use 10-Mbps Ethernet for a long time, particularly if an ATM hub or local switch can put the workstation on a segment by itself so that it has full access to the 10 Mbps of bandwidth, for example, through the use of Ethernet switching.

11.4.3 Collapsed Backbone

Another key use of ATM will likely be that of a collapsed backbone interconnecting many smaller LANs and access-level routers and bridges at a building or enterprise level. Key functions are automatic configuration, fault recovery, address resolution, address assignment and network management for a collapsed backbone. Indeed, a major benefit of ATM may be the consolidation and unification of network management in this environment [3]. For more details on the collapsed backbone architecture see [13].

A single, shared medium segment runs out of bandwidth quickly as more users require interconnections to shared resources such as servers, or communication increases due to the re-engineering of the corporation. The bridging of LAN segments does not scale well beyond a certain point. An architecture employed by many users confronted with this problem was to effectively implement a hierarchy or star network collapsed on a router as shown in Figure 11.11a. As the demand for speed and connectivity increases, ATM is considered a strong candidate to provide the increased bandwidth requirements in the hierarchical collapsed backbone architecture, as shown in Figure 11.11b. The ATM architecture also has the potential to move away from this hierarchy to a more distributed backbone architecture that can better between serve communities of interest with large amounts of traffic.

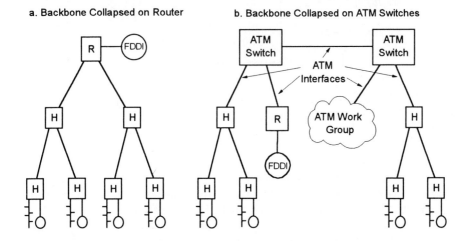

Figure 11.11 Use of ATM in the Enterprise's Collapsed Backbone

11.4.4 ATM Workgroups

An early use of ATM switches will be for the support of ATM workgroups, typically a set of high-performance computing systems that can utilize the high speed and flexibility of ATM. Also, certain workstations will be able to initially justify the performance and cost of an ATM interface. For the high-end segment of the workstation market, FDDI is already too slow for visually-intensive applications. These types of users often work in smaller workgroups in leading-edge segments of the industry. An early role of ATM will likely be to empower these smaller groups through the use of ATM switching as illustrated in Figure 11.12. As discussed in Chapter 1, desktop bandwidth requirements are exploding and ATM is prepared to meet that explosion. These leading-edge work groups can be interconnected to the older, lower-speed legacy LAN networks through virtual networking and a collapsed backbone architecture. The trend — ATM moving full force into the workplace!

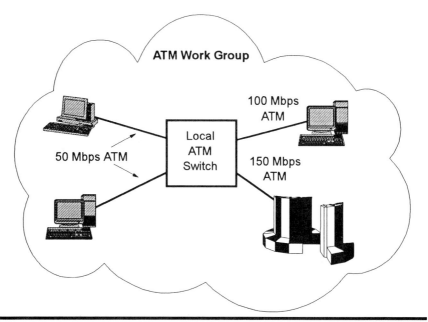

Figure 11.12 ATM Support for a Leading-Edge Work Group

11.4.5   Seamless Networking

An attractive application of ATM is that if it is used from the desktop, throughout all hubs, bridges, routers, and switches in a building, used to interconnect a campus, provide access in a metropolitan area, and extend across national and even international boundaries, then great benefits must arise based on the same underlying technology employed end-to-end.   ATM is but a small part of the overall picture, with many additional, higher-layer protocol interworking required before this vision is realized.

There are also some fundamental, inescapable limits to this vision. In order to achieve seamless ATM networking, the entire suite of ATM-based protocols, interfaces, and procedures must be scaleable in distance, speed, and number of users.   Witness the IBM plan to offer ATM products from CO-based switches down to workstation interface cards.

Propagation delay increases with distance — period.   There is no technology that can change this, so far.   A client-server application that operates well with a 10-μs round-trip delay in the same building may perform horribly when confronted with a hundred times greater delay that is encountered going to the next major office miles away. Careful attention to higher layer protocol design to provide scalability in distance will be an important challenge in achieving seamless ATM networking.

As introduced earlier, servers need high bandwidth to keep latency low.   Servers may be connected by lower speeds when scaled over a wider area — where bandwidth comes at a higher price — thus increasing latency.   There comes a tradeoff between latency and delay in networks that limits scalability in speed.   This tradeoff will be explored further in Chapter 16.

An ATM LAN solution that works well for an ATM workgroup of a dozen people may not scale to a collaborative research group of 100 individuals.   The software and protocol infrastructure designed on top of ATM must be planned for success.   Address and capacity plans should be made larger than any conceivable set of applications could currently use, because it will only be a matter of time until applications that exceed the address space will be required.

See Ref. 14 for further information and other viewpoints on seamless ATM networking.

11.5   FUTURE DIRECTIONS

There are two major standards activities underway to address some of the software- and application-oriented issues raised in this chapter: ATM Forum and IETF activities.

11.5.1   ATM Forum Activities

The ATM Forum is adopting a structure to get early input from users in the protocol and interface specification process using the Enterprise Network Roundtable (ENR). This is a great experiment that will hopefully deliver needed capabilities to users sooner, and provide a high degree of interoperability.

The ATM Forum has established a number of technical workgroups to directly address some key issues through interoperable protocol and interface specifications in the following areas:

Private-Network-Network Interface (P-NNI): This group is focusing on defining address resolution, link state advertisement, signaling, and congestion control between private networks.

LAN Emulation Service (LES): This group is working on the interfaces and protocols that will allow LAN-like features to be emulated by interfaces and protocols built on top of ATM.

System Aspects and Applications Application Program Interface (API): This group is focused on defining an API with characteristics described earlier in this chapter.

SAA Video Coding: The ATM Forum is reviewing a method to achieve early interoperability of MPEG video coding over ATM.

11.5.2   IETF Activities

The IETF has several working groups focusing on the following aspects of ATM:

Classical IP over ATM: This group is chartered with defining how current IP implementations can take the first step to operation over ATM. The current state of work in this group was described earlier in this chapter.

Routing Over Large Clouds (ROLC): This group is chartered with defining how IP is routed over large data networks in the long term, such as future ATM, SMDS, and Frame Relay public network services. The primary issue is that in a large network all routers cannot be aware of each other because the routing tables will become too large.

IP Maximum Transfer Unit (MTU): This document details the determination and long-term direction of how the Maximum Transfer Unit (MTU) is negotiated in IP over ATM.

11.6   REVIEW

This chapter began by looking at the difficult requirements of networking, addressing, and routing in the context of ATM. Two specific examples of IP networking over ATM were detailed: an early ATM implementation by Fore Systems, and the approach specified by the IETF RFC 1577 classical IP over ATM standard.   Issues regarding ATM end system hardware and software were discussed. The impact of hardware adapter design on throughput and processing power available to run other applications was identified. Next, some open issues in operating systems and the interfaces with network protocols such as IP were described.   The premier applications proposed for ATM were described and illustrated: multimedia, client-server architecture support, collapsed backbones, high-performance workgroups, and seamless networking. In order to meet the objectives outlined in this chapter, significant additional standards work must be completed, particularly at the higher layer protocols and in defining how older legacy protocols interwork with the new ATM protocols. The chapter concluded with a summary of the standards activities in progress at the ATM Forum and the IETF toward addressing these issues.

11.7 REFERENCES

[1]   E. Biagioni , E. Cooper, R. Sansom, "Designing a Practical ATM LAN," *IEEE Network*, March 1993.

[2]   M. Laubach, "IETF Draft RFC - Classical IP and ARP over ATM," Version 6.0, December 20, 1993.

[3]   N. Lippis, "ATM's Biggest Benefit: Better Net Management," *Data Communications*, December 1992.

[4]   T. Bradley, C. Brown, "Inverse Address Resolution Protocol", RFC1293, USC/Information Sciences Institute, January 1992.

[5]   R. Perlman, *Interconnections*, Addison-Wesley, 1992.

[6]   U. Black, *TCP/IP and Related Protocols*, McGraw-Hill, 1992.

[7]   D. L. Comer, Interworking with TCP/IP - Volume I Principles, Protocols and Architecture, Prentice Hall, 1991.

[8]   H. E. Meleis, D. N. Serpanos, "Designing Communication Subsystems for High-Speed Networks," *IEEE Network*, July 1992.

[9]   C. B. S. Traw, J. M. Smith, "Hardware/Software Organization of a High-Performance ATM Host Interface," *IEEE JSAC*, February 1993.

[10]   L. Zhang, S. Deering, D. Estrin, S. Shenker, D. Zappalla, "RSVP: A New Resource ReSerVation Protocol," *IEEE Network*, September 1993.

[11]   H. Armbruster, K. Wimmer, "Broadband Multimedia Applications Using ATM Networks: High-Performance Computing, High-Capacity Storage, and High-Speed Communication," *IEEE JSAC*, December 1992.

[12]   N. Lippis, "Multimedia Networking," *Data Communications*, February 1993.

[13]   M. Zeile, "Expanding the Enterprise by Collapsing the Backbone," *Data Communications*, November 21, 1992.

[14]   G. Smith, J. P. van Steerteghem, "Seamless ATM Networks," *Telecommunications*, September 1993.

[15]   P. Sevcik, "Arguments and Architectures for ATM," *Business Communications Review*, October 1993.

[16]   ATM Forum, "53 Bytes Newsletter," February 1994.

ATM Traffic Management, Congestion Control, and Traffic Engineering

This section is designed to provide the reader with an application-oriented view of the ATM traffic contract, traffic control, congestion control, traffic engineering, and design considerations. First, standards and theoretical traffic modeling terminology are defined. Each concept is explained using analogies and several different viewpoints. This simplifies as much as possible a subject that is usually viewed as very complex. References to more detailed discussions are given for the more technically oriented reader. Next, the theory is applied to the real-world business problems defined in the previous sections. When basic mathematical theories are used, they are introduced with sufficient background for the reader to understand the results. Any formulas presented are capable of being "programmed" onto a spreadsheet. Examples are given of how the formulas can be used in network design and comparison studies.

12

The Traffic Contract

This chapter explains the formal concept of a traffic contract from ITU-T Recommendation I.371 and the ATM Forum UNI specification version 3.0 in an application-oriented manner. Some specific terms and concepts, such as traffic parameters, conformance checking, and Quality of Service (QoS) parameters are introduced and explained through the use of examples and analogies. Finally, some considerations in choosing traffic parameters and tolerances are given.

12.1 THE TRAFFIC CONTRACT

In essence, a separate traffic contract exists for every Virtual Path Connection (VPC) or Virtual Channel Connection (VCC). The traffic contract is an agreement between a user and a network across a User-Network Interface (UNI) regarding the following interrelated aspects of any VPC or VCC ATM cell flow:

☞ The Quality of Service (QoS) that a network is expected to provide
☞ The traffic parameters that specify characteristics of the cell flow
☞ The conformance checking rule used to interpret the traffic parameters
☞ The network's definition of a compliant connection

The definition of a compliant connection allows some latitude in the realization of checking conformance of the user's cell flow. A compliant connection can identify some portion of cells to be non-conforming, but no more than the portion which the ideal conformance checking rule would identify as nonconforming.

The next sections of this chapter define a reference configuration, Quality of Service (QoS) parameters, traffic parameters, and the leaky bucket conformance checking rule. Also some practical guidelines are provided for the use of these traffic parameters and how tolerances should be allocated for them.

12.2 REFERENCE MODEL

The basis of the traffic contract is a reference configuration, which in the standards is called an *equivalent terminal* reference model illustrated in Figure 12.1.

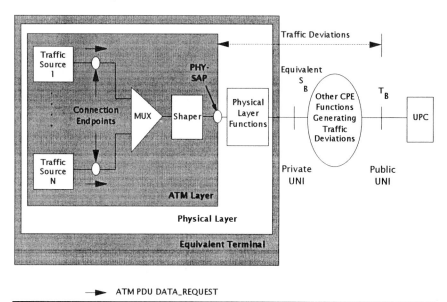

Figure 12.1 Equivalent Terminal Reference Model

An equivalent terminal need not be a real device, and indeed may be a collection of devices. The ATM cell traffic is generated by a number of cell sources, for example, a number of workstations, which each have either a VPC or VCC connection endpoint. These are all connected to a cell multiplexer – which in a distributed implementation could be a local ATM switch, router, or hub.

Associated with the multiplexing function is a traffic shaper, which assures that the cell stream conforms to a set of traffic parameters defined by a particular conformance-checking algorithm. The output of the shaper is the physical (PHY) layer Service Access Point (SAP) in the OSI layered model of ATM.

After the shaper function, some physical layer (and other) functions may change the actual cell flow emitted over a private ATM UNI (or S_B reference point) so that it no longer conforms to the traffic parameters. This ATM cell stream may then be switched through other CPE, such as a collapsed ATM backbone, before it is delivered to the public ATM UNI (or T_B reference point).

The end-to-end QoS reference model may contain one or more intervening networks, each with multiple nodes as depicted in Figure 12.2. Each of these intervening networks may introduce additional fluctuations in the cell flow due to multiplexing and switching, thereby impacting QoS. In principle, the user should not have to be concerned about how many intervening networks there are and/or what characteristics they have, but should always be provided the guaranteed end-to-end QoS for all configurations. However, this principle and reality have not yet been aligned in standards or interworking of multiple networks.

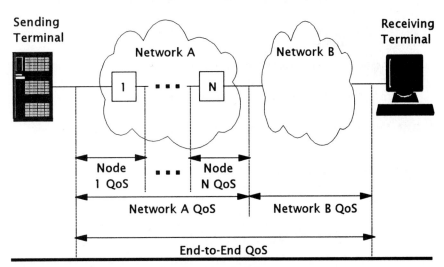

Figure 12.2 End-to-End QoS Reference Model

12.3 QUALITY OF SERVICE

Quality of Service (QoS) is defined by specific parameters for cells that are conforming to the traffic contract. In order to simplify a user's request for a certain QoS, certain classes are defined.

12.3.1 Quality of Service (QoS) Parameters

Quality of Service (QoS) is defined on an end-to-end basis – a perspective that is actually meaningful to an end user. The definition of *end* can be the end workstation, a customer premises network, a private ATM UNI, or a public ATM UNI.

QoS is defined in terms of the one of the following measurement outcomes, which Chapter 21 defines in detail. The measurement is done with respect to cells sent from an originating user to a destination user.

- ➢ A **Transmitted Cell** from the originating user
- ➢ A **Successfully Transferred Cell** to the destination user
- ➢ A **Lost Cell** which does not reach the destination user
- ➢ An **Errored Cell** which arrives at the destination but has errors in the payload.
- ➢ A **Misinserted Cell** which arrives at the destination but was not sent by the originator. This can occur due to an undetected cell header error or a configuration error.

The QoS parameters are defined in terms of the above outcomes by the following definitions:

$$\text{Cell Loss Ratio} = \frac{\text{Lost Cells}}{\text{Transmitted Cells}}$$

$$\text{Cell Error Ratio} = \frac{\text{Errored Cells}}{\text{Successfully Transferred Cells} + \text{Errored Cells}}$$

$$\text{Severely Errored Cell Block Ratio} = \frac{\text{Severely Errored Cell Blocks}}{\text{Total Transmitted Cell Blocks}}$$

A severely errored cell block is defined as the case where more than M out of N cells are in error, lost, or misinserted.

$$\text{Cell Misinsertion Rate} = \frac{\text{Misinserted Cells}}{\text{Time Interval}}$$

The Cell Transfer Delay is comprised of the following components, illustrated in Figure 12.3:

➤ T1   = Coding and Decoding Delay:
    ▫ T11   = Coding delay
    ▫ T12   = Decoding delay
➤ T2 = Segmentation and Reassembly Delay:
    ▫ T21   = Sending-side AAL segmentation delay
    ▫ T22   = Receiving-side AAL reassembly/smoothing delay
➤ T3 = Cell Transfer Delay (End-to-End):
    ▫ T31   = Inter-ATM node transmission delay
    ▫ T32   = Total ATM node processing delay (due to queueing, switching, routing, etc.)

Delay can occur on the sending and receiving sides of the end terminal, in intermediate ATM nodes, and on the transmission links connecting ATM nodes.

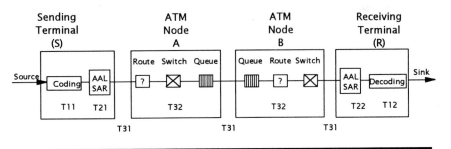

Figure 12.3 Illustration of Sources of Delay

Mean cell transfer delay is the average of the random component T32 and the fixed propagation delay T31. We will investigate the effects of queueing and switching delays in later chapters. The methods for measuring mean delay are covered in Chapter 20.

Cell Delay Variation (CDV) is currently defined as a measure of cell clumping, which is heuristically how much more closely the cells are spaced than the nominal interval. CDV can be computed at a single point against the nominal intercell spacing, or from an entry to exit point. Details on computing CDV and its interpretation are covered in ITU-T Recommendation I.356 and the ATM UNI specifcation 3.0. Cell clumping is of concern because if too many cells arrive too closely together, then cell buffers may overflow, as covered in the example of Chapter 16. Quality of Service (QoS) classes are defined primarily in terms of the following parameters defined by CCITT I.350 for each ATM VPC or VCC:

✿ Average delay
✿ Cell delay variation
✿ Loss on CLP=0 cells for ATM
✿ Loss on CLP=1 cells for ATM
✿ Error rate

The error rate is principally determined by fiber optic error transmission characteristics and is common to all QoS classes. Average delay is largely impacted by the propagation delay in the Wide Area Network (WAN), and average queueing behavior. A lower bound on loss is determined by fiber optic error characteristics, with higher values of loss dominated by the effects of queueing strategy and buffer sizes. Delay, delay variation, and loss are impacted by buffer size and buffering strategy. For a single, shared buffer a larger buffer results in lower loss, but greater average delay and delay variation. With multiple buffers more flexibility can be achieved in the tradeoff between delay and loss as explored in Chapter 13.

For those connections which do not (or cannot) specify traffic parameters and a QoS class, there is a capability defined by the ATM Forum as *best effort* where no QoS guarantees are made and no specific traffic parameters need be stated. This traffic can also be viewed as "at risk" since there are no performance guarantees. In this case, the network admits this traffic and allows it to utilize capacity unused by connections which have specified traffic parameters and have requested a QoS class. It is assumed that connections utilizing the best effort capability can determine the available capacity on the route allocated by the network.

12.3.2 QoS Classes

In order to make things simpler on users, a small number of pre-defined QoS classes are defined, with particular values of parameters (defined above) pre-specified by a network in each of a few QoS Classes. The ATM Forum UNI specification version 3.0 defines the five numbered QoS classes and example applications summarized in Table 12.1.

Table 12.1 ATM Forum QoS Classes

QoS Class	QoS Parameters	Application
0	Unspecified	"Best Effort" "At Risk"
1	Specified	Circuit Emulation, CBR
2	Specified	VBR Video/Audio
3	Specified	Connection-Oriented Data
4	Specified	Connectionless Data

A QoS class is defined by at least the following parameters:

• Cell loss ratio for the CLP=0 flow
• Cell loss ratio for the CLP=1 flow
• Cell delay variation for the aggregate CLP=0+1 flow
• Average delay for the aggregate CLP=0+1 flow

The CLP=0 flow refers to only cells which have the CLP header field set to 0, while the CLP=1 flow refers to only cells which have the CLP header field set to 1. The aggregate CLP=0+1 flow refers to all cells in the virtual connection.

12.3.3   Specified QoS Classes

A specified QoS class provides performance to a ATM virtual connection (VCC or VPC) as specified by a subset of the ATM performance parameters. For each specified QoS class, there is one specified objective value for each performance parameter, where a particular parameter may be essentially unspecified – for example, a loss probability of 1. Initially, each network provider should define the ATM performance parameters for at least the following service classes from ITU-T Recommendation I.362 in a reference configuration that may depend on mileage and other factors:

• Service Class A: circuit emulation, constant bit rate video
• Service Class B: variable bit rate audio and video
• Service Class C: connection-oriented data transfer
• Service Class D: connectionless data transfer

Chapter 8 defined how these service classes are related to AALs. In the future, more QoS classes may be defined for a given service class. The following specified QoS Classes are currently defined by the ATM Forum:

Specified QoS Class 1 supports a QoS that meets service class A performance requirements. This class should yield performance comparable to current digital private line performance.

Specified QoS Class 2 supports a QoS that meets service class B performance requirements. This class is intended for packetized video and audio in teleconferencing and multimedia applications.

Specified QoS Class 3 supports a QoS that meets service class C performance requirements. This class is intended for interoperation of connection-oriented protocols, such as Frame Relay.

Specified QoS Class 4 supports a QoS that meets service class D performance requirements. This class is intended for interoperation of connectionless protocols, such as IP or SMDS.

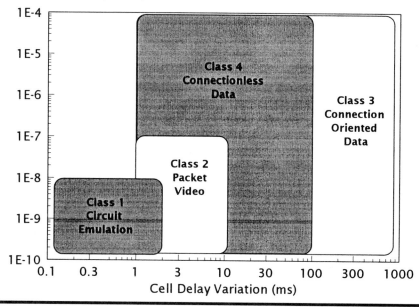

Figure 12.4 Example of QoS Class Value Assignments

Figure 12.4 gives a concrete example of how the QoS parameters for cell loss ratio for the CLP=0 flow and cell delay variation might be assigned for the four specified QoS classes. See Ref. 1 for another example. A network operator may provide the same performance for all or a subset of specified QoS classes, subject to the constraint that the requirements of the most stringent service class are met. Note that class 1 requirements meet the requirements of all of the other classes, a requirement stated in Appendix A of the ATM Forum UNI version 3.0 specification.

12.3.4 Unspecified QoS

In the unspecified QoS class, no objective is specified by the network operator for the performance parameters (however, an internal objective may be set). Services using the unspecified QoS class may have explicitly specified traffic parameters.

An example application of the unspecified QoS class is the support of *best effort* service, where effectively no traffic parameters are specified. For this type of "best effort" service, the user does not effectively specify any traffic parameters and does not expect a performance commitment from the network. One component of the "best effort" service is that the user application is expected to adapt to the time-variable, available network resources. The interpretation and clearer definition of the best effort service is an ongoing activity in the ATM Forum. The current name for this type of service is the Unspecified Bit Rate (UBR). An adaptive, flow-controlled service is currently being defined by the ATM Forum as the Available Bit Rate (ABR), which is similar to the ANSI Class Y service (see Ref. 2).

12.4 APPLICATION QoS REQUIREMENTS

Various applications have different QoS requirements. For example, some well-known requirements exist for voice after 30 years of experience in telephony. If voice has greater than about 15 ms of delay, then echo cancellation is usually required. Packetized voice can accept almost a 1% cell loss rate without being objectionable to most listeners. Newer applications do not have such a basis, or well-defined requirements; however, there are some general requirements that the following paragraphs summarize.

Video application requirements depend upon several factors, including the video coding algorithm, the degree of motion required in the image sequence, and the resolution required in the image. Loss generally causes some image degradation, ranging from distorted portions of an image to loss of an entire frame, depending upon the extent of the loss and the sensitivity of the video coding algorithm. Also, variations in delay of greater than 20 to 40 ms can cause perceivable jerkiness in the video playback.

Combined video and audio is very sensitive to differential delays. Human perception is highly attuned to the correct correlation of audio and video, which is apparent in some foreign language dubbed films. File transfer applications are also sensitive to loss and variations in delay, which result in retransmissions and consequent reduction is usable throughput. The sensitivity is related to the time-out and loss identification algorithm and the retransmission strategy of the application.

Users of interactive applications are also sensitive to loss and variations in delay due to retransmissions, and inconsistent response time, which can decrease productivity. Consistent response time (or the lack thereof) can effect how users perceive data service quality.

Distributed computing and database applications can be very sensitive to absolute delay, loss, and variations in delay. The ideal for these types of applications is infinite bandwidth with latency close to that of the speed of light in fiber. A practical model is that of performance comparable to a locally attached disk drive or CD-ROM, which ranges from 10 to 100 ms.

12.5 TRAFFIC DESCRIPTOR

The traffic descriptor is a list of parameters which captures intrinsic source traffic characteristics. It must be understandable and enforceable. This section describes the following traffic parameters defined by the ATM Forum UNI specification version 3.0:

- A mandatory Peak Cell Rate (PCR) in cells/second in conjunction with a CDV tolerance in seconds.
- An optional Sustainable Cell Rate (SCR) in cells/second (always less than or equal to PCR) in conjunction with a Maximum Burst Size (MBS) in cells

The following statements summarize the current state of the ATM Forum agreements. Figure 12.5 illustrates the following key traffic contract parameters:

◎ Peak Cell Rate (PCR) = 1/T in units of cells/second, where T is the minimum intercell spacing in seconds (i.e., the time interval from the first bit of one cell to the first bit of the next cell).

◎ Cell Delay Variation (CDV) Tolerance = τ in seconds. This traffic parameter normally cannot be specified by the user, but is set instead by the network. The number of cells that can be sent back-to-back at the access line rate is $\tau/T+1$ as shown in the figure.

◎ Sustainable Cell Rate (SCR) is the maximum average rate that a bursty, on-off traffic source that can be sent at the peak rate, such as that depicted in Figure 12.5.

◎ Maximum Burst Size (MBS) is the maximum number of cells that can be sent at the peak rate.

Figure 12.5 also depicts the minimum burst interarrival time as Ti, which is related to the SCR and MBS as defined by the equations at the bottom of the figure. The maximum burst duration in seconds is given by Tb, which is also defined at the bottom of the figure. These definitions may be helpful in understanding the traffic parameters but are not part of the formal traffic contract.

ITU-T Recommendation I.371 specifies only the peak cell rate. The ATM Forum added the specification of the sustainable cell rate and maximum burst size in a manner patterned after the PCR definition to better model bursty data traffic. This modeling of the peak, average, and burst length characteristics enables ATM networks to achieve statistical multiplex gain within a specified loss rate, as described in Chapter 15.

Figure 12.5, however, does not represent a rigorous definition of the traffic parameters. A more formal, rigorous definition has been defined in standards and industry specifications. The formal algorithm (which has the informal name of leaky bucket), is not described in this text, but is instead illustrated through simple analogies and examples.

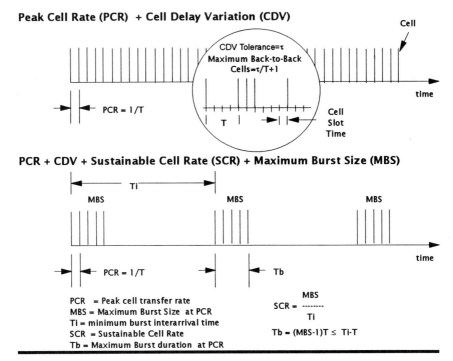

Figure 12.5 Illustration of Principal ATM Traffic Parameters

12.6 LEAKY BUCKET ALGORITHM

The *leaky bucket* algorithm is key to defining the meaning of conformance checking for an arriving cell stream against the traffic parameters in the traffic contract. This section defines the leaky bucket method using several analogies. A formal definition of the leaky bucket algorithm may be found in the ATM Forum UNI specification or CCITT Recommendation I.371.

The leaky bucket analogy refers to a bucket with a hole in the bottom that causes it to *leak* at a certain rate corresponding to a traffic cell rate parameter. The *depth* of the bucket corresponds to a traffic parameter or a tolerance parameter. These tolerance and traffic parameters are defined in a subsequent section. Each cell arrival creates a *cup* of fluid flow that is *poured* into one or more buckets for use in conformance checking. The funneling of cell arrival fluid into buckets is controlled by the Cell Loss Priority (CLP) bit in the ATM cell header. First, the case of a single bucket is described, and is then extended to the case of two buckets.

In the leaky bucket analogy, the cells do not actually flow through the bucket; only the check for conforming admission does. Note that one implementation of traffic shaping is to actually have the cells flow through the bucket; we cover this in the next chapter. The operation of the leaky bucket is described with reference to the following figures for examples of a conforming and non-conforming cell flow. In all of the following examples, the nominal cell interarrival time is four cell times, which is the bucket increment, and the bucket depth is six cell times in both examples. The notion commonly employed in queueing theory of a fictional "gremlin" performing some action is used to explain each of the examples.

When a cell arrives, a gremlin checks to see if the entire bucket increment for a cell can be added to the current bucket contents without overflowing. If the bucket would not overflow, then the cell is *conforming*; otherwise it is *nonconforming*. The gremlin pours the fluid for nonconforming cells on the floor. Fluid from a cell arrival is added to the bucket only if the cell is conforming, otherwise accumulated fluid from nonconforming cells might cause later cells to be identified as nonconforming.

Each cell time the bucket drains one increment. Each cell arrival adds a number of units specified by the increment. In other words, upon arrival of a cell, its fluid is completely drained out after a number of cell times given by the leaky bucket increment.

For the conforming cell flow example shown in Figure 12.6, the first cell arrival finds an empty bucket, and fills it to a depth of four units. At the third cell time two units have drained from the bucket, and a cell arrives. The gremlin determines that the fluid from this cell would fill the bucket to the brim (i.e., to a depth of six); therefore, it is conforming so that its increment is added to the bucket. Now the next earliest conforming cell arrival time would be four cell times later (i.e., cell time 7), since four increments must be drained from the bucket in order for a cell arrival to not cause the bucket of depth equal to 6 units to overflow. This worst case arrival of cells continues in cell times 15 and 19 in the example.

In the nonconforming cell flow example of Figure 12.7, the first cell arrival at cell time 1 finds an empty bucket, and fills it to a depth of four units. Over the next four cell times the bucket drains completely – one unit per cell time. On the fifth cell time another cell arrives and fills the empty bucket with four increments of fluid. At the sixth cell time a cell arrives, and the gremlin determines that the bucket would overflow if the arriving cell's fluid were to be added, therefore this definition determines that this cell is nonconforming. The gremlin then pours the fluid for this cell onto the floor, bypassing the bucket. Since the fluid for the nonconforming cell was not added to the bucket, the next conforming cell can arrive at cell time 7, completely filling the bucket. The next cell arrives at cell time 13, and fills the empty bucket with four increments. The next cell arrival at cell time 15 fills the bucket to the brim. Finally the cell arrival at cell time 17 would cause the bucket to overflow; hence it is by definition nonconforming, and the gremlin pours the non-conforming cell's fluid on the floor, bypassing the bucket again.

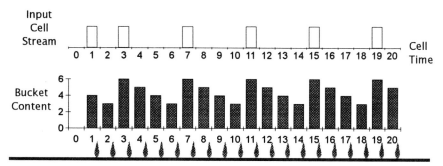

Figure 12.6 Illustration of a Conforming Cell Flow

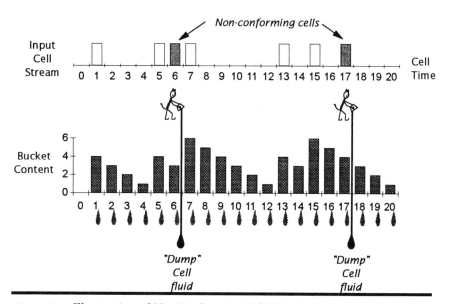

Figure 12.7 Illustration of NonConforming Cell Flow

12.7 TRAFFIC DESCRIPTORS AND TOLERANCES

The Peak Cell Rate (PCR) and Sustainable Cell Rate (SCR) traffic parameters are formally defined in terms of a virtual scheduling algorithm (which is equivalent to a leaky bucket algorithm) in ITU/CCITT Recommendation I.371 and the ATM Forum UNI specification. These parameters are specified either in a signaling message or at subscription time or are implicitly defined by the network according to default rules.

The Peak Cell Rate (PCR) is modeled as a leaky bucket drain rate, and the Cell Delay Variation (CDV) Tolerance defines the bucket depth for Peak Rate conformance checking on either the CLP=0 or the combined CLP=0+1 flows.

The Sustainable Cell Rate (SCR) is modeled as a leaky bucket drain rate, and the burst tolerance (which is proportional to the maximum burst size) defines the bucket depth for sustainable rate conformance checking on either the CLP=0, CLP=1, or CLP=0+1 flows. The burst tolerance, which defines the SCR bucket depth, is defined by the following formula in the ATM Forum UNI version 3.0 specification:

$$\text{Burst Tolerance} \quad = \quad (\text{MBS} - 1) \left(\frac{1}{\text{SCR}} + \frac{1}{\text{PCR}} \right)$$

The burst tolerance, or bucket depth, for the SCR is not simply the MBS because the sustainable rate bucket is draining at the rate given by SCR. The formula above is determined by calculating the bucket depth required for MBS cells arriving a rate given by the PCR, while draining out at the rate given by SCR.

The following combinations of two leaky buckets and associated traffic parameters are specified by the ATM Forum for interoperability. These traffic parameters may be either explicitly specified, for example, in a signaling message; or implicitly specified, for example, by a default rule in the network.

1. PCR on CLP=0+1
2. PCR on CLP=0+1, and PCR on CLP=0 without tagging
3. PCR on CLP=0+1, and PCR on CLP=0 with tagging
4. PCR on CLP=0+1, and SCR+MBS on CLP=0+1 without tagging

5. PCR on CLP=0+1, and SCR+MBS on CLP=0 without tagging
6. PCR on CLP=0+1, and SCR+MBS on CLP=0 with tagging

Figure 12.8 illustrates these interoperable ATM Forum Leaky Bucket configurations. When there are two buckets, an equal amount of cell "fluid" is directed into both buckets when the figures show a diagonal line and a two-headed arrow. The analogy for the diagonal line upon which the fluid is poured for direction into the buckets is a "rough board" which creates enough turbulence in the fluid such that a single cup of fluid from an arriving cell fills both buckets to the same depth as if the fluid from a single cell was smoothly poured into one bucket.

Tagging is the action of changing the CLP=0 cell header field to CLP=1 when a nonconforming cell is detected. This controls how a bucket acting on CLP=0 only interacts with a bucket that operates on both CLP=0 and CLP=1 flows, referred to as CLP=0+1. Tagging will be covered in more detail in the next chapter.

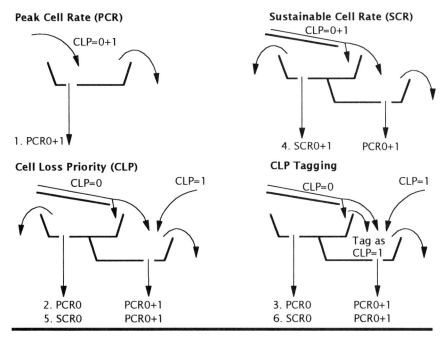

Figure 12.8 ATM Forum Leaky Bucket Configurations

All of these configurations contain the Peak Cell Rate (PCR) on the aggregate CLP=0+1 cell flow to achieve interoperability with the minimum requirement for ITU-T Recommendation I.371.

Configuration 1 provides the check on PCR for the aggregate CLP=0+1 cell flow as the minimum, required configuration. All others are optional. Configuration 2 checks PCR conformance on the CLP=0 and aggregate CLP=0+1 cell flows separately, as shown in Figure 12.8. Configuration 3 is similar to configuration 2 but changes the CLP field from 0 to 1 for cells in the arriving CLP=0 flow that exceed the PCR. An example application for configurations 2 and 3 would be dual-level coded video where the critical information is coded as CLP=0, and the noncritical information is coded as CLP=1. Configuration 4 checks the peak and sustainable rate on the aggregate CLP=0+1 flow, discarding any cells which violate either check. Configurations 5 and 6 are similar in form to configurations 2 and 3 as shown in the lower part of Figure 12.8; however, the check is on peak and sustainable rate. Appendix B of the ATM Forum UNI specification version 3.0 gives an example of how configuration 5 can be used to emulate the SMDS access class credit manage, and how configuration 6 can be used to emulate the Frame Relay Committed and Excess Information Rate (CIR and EIR) traffic parameters.

12.8 ALLOCATION OF TOLERANCES

There are several considerations involved in setting the leaky bucket depths (i.e., traffic parameter tolerances). These are different for the peak rate and the sustainable rate.

For the peak rate the bucket depth should not be much greater than that of a single cell; otherwise cells may arrive too closely together. For a single bucket depth of D cells and a nominal spacing S, note that approximately D/S cells can arrive back-to-back.

For the sustainable rate the bucket depth should be set to a value greater than the greatest burst expected from the user. The burst length is at least as long as the number of cells corresponding to the longest higher layer Protocol Data Unit (PDU) originated by the user. Furthermore, some transport protocols, such as TCP, may generate bursts that are many PDUs in length. Some additional tolerance should be added to allow for the effect of multiplexing and intermediate networks prior to the point where traffic conformance is checked.

12.9 REVIEW

This chapter introduced several key concepts and some basic terminology that will be used in the remainder of Part 5. A traffic contract is an agreement between a user and a network regarding the Quality of Service (QoS) that a cell flow is guaranteed – if the cell flow conforms to a set of traffic parameters defined by the leaky bucket rule. The principal QoS parameters are: average delay, variation in delay, and loss ratio. The traffic parameters define at least the Peak Cell Rate (PCR), and may optionally define a Sustainable Cell Rate (SCR) and Maximum Burst Size (MBS). A CDV tolerance parameter is also associated with the peak rate, but is not usually specified by the user. A leaky bucket algorithm in the network checks conformance of a cell flow from the user by pouring a cup of fluid for each cell into a set of buckets leaking at rates corresponding to the PCR, and optionally the SCR. If the addition of any cup of cell fluid would cause a bucket to overflow, then the cell arrival is considered *nonconforming*, and its fluid is not added to the bucket. Additional considerations in setting the depth of the leaky buckets to account for tolerances in the traffic parameters are also described.

12.10 REFERENCES

[1]   G. Woodruff, R. Kositpaiboon, "Multimedia Traffic Management Principles for Guaranteed ATM Network Performance," IEEE JSAC, April 1990
[2]   ATM Forum, "53 Bytes Newsletter," January 1994, Volume 2, Issue 1

13

Traffic Control

This chapter introduces, compares, and contrasts some basic concepts from standards groups and the industry regarding traffic control. Traffic control provides the means that allow a user to ensure that the offered cell flows meet the rate specified in the traffic contract, and the means for networks to ensure that the traffic contract rates are enforced such that the QoS performance is achieved across all users. This chapter describes several types of Usage Parameter Control (UPC) and traffic shaping implementations. One standardized means for handling priority is that of selective cell discard based upon the Cell Loss Priority (CLP) bit in the ATM cell header. Another is that of Explicit Forward Congestion Indication (EFCI). Finally, concepts are introduced from fast resource management, which is an area of current standards activity. Other schemes that have been introduced to perform traffic control functions are also summarized.

13.1 TRAFFIC AND CONGESTION CONTROL

This section provides an overview of traffic and congestion control as an introduction to this chapter and Chapter 14 on congestion control. A generic reference model derived from I.371 as shown in Figure 13.1 illustrates the placement of various traffic and congestion control functions. Shaping of cell flows in order to conform with traffic parameters can be performed in the terminal equipment. It is then required that conformance to these traffic parameters be checked, or "policed," by the Usage Parameter Control (UPC) function by the network at the User Network Interface (UNI). In a similar manner networks may check the arriving cell flows from a prior network using Network Parameter Control (NPC). Within networks functions such as Connection Admission Control (CAC), resource management, priority control, and others may be employed.

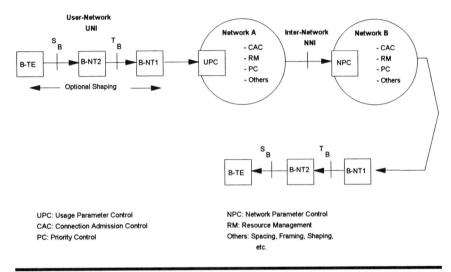

Figure 13.1 Overview of Traffic/Congestion Control Functions

In traffic and congestion control the time scale over which a particular control is applicable is usually very important. This is illustrated in Figure 13.2. The minimum time scale that traffic or congestion control can act on is a cell time (which is approximately 10 μs on a DS3, and 3 μs on an STS-3c), which is covered in this chapter in detail. The next time scale that makes sense is that of round-trip propagation time, which is covered in detail in the section on feedback congestion controls in the next chapter. The next major event that changes over time is the arrival of either PVC provisioning or SVC signaling that creates requests to either establish or relinquish a virtual connection. This time scale can be greater than the round-trip delay. Finally, there are long-term network engineering and network management procedures that operate on much longer time scales, on the order of hours to days and even months and years for long-range network planning.

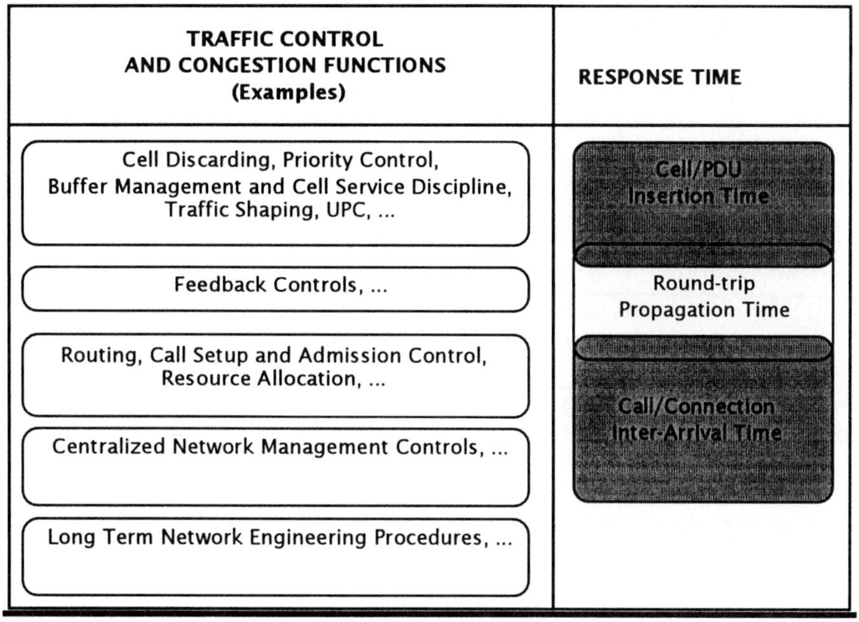

TRAFFIC CONTROL AND CONGESTION FUNCTIONS (Examples)	RESPONSE TIME
Cell Discarding, Priority Control, Buffer Management and Cell Service Discipline, Traffic Shaping, UPC, ...	Cell/PDU Insertion Time
Feedback Controls, ...	Round-trip Propagation Time
Routing, Call Setup and Admission Control, Resource Allocation, ...	Call/Connection Inter-Arrival Time
Centralized Network Management Controls, ...	
Long Term Network Engineering Procedures, ...	

Figure 13.2 Time Scales of Traffic/Congestion Control

13.2 USAGE/NETWORK PARAMETER CONTROL

The actions taken by the network with regard to the conformance of cell flows from a user, or another network, are called Usage Parameter Control (UPC) and Network Parameter Control (NPC), respectively. Another commonly used name for UPC/NPC is *policing* [2]. This is a good analogy, because UPC/NPC perform a role similar to the police in society. Police are supposed to enforce the law, to ensure fair treatment for all people. UPC/NPC do this by ensuring that bandwidth and buffering resources are fairly allocated among the users according to their traffic contracts. Without UPC/NPC, situations that would be unfair, or where a single user could "hog" resources, could occur. Most of the functions and discussion in the following apply equally to UPC and NPC, with any differences identified explicitly.

Standards do not specify exactly how the UPC and NPC functions are to be implemented; instead the performance of any UPC/NPC implementation is specified in relation to the leaky bucket conformance checking algorithm. Indeed, the compliant connection definition part of the traffic contract identifies how much slack exists between the network's UPC implementation and that of the ideal conformance checking rule, in other words, the looseness of the UPC. The other requirement is that the UPC should not take a policing action (i.e., tag or discard) on more than the fraction of cells which are conforming to the leaky bucket rule; or in other words, the UPC cannot be too tight.

Also note that the UPC may police different cells than an ideal leaky bucket due to inevitable differences in initialization, the latitude defined for a compliant connection, or the fact that the UPC is not implemented using the leaky bucket.

The following sections give three examples of UPC/NPC implementations; one using the leaky bucket algorithm, and two windowing schemes to illustrate differences in how cell flows are compliance-checked for different algorithms using the same traffic parameters.

13.2.1 Leaky Bucket UPC/NPC

The leaky bucket example of Figure 13.3 uses the same nonconforming cell flow as input as the shaping example of Figure 13.8. Three more gremlins: *Tag*, *Discard*, and *Monitor* join the *Dump* gremlin from Chapter 12 in this example to illustrate the possible UPC actions. The sequence of cell arrivals is illustrated along the horizontal axis at the top of Figure 13.3, with the *Dump* gremlin causing the fluid from nonconforming cells (shaded in the figure) to bypass the leaky buckets. *Tag*, *Discard*, and *Monitor* all operate in conjunction with *Drops* identification of a nonconforming cell. *Tag* sets the CLP bit to 1 (regardless of its input value) and lets the cell pass through, with its position unchanged. *Discard* simply does not allow the cell to be transmitted. *Monitor* simply keeps track of how many cells were nonconforming on his notepad. There is a fourth possible UPC action in the standard, namely, do nothing, which corresponds to all of the gremlins being out to lunch in this example.

13.2.2 Sliding and Jumping Window UPC

Next, two windowing UPC mechanisms are compared to the leaky bucket mechanism to illustrate how different UPC implementations designed to police the same portion of nonconforming cells can police different cells, and even police at different portions of nonconforming cells! From Figure 13.4 observe that the worst-case conforming cell flow can have at most 3 cells in 10 cell times. For any set of single leaky bucket parameters, an equivalent relationship for a window of M cells in N cell times can be derived. In the examples, M=3 and N=10. Two types of windowing UPCs are considered: a sliding window and a jumping window method. Two more gremlins are used to describe these UPCs: "Slide" and "Jump."

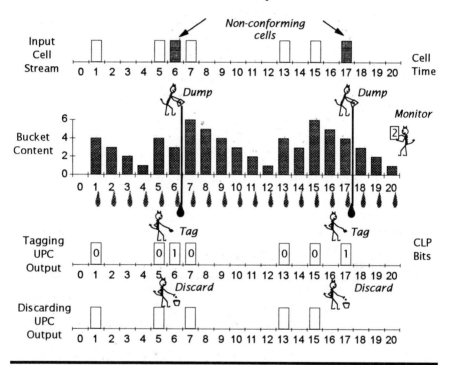

Figure 13.3 UPC via Leaky Bucket Example

For reference, the nonconforming cell flow of Chapter 12 is shown at the top Figure 13.4. In the sliding window protocol a window of N cell times is moved to the right by "Slide" each cell time. If there are less than M cells in the N cell times, then no UPC action is performed. However, if by sliding the window one unit to the right M (or more) cells are included in the window, then UPC is taken on all cells after the Mth cell in the window. This process is repeated each cell time by "Slide." Note that in the example the same number of cells are not acted on by the different UPCs. In the sliding window and leaky bucket UPCs there are the same number of cells acted upon; however, they are different cells! In the jumping window scheme "Jump" moves the window N units to the right every N cell times. The same M out of N count rule is applied, but note that fewer cells are acted on by the UPC. In general a jumping window is more permissive than a sliding window UPC.

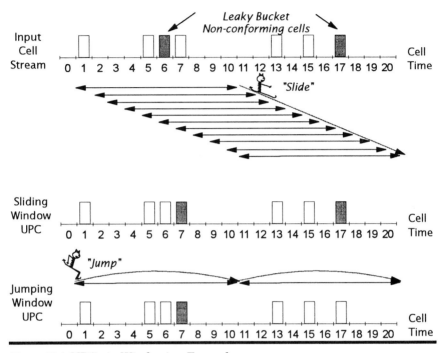

Figure 13.4 UPC via Windowing Example

In the example of Figure 13.4, the jumping window UPC was looser than the leaky bucket and sliding window UPC methods. The fact that different UPC algorithms, or even the same algorithm, may police different cells is called measurement skew and is the reason that UPC performance is stated in terms of the fraction of cells that are conforming. Figure 13.5 illustrates a pathological case where the difference between a leaky bucket, sliding window, and jumping window UPC is even more pronounced.

Each algorithm has parameters chosen to admit one cell every three cell times on the average, and allow at most two back-to-back cells. The leaky bucket UPC has an increment of three cells and a bucket depth of five cells, while the window algorithms have parameters M=2 and N=6. Figure 13.5 illustrates the arrival of 10 cells. The leaky bucket UPC identifies 20% as noncompliant, the sliding wind identifies 40% as noncompliant, and the jumping window identifies 30% as being noncompliant.

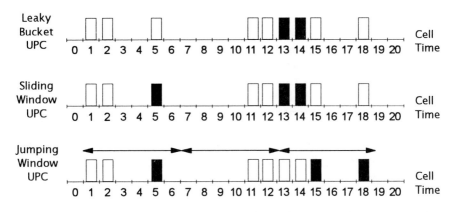

Figure 13.5 Pathological Example of UPC Differences

13.3 PRIORITY CONTROL

Priority control can help achieve the full range of QoS loss and delay parameters required by the range of high-performance applications. This can be accomplished by what is also called priority queueing, service scheduling, or fair queueing [4], [5], [3]. Basically, multiple queues are implemented in the switch, such that traffic on certain VPC/VCCs that are not tolerant of delay can "jump ahead" of those that are more tolerant of delay .

Priority queueing is defined between different VPCs and VCCs in order to meet different delay and loss priorities simultaneously. It is described by example with reference to the block diagram of Figure 13.6. In this example the priority queueing function occurs on the output side of an output buffered switch. Arriving cell streams (on multiple input ports) look up an internal priority value, and are directed to one of several corresponding queues for the output port. The output ATM port serves each of the queues according to a particular scheduling function.

An example scheduling function is one where the highest priority, nonempty queue is always served next. This scheduling function ensures that the highest priority buffer has the least loss and delay. Other scheduling functions can be chosen to spread out the variation in delay across the multiple queues. A scheduling function could send cells just before the maximum delay is reached for cells in each of the queues. This can make the variation in delay less for the lower priority queues. This is of particular interest at lower speeds with frame based protocols, such as Frame Relay.

Combinations of thresholded discard operating on each of the priority queues result in the capability to provide a range of loss and delay priority to each priority class, as well as a means to perform the congestion control that is discussed in depth in the next chapter.

A dedicated set of buffers may be allocated to each output port or split between the input and output ports, or a memory may be shared between multiple ports and/or the multiple priorities. If a shared memory approach is used, then limits on the individual queue sizes may be configurable under software control.

13.4 GENERIC FLOW CONTROL

The concept of Generic Flow Control (GFC) has seen a considerable evolution over the past several years in the standardization process. Initially, it was viewed as a means to implement a function very similar to the Distributed Queue Dual Bus (DQDB) protocol with a shared access medium. What appears likely to be standardized by 1994 or 1995 is a point-to-point configuration that allows a multiplexer to control contention for a shared trunk resource through use of traffic-type selective controls.

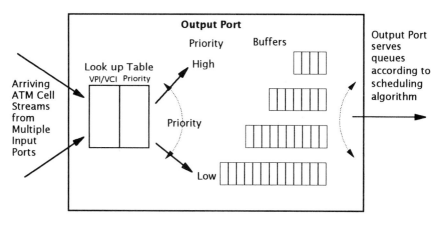

Figure 13.6. Illustration of Priority Queueing Operation

There are four bits allocated in the ATM cell header for the purpose of Generic Flow Control (GFC) at the ATM UNI. Since the GFC is part of the cell header, no additional bandwidth of VPI/VCI allocations is needed for a multiplexer to control a terminal. The first of these two bits are for controls and the last two are for parameters. The four bits of GFC represent almost 1% of the available ATM cell payload rate, and are therefore an important resource. Figure 13.7 illustrates the multiplexer configuration envisioned for GFC along with the controls and parameters used between the multiplexer and terminals. The default coding of the GFC is *null*, or all zeroes, which indicates that the interface is not under GFC, which is called *uncontrolled*. ATM terminals may have either a single traffic type or multiple traffic types. A terminal with multiple traffic types may be viewed as effectively having multiple, prioritized queues as indicated in the figure. A terminal with only a single traffic type can be viewed as having only a single queue. The protocol between the multiplexer and the terminals is asymmetric; the multiplexer controls the terminals, and the terminals only respond or convey information. The multiplexer may command a terminal to either start or stop traffic selectively by type if the shared trunk resource is becoming too congested. The terminal echoes the nonzero control field, which indicates to the multiplexer that the terminal understands GFC, which in standards terminology means that it is called a *controlled* terminal. The terminal also indicates the traffic type for a particular VPI/VCI in the function field. The start control can be used by the terminal to load an allocation counter, which counts down for each cell transmitted. Therefore, as long as the multiplexer sends a start command to the terminal periodically, it can continue to send data.

GFC may also be employed in a unidirectional ring using this protocol. There is also a possibility that further information could be multiplexed into the GFC field for more sophisticated controls.

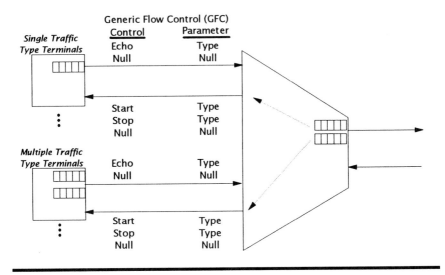

Figure 13.7 Illustration of GFC Configuration and Function

13.5 TRAFFIC SHAPING

A key element of the traffic contract from the user perspective is the sequence of cells that can be sent to the network and still be compliant with the traffic parameters in the traffic contract. The method specified in standards is called "shaping." In other words, the user equipment can process the source cell stream such that the resultant output toward the network is conforming to the traffic parameters according to the leaky bucket algorithm configuration in the traffic contract. This function is identified as being optional in standards, but if it is not done, then a cell flow may not be conforming, and the network need not guarantee QoS performance. Also, a network may use shaping when transferring a cell flow to another network in order to meet the rate conditions of a network-to-network traffic contract, or in order to ensure that the receiving user application will operate in an acceptable way.

A real-life example of the need for such shaping is the case where a receiving switch or user application may accept only a certain degree of cell clumping (i.e., CDV tolerance); otherwise its buffers would overflow.

13.5.1 Overview of Possible Shaping Methods

A list of possible implementations of traffic shaping as proposed in the standards and literature is summarized below:

- Buffering
- Spacing
- Peak Cell Rate Reduction
- Scheduling
- Burst Length Limiting
- Source Rate Limitation
- Priority Queueing
- Framing

A brief summary will be given for each of the proposals listed above, with reference to further detail given for the interested reader. Detailed examples for buffering, spacing, framing, and priority queueing will then be given.

Buffering operates in conjunction with the leaky bucket algorithm to ensure that cells will not violate the traffic parameters of the contract by buffering cells until the leaky bucket would admit them.

Spacing involves the end terminal holding cells from multiple virtual connections in a queue, scheduling their departures such that the traffic parameters are not violated, and that some measure of delay variation is minimized.

Peak cell rate reduction can be achieved by operating the sending terminal at a peak rate less than that in the traffic contract, reducing the possibility of conformance violation.

Burst length limiting is similar to peak rate reduction in that the source limit on burst length is set less than the Maximum Burst Size (MBS) in the traffic contract.

Source rate limitation is an implicit form of shaping that occurs when the actual source rate is limited in some other way; for example, in DS1 circuit emulation the source rate is inherently limited.

Framing superimposes a TDM-like structure onto the sequences of ATM cells, and utilizes this frame structure to schedule those cell streams that require controlled delay variation into the next frame time.

13.5.2   Leaky Bucket Traffic Shaping

A detailed example of traffic shaping using buffering and a leaky bucket implementation will be given. The example will show how the nonconforming cell flow of the previous chapter is converted into a conforming cell flow using buffering. This example uses the same notation as in the previous chapter for cell arrivals in cell time along the horizontal axis, the same nominal interarrival time of 4 and the same leaky bucket with a depth of 6.   Two new gremlins are introduced in this example, *Stop* and *Go,* to illustrate the buffering operation.

The gremlin *Stop* replaces *Dump* in the nonconforming example. *Stop* causes a cell to be buffered if its fluid flow would cause the bucket to overflow, and *Go* allows a cell to be transmitted and consequently have its fluid added to the bucket, as soon as the bucket drains far enough so that it would not overflow. Figure 13.8 illustrates this operation with the individual cells labeled A through G, and the nonconforming cells from the previous example indicated by shading.

Cell arrivals A and B are conforming, and leave the bucket in a state such that arrival C at cell time 6 is nonconforming, and hence *Stop* causes cell C to be buffered. D arrives immediately after C, so the gremlin *Stop* also causes D to be buffered. In the same cell time the bucket empties enough so that *Stop's* partner *Go* allows cell C to be transmitted and its flow added to the bucket. At cell time 11,*Go* allows cell D to be sent and fills the bucket. At cell time 13 cell arrival E would cause the bucket to overflow; hence *Stop* causes it to be buffered.   Cells F and G are similarly buffered by *Stop* and transmitted at the earliest conforming time by *Go* as illustrated in the figure. Cell G would be transmitted at cell time 23 (not shown). Note that the output cell flow from this process is conforming, as can be checked from the conformance test of the leaky bucket defined in the previous chapter.

The leaky bucket shaper smoothes out the input stream completely and does not drop any cells. A leaky bucket policing algorithm would find this output stream to be compliant.

13.5.3   Spacing

One proposal for a UPC/NPC traffic control function is that of spacing [1]. In this UPC/NPC implementation the resultant output never violates the nominal intercell spacing, but may discard additional cells. In order to describe this function, yet another gremlin called *Space*, is employed. *Space* implements a virtual scheduling algorithm which computes a Theoretical Reemission Time (TRT) such that the output never violates the nominal cell spacing, and discards any input bursts that cannot be spaced out within a tolerance specified in a manner analogous to the leaky bucket. Figure 13.9 illustrates the spacing function with the same nonconforming cell flow example of Chapter 12. Cells arrive as shown in the upper axis, with nonconforming cells shown shaded as before. The gremlin *Space* is shown wherever a cell is rescheduled, or spaced, showing the TRT on the card in his hand.

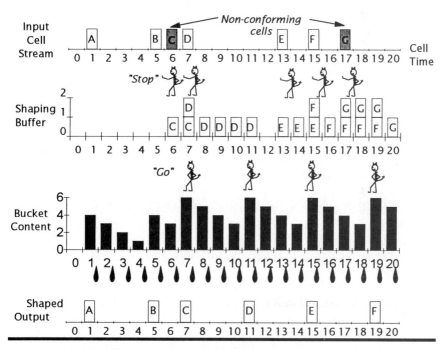

Figure 13.8 Traffic Shaping Example (Buffering)

Note that the output of the spacer is very regular, and never violates the nominal cell emission interval. This has advantages in terms of controlling the delay variation that will be observed by subsequent networks or the receiving terminal. This is important for functions such as video, audio, and circuit emulation where the application expects to receive cells at a regular rate. A disadvantage is that cells from a burst may be discarded, disrupting such higher layer functions as illustrated at cell times 6 and 17 in the example.

It is possible to not discard cells, only tagging them; for example, however, the function of taking this into account would add additional complexity to the spacing function.

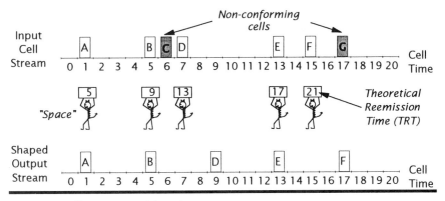

Figure 13.9 Illustration of Spacing

13.5.4 Framing

Another method of controlling delay variation for delay-sensitive virtual connections is that of framing [6]. Framing basically overlays a synchronous structure of frame boundaries on the asynchronous cell stream. Cells from flows that require a tightly controlled delay variation are scheduled to depart from a switching node in the next frame in a priority manner, similar to catching the next train. If the arrival time between frames is small enough, then the variation can be controlled with a limit equal to the number of nodes times the frame length. This concept is illustrated in Figure 13.10.

This type of function would be well suited to AAL1 circuit emulation over ATM, where the receiving connection endpoints have only a limited amount of cell buffer for absorption of the jitter accumulated across the network.

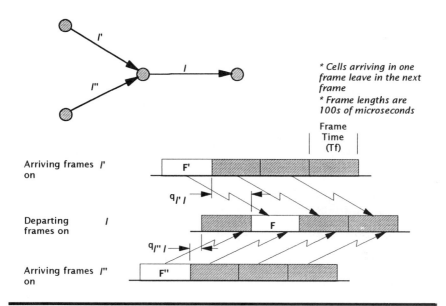

Figure 13.10 Illustration of Framing

13.6   CONNECTION ADMISSION CONTROL (CAC)

Connection Admission Control (CAC) is a software function in a switch that is responsible for determining whether a connection request is admitted or denied.   A connection request defines the source traffic parameters and the requested QoS Class.   CAC determines whether the connection request can be accepted at PVC provisioning time or SVC call origination time.   CAC can accept the request only if the QoS for all existing connections would still be met if the request was accepted.   CAC could be done on a node-by-node basis, or in a centralized system.   For accepted requests, CAC determines UPC/NPC parameters, routing, and resource allocation. Resources allocated include trunk bandwidth, buffer space, and internal switch resources.

CAC must be simple and rapid for high switching rates.   CAC complexity is related to the traffic descriptor.   The simplest CAC algorithm is for peak rate allocation, where a connection request cannot be admitted if the sum of the peak rates would exceed the trunk bandwidth.   This is illustrated in Figure 13.11, where a request for bandwidth R is submitted to peak rate Connection Admission Control (CAC) logic.   The "pipe" bandwidth is P, of which a certain portion A is already assigned. If the request R exceeds the available bandwidth (P-A), then the request is denied; otherwise, it is accepted. Actually the admission threshold may be somewhat less due to the slack implied by the compliant connection definition, the CDV tolerance parameter, and the buffer size available for a certain cell loss and delay QoS objective. CAC may also permit a certain amount of resource "overbooking" in order to increase statistical multiplex gain, which will be covered in Chapter 15.

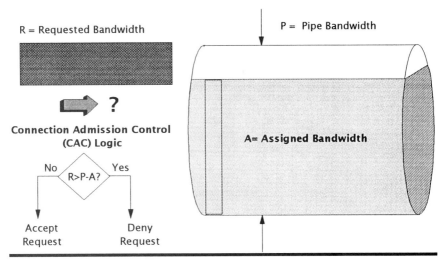

Figure 13.11 Illustration of Peak Rate Connection Admission Control

13.7 RESOURCE MANAGEMENT

There is a need to manage critical resources in the nodes of an ATM network. Two critical resources are buffer space and trunk bandwidth.

One way of simplifying the management of the trunk bandwidth is through the use of Virtual Paths (VPs). Recall from Chapter 8 that a VP can contain many VCs, and that VP cell relaying only operates on the VPI portion of the cell header. If every node in a network is interconnected by a VPC, then only the total available entry-to-exit VPC bandwidth need be considered in CAC decisions. A VPC is easier to manage as a larger aggregate than multiple, individual VCCs.

Note that QoS is determined by the VCC with the most stringent QoS requirement in a VPC. One could envision a network of nodes interconnected by a VPC for each QoS class; however, this could quickly exhaust the VPC address space if there are more than a few QoS classes. The section on ATM network design considerations elaborates further upon this point.

The complexity and number of changes required to implement routing, restoration, and measurement also are reduced by VPCs as compared to VCCs.

13.8 FAST RESOURCE MANAGEMENT

This section presents several additional proposals generically called Fast Resource Management (FRM) for managing bandwidth and buffering resources dynamically. The ITU-TSS has standardized a Payload Type (PT) code point of 110 in the ATM cell header and a FRM cell format shown in Figure 13.12. Specific FRM information field formats and protocols have not been specified yet.

This section briefly describes the two generic categories of FRM that have been proposed to date, bandwidth reservation and buffer reservation, and then compare and contrast their advantages and disadvantages.

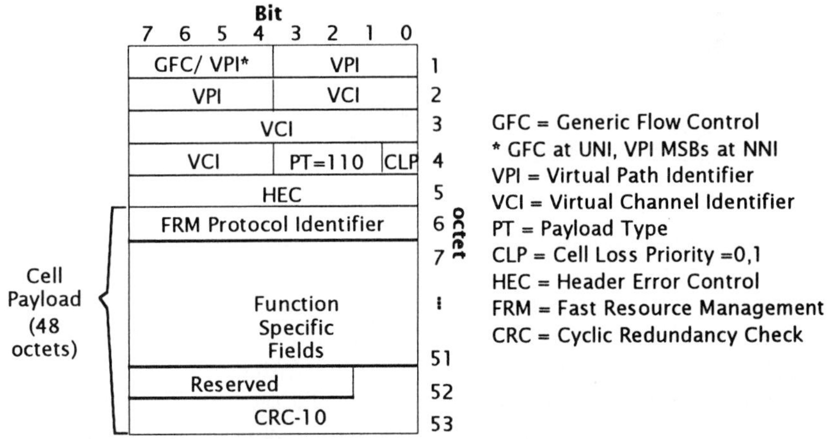

Figure 13.12 Fast Resource Management (FRM) Cell Format

13.8.1   Fast Bandwidth Reservation

Fast bandwidth reservation basically involves reserving bandwidth at each node along an end-to-end route. There are two possible modes for bandwidth reservation: immediate unguaranteed and delayed guaranteed.

The delayed bandwidth reservation method guarantees that bandwidth is reserved end-to-end. This is accomplished by the method illustrated in Figure 13.13.

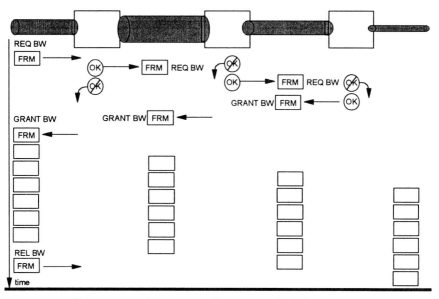

Figure 13.13 Illustration of Guaranteed Fast Bandwidth Reservation

A burst is preceded by an FRM cell indicating the request to reserve the bandwidth (REQ BW) for each intermediate node on the end-to-end route. This reservation request progresses through each intermediate node, either reserving the requested bandwidth and forwarding the request (OK), or in the event of failure discarding the request (NOT OK) or even returning a failure message to the origination. In the successful case illustrated in the figure the bandwidth is reserved at all the intermediate nodes, the bandwidth granting response (GRANT BW) is returned to the source, and the burst begins. Upon completion of the burst the source sends a release bandwidth reservation message which is propagated to all of the intermediate nodes. A reliable message transport protocol or time-outs must be employed to ensure that reservation releases are not disrupted by lost or errored cells. This scheme works well if the burst length is much greater than the round trip propagation delay, in that it is equivalent to fast call setup and tear down.

The immediate mode involves a user sending an FRM cell indicating the traffic parameters that are required for the subsequent burst, immediately followed by the actual burst. The entry node and intermediate nodes can then make use of this information to determine whether the entire burst, or none of the burst, can be handled. Therefore, the delivery of the burst is not guaranteed since congestion may occur at an intermediate node, resulting in the entire burst, or portions of it, being discarded.

13.8.2   Fast Buffer Reservation

The other critical resource that can be managed is that of buffer space in the intermediate nodes [7]. In this type of protocol a burst requests that buffer space be reserved in an intermediate node before it is admitted, as illustrated in Figure 13.14.

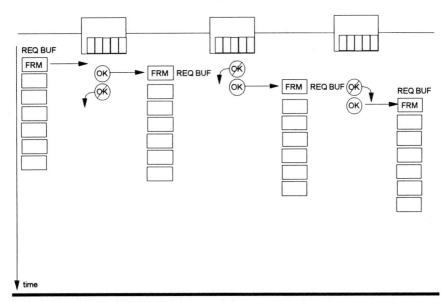

Figure 13.14 Illustration of Fast Buffer Resource Management

This can be done by an FRM cell at the beginning of the burst, by an AAL field, or by a predefined maximum burst length defined at subscription time on a per VPI/VCI basis. The basic operation is that when a burst arrives at a node, a check is performed against the requested buffer (REQ BUF) size and the available buffer space at the node. If the requested buffer size is greater than the available buffer space, then the entire burst is not admitted; otherwise the entire burst is admitted. By ensuring that the entire burst can be buffered at a node, it can be guaranteed that no burst is ever lost, except through errored or misinserted cells (unlikely events under normal circumstances).

A higher layer protocol must detect if a burst was dropped, and retransmit it.

13.9 OTHER TRAFFIC CONTROL METHODS

Other methods of traffic control are also possible. Further definition of traffic parameters, improved shaping and policing functions, and priority queueing are being developed and included in emerging ATM products. Also, some methods covered in the next chapter on congestion control can also be used for traffic control purposes as well. The use of selective cell discard based upon the Cell Loss Priority (CLP) bit allows traffic control to be selectively applied. The feedback and flow control techniques described in Chapter 14 could also be used to adaptively control UPC parameters.

13.10 REVIEW

This chapter introduced the concepts of traffic and congestion control and indicated where these would occur in user equipment or networks. The time scale on which these traffic and congestion schemes are applicable was also summarized. The Usage/Network Parameter Control (UPC/NPC) actions of Cell Loss Priority (CLP) bit tagging, cell discarding, and monitoring were then introduced using the same leaky bucket example. Two different windowing-based UPC implementations were compared to the leaky bucket method to show how different cells, and even different proportions of cells, are policed by different implementations. The discussion then introduced the concept of priority queueing and showed how it could help support multiple QoS classes. The application of the Generic Flow Control (GFC) field in the ATM cell header to implement multiplexing was reviewed. First introduced was the user equipment traffic control function called "shaping," which attempts to change the traffic parameters of a received cell flow to conform with the traffic contract. A description of how shaping could be implemented using the leaky bucket concept from the previous chapter or other methods was discussed. The role of Connection Admission Control (CAC) operating at the call level time scale in ensuring that QoS guarantees are met was then covered. The treatment then moved to the role of resource management, and in particular the use of virtual paths in simplifying this problem. Finally, the areas of Fast Resource Management (FRM) bandwidth and buffer reservation techniques that are in the process of standardization were surveyed.

13.11 REFERENCES

[1] P. Boyer, F. Guillemin, M. Servel, J-P. Coudreuse, "Spacing Cells Protects and Enhances Utilization of ATM Network Links," *IEEE Network*, September 1992
[2] E. Rathgeb, "Modeling and Performance Comparison of Policing Mechanisms for ATM Networks," *IEEE JSAC*, April 1991
[3] D. Hong, T. Suda, "Congestion Control and Prevention in ATM Networks," *IEEE Network*, July 1991
[4] M. Katevenis, S. Sidiropoulos, C. Courcoubetis, "Weighted Round-Robin Cell Multiplexing in a General-Purpose ATM Switch Chip", *IEEE JSAC*, October 1991
[5] H. Kröner, G. Hébuterne, P. Boyer, A. Gravey, "Priority Management in ATM Switching Nodes," *IEEE JSAC*, April 1991
[6] L. Trajkovic', S. Golestani, "Congestion Control for Multimedia Services," *IEEE Network*, September 1992
[7] J. Turner, "Managing Bandwidth in ATM Networks with Bursty Traffic," *IEEE Network*, September 1992

14

Congestion Control

This chapter covers the topic of congestion control introduced in the previous chapter. Congestion is an abnormal situation, which should be avoided if possible, but may warrant different levels of response depending upon its severity or duration. First, this chapter introduces the levels of congestion that can occur, and the types of responses, or controls, that exist. The standard method of selective cell discard is described in detail. The longer time scale methods of congestion control, namely resource allocation, network engineering, and network management controls are then discussed. Finally, other alternatives proposed by the industry to adaptively control the flow of user traffic into the network using either flow control or backward congestion notification are reviewed.

14.1 WHAT IS CONGESTION CONTROL?

This section introduces the definition of congestion through everyday examples, and then presents terminology more specific to ATM. Next, the factors that categorize congestion control schemes are defined. The treatment then moves to a definition of the metrics commonly used in comparing the relative performance of congestion control schemes. Finally, the classes of congestion control schemes are introduced.

14.1.1   Congestion Defined

Many of us experience congestion daily in the form of traffic jams, long checkout lines at stores, movie ticket lines, or just waiting for some form of service.   Congestion is generally defined as the condition reached when the demand for resources exceeds the available resources, over a certain time interval. Take the real-life example of a traffic jam. Congestion occurs because the number of vehicles wishing to use a road (demand) exceeds the number of vehicles that can travel on that road (available resources) during a rush hour (a time interval).

More specific to ATM, congestion is defined as the condition where the offered load (demand) from the user to the network is approaching, or exceeds, the network design limits for guaranteeing the Quality of Service (QoS) specified in the traffic contract. This demand may exceed the resource design limit because the resources were overbooked, or because of failures within the network.

In ATM networks the resources that can become congested include switch ports, buffers, transmission links, ATM Adaptation Layer (AAL) processors, and Connection Admission Control (CAC) processors. The resource where demand exceeds capacity is called the bottleneck, congestion point, or constraint.

14.1.2   Determining the Impact of Congestion

A number of application characteristics determine the impact of congestion, such as: connection mode, retransmission policy, acknowledgment policy, responsiveness, and flow control. In concert with the application characteristics, certain network characteristics also determine the response to congestion, such as: queueing strategy, service scheduling policy, discard strategy, route selection, propagation delay, processing delay, and connection mode.

Congestion can also occur on several levels as discussed in the last chapter. Congestion can occur in time at the cell level, the burst level, or the call level. See Refs, 1, 2, and 8 for more details on categorization of congestion time scales. Congestion can occur in space within a single resource or multiple resources. The detection of congestion is termed an indication, feedback, or notification.

The reaction to congestion can occur in either time or space. The reaction in *time* can be on a cell-by-cell basis, on a burst basis, or at the call level. In *space*, the reaction can be at a single node, at the source, at the receiver, or at multiple nodes. The reaction to the detection of congestion control, response, or action.

Definition of the congestion control problem is difficult because of the large number of combinations of application characteristics, network characteristics, and levels of congestion detection and reaction. One congestion control scheme that works well for certain applications and network characteristics at a certain level may work poorly for different characteristics and/or a different level. Congestion control in broadband networks has been the subject of intensive research and many papers.

14.1.3 Congestion Control Performance

There are two basic measurements to be considered in studying congestion — useful throughput and effective delay. We define *useful throughput* as the throughput that is actually achieved by the end application. For example, in a file transfer application if a packet is lost, then that packet (and possibly the packets sent after it) must be retransmitted. Although the ATM network transferred at least some of the cells corresponding to the lost packet (and possibly those after it), this was not useful to the end application since all cells were retransmitted. In a similar manner, the effective delay was not the delay required to send the packet unsuccessfully the first time, but the interval from the first transmission until the final successful reception at the destination. The useful throughput and effective delay for some applications can be identical to that of the underlying ATM network. For example, voice or video that is coded to operate acceptably even under loss is not retransmitted, and hence usable throughput and effective delay are the same as the ATM layer. In practice only loss or delay up to a critical value is acceptable; then the application performance, subjective perception of the image, or audio playback becomes unacceptable.

These two examples, a file transfer and voice and video traffic, represent extremes of application sensitivity to loss, and will be used throughout the remainder of this chapter to illustrate the relative performance of various congestion control schemes.

When congestion occurs, a critical phenomenon called *congestion collapse* can occur as illustrated in Figure 14.1. As offered load increases into the mild congestion region, the actual carried load is limited by the bandwidth and buffering resources up to maximum values. As offered load increases further into the severe congestion region, the carried load can actually *decrease* due to user application retransmissions caused by loss or excessive delay. The degree to which carried load decreases in the severe congestion region is known as the *congestion collapse* phenomenon. The collapse is determined by both the application and network characteristics.

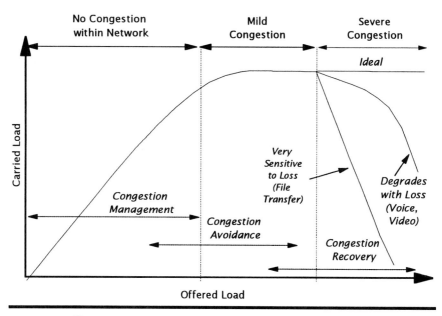

Figure 14.1 Illustration of Congestion Regions and Collapse

The simple file transfer example tends to have a congestion collapse that is load-sensitive. Thus we see a *cliff* of declining useful throughput when severe congestion is reached because of unproductive retransmissions which, in turn, actually increases the offered load-making the congestion even worse. The other example of voice or video coding is robust to loss and extends this cliff even further until the offered load is some fraction greater than the bottleneck resource; in other words, it is loss-limited. An ideal congestion control scheme is one where there is no congestion collapse, and the carried load increases to the available capacity of the bottleneck resource, and stays constant.

In the severe congestion region, the principle QoS degradation that can occur is either markedly increased delay or loss. The effectiveness of a particular congestion control scheme can also be measured by how much delay or loss occurs under the scenario of an offered load in excess of the design limit, as illustrated in Figure 14.2. A very load-sensitive application has delay and loss that increase markedly as severe congestion occurs. A loss-limited application can still achieve acceptable performance until the threshold between mild and severe congestion is reached. The ideal congestion-controlled application has bounded delay and loss at all values of offered load. Other applications may have delay and loss that are less sensitive to load. Figure 14.2 illustrates only the general trend of delay and loss versus load. In general, the curves would look markedly different when we plot delay and loss versus load, as will be shown in examples in Chapter 15. Normally delay is plotted on a linear scale, while loss is plotted on a logarithmic scale.

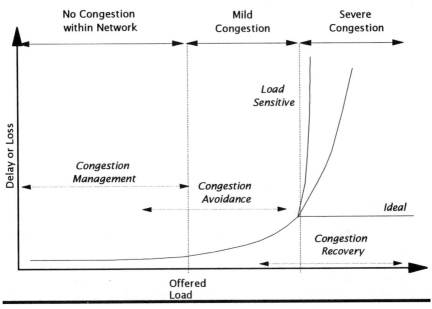

Figure 14.2 Effect of Congestion on Delay

14.1.4 Categories of Congestion Control

The terminology for the categorization of traffic and congestion control varies. For other categorizations of congestion see Refs. 1, 10, and 8. We have chosen the following categories as a means to structure the presentation of congestion control techniques in this chapter. The categories of response to congestion are: management, avoidance, and recovery. Each of these may operate at the cell level, the burst level, or the call level as illustrated in Table 14.1, which is essentially a road map to the remainder of the chapter.

Table 14.1 Congestion Control Categories and Levels

Category	Cell Level	Burst Level	Call Level
Management	UPC Discard	Resource Allocation	Network Engineering
Avoidance	EFCI, UPC Tagging	Window, Rate, or Credit Flow Control	"Overbooked" CAC, Call Blocking
Recovery	Selective Cell Discard, Dynamic UPC	Loss Feedback	Call Disconnection, Operations Procedures

Congestion management operates in the region of no congestion with the objective of attempting to ensure that the congested regions as illustrated in Figures 14.1 and 14.2 are never entered. This includes allocation of resources, discarding Usage Parameter Control (UPC), fully booked, or absolutely guaranteed, bandwidth, Connection Admission Control (CAC), and network engineering.

Congestion avoidance is a set of real-time mechanisms to prevent and recover from congestion during periods of coincident peak traffic demands or network overloads. One example of its use is when nodes and/or links have failed. Congestion avoidance procedures usually operate around the point between no congestion and mild congestion, and throughout the mildly congested region as illustrated in Figures 14.1 and 14.2. Congestion avoidance includes Explicit Forward Congestion Indication (EFCI), Usage Parameter Control (UPC) tagging using the Cell Loss Priority (CLP) bit, Overbooked Connection Admission Control (CAC), Blocking CAC, and either window-, rate-, or credit-based flow control.

Congestion recovery procedures are initiated to prevent congestion from severely degrading the end user perceived Quality of Service (QoS) delivered by the network. These procedures are typically initiated when the network has begun to experience loss or markedly increased delay due to congestion. Congestion Recovery includes selective cell discard, dynamic setting of UPC parameters, feedback driven by actual loss or disconnection, and operations procedures.

14.2 CONGESTION MANAGEMENT

Congestion management attempts to ensure that congestion is never experienced. For example, our form of congestion management is to not normally travel in rush hour traffic, or to wait until we know that there will be short lines in waiting for a particular service. This section covers the following congestion management methods:

☺ Resource Allocation
☺ Usage Parameter Control (UPC) Discard
☺ Fully Booked Connection Admission Control (CAC)
☺ Network Engineering

14.2.1   Resource Allocation

Of course, one way to control congestion is to avoid it entirely. This can be done by proper resource allocation. Resources subject to allocation include:

☞ Trunk Capacity
☞ Buffer Space
☞ UPC/NPC Parameters
☞ Virtual Path Connection (VPC) Parameters

If the UPC (policing) action is set to discard cells in excess of the peak rate and all trunk and buffer resources are allocated for the peak rate, then congestion simply cannot occur. This design can be extended to handle the worst-case failure scenario so that, under normal conditions, the network is well below full loading since some reserve capacity is allocated for restoration. Although this approach avoids congestion completely, the resulting utilization of the network may be very low, making this a potentially expensive proposition. This may be a viable approach in a local area where transmission and ports are relatively inexpensive. This approach is also referred to as "overengineering" the network.

The manner in which resources are allocated to meet a guaranteed QoS is, of course, a network decision. For example, it is optional whether any resources are allocated to CLP=1 flows. There should be sufficient resources allocated in order to meet the performance requirements for the expected mix of QoS class traffic. The Connection Admission Control (CAC) function makes a call-by-call decision on whether a connection request can be admitted based upon available resources. The resources for all QoS classes may be in a single shared pool, or placed in separate pools in order to achieve isolation between the classes.

14.2.2   Usage Parameter Control (UPC) Discard

Usage Parameter Control (UPC) acts as the traffic cop at the network ingress point. UPC with a discard capability manages to ensure that congestion cannot occur if resources are fully allocated as described in the previous section. The first node in the network implements UPC, and hence traffic that could congest the network is not admitted to the network to possibly congest other downstream nodes. UPC discard may also be used in conjunction with overbooking, in which case congestion internal to the network can occur.

14.2.3   Fully Booked Connection Admission Control (CAC)

The peak cell rate, sustainable cell rate, and maximum burst size for the Cell Loss Priority (CLP) flows as defined in the traffic contract can be used to fully book the buffer, trunk, and switch resources. This ensures that even if all sources were sending the worst-case, conforming cell streams, the specified Quality of Service (QoS) would still be achieved. This is called "fully booked" Connection Admission Control (CAC) because it admits only calls which have traffic parameters that would not cause the QoS of other connections to be violated. These traffic parameters must be enforced by Usage Parameter Control (UPC).

14.2.4   Network Engineering

One method for efficiently allocating resources is to base such decisions upon long-term, historical trending and projections. This is the method that is used in most large private and public networks today. These types of decisions include a determination of when and where to install or upgrade switches or trunk capacity. Various statistical measurements of the traffic and actual performance may be collected in order to accurately model the traffic sources for use in network planning algorithms. We will discuss several basic traffic source models, resource models, and traffic engineering methods in the next chapter. There is essentially no standardization in this area, and hence it is a network provider decision.

14.3 CONGESTION AVOIDANCE

Congestion avoidance attempts just that — to avoid severe congestion — but continues to push the offered load into the mildly congested region. In other words, it attempts to operate at the "knee" of the throughput versus load curve. This is analogous to life in the fast-paced modern world where we try to travel either just before, or just after, rush hour. Or when we try to arrive at the airport with as little time to spare as possible.

This section covers the following congestion avoidance methods:

☺ Explicit Forward Congestion Indication (EFCI)
☺ Usage Parameter Control (UPC) Tagging
☺ Connection Admission Control (CAC) "Overbooking"
☺ Call Blocking
☺ Window, Rate, or Credit Flow Control

14.3.1 Explicit Forward Congestion Indication (EFCI)

A network element in a congested state may set the Explicit Forward Congestion Indication (EFCI) payload type codepoint in the cell header for use by other nodes in the network or by the destination equipment. This could be done by setting EFCI when a threshold in a buffer is exceeded. Note that since the EFCI payload type is set in the cell, the cell is not discarded, and hence is appropriate to congestion avoidance. A network element not in a congested state should not modify EFCI, since it is used to communicate the existence of congestion from any intermediate point to the receiving end. Usage of EFCI by higher-layer protocols is currently not standardized.

As introduced in Chapter 5, Frame Relay (FR) has a similar congestion indication called Forward Explicit Congestion Notification (FECN). Additionally, frame relay also has a Backward Explicit Congestion Notification (BECN) for which there is no analogy in ATM. One reason that backward congestion notification was not standardized in ATM (as it was in frame relay), was that it was believed that the destination application protocol should communicate to the source destination protocol the command to slow down transmissions if network congestion was being experienced. To a large extent, the analogous congestion indications from FR (FECN and BECN) have not been utilized by end systems or higher layer protocols such as Transmission Control Protocol (TCP). A similar situation may occur in ATM with the EFCI bit not being utilized by many end systems or higher layer protocols and applications. There is also a problem for intermediate equipment, such as routers, in utilizing congestion indication information. If they slow down and the source application protocol does not, then loss will occur anyway in this intermediate equipment. Most routers do collect statistics on the number of congestion messages received, which is useful in network planning.

Networks may use ATM EFCI and a currently nonstandard internal backward congestion notification like FR BECN in a feedback loop, such as that illustrated in Figure 14.3. Basically, if congestion is detected anywhere along the route, including congestion for the outgoing link as shown in the figure, a feedback congestion message is sent to the originating node. The originating node may slow down service for that connection, selectively discard cells, or a combination of both. In either case, the source that is causing the congestion is throttled back to recover from the congestion state. The expected result is that useful throughput, for example in a file transfer application, does not suffer from congestion collapse, as indicated in the graph in the figure. See References 3 and 4 for further examples of this use of feedback control and simulation results.

This method can work only if the congestion interval is substantially greater than the round-trip delay; otherwise the congestion would have abated by the time the feedback control can act. The worst-case scenario for such a feedback scheme would be that of periodic input traffic, with a period approximately equal to the round-trip time. A realistic scenario that can result in long-term overload would be that of major trunk and/or nodal failures in a network during a busy interval. This will likely result in congestion that persists for the duration of the failure, in which case feedback control can be an effective technique for recovering from congestion and splitting the impairment fairly across different sources.

Several closed-loop flow control techniques, which operate in a similar, closed-loop manner, are covered later in this chapter.

14.3.2   Usage Parameter Control (UPC) Tagging

Usage Parameter Control (UPC) is primarily involved during traffic control, but can also be involved in congestion avoidance. One example is the UPC tagging cells by changing the Cell Loss Priority

(CLP) bit to indicate the nonconforming cells (CLP=1). This allows traffic in excess of the traffic parameters to be admitted to the network, which may cause congestion to occur. If this technique of congestion avoidance is used, then a corresponding technique such as selective cell discard or dynamic UPC must be used to recover if severe congestion is experienced.

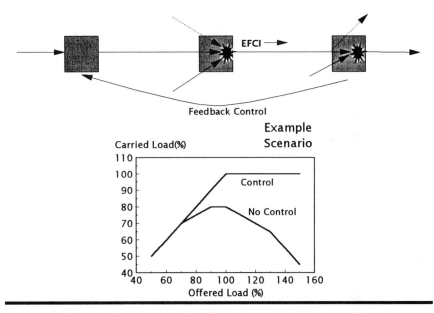

Figure 14.3 Feedback Control Example

14.3.3   Connection Admission Control (CAC) "Overbooking"

A more aggressive form of Connection Admission Control (CAC) than that of "full booking" described earlier is where a certain degree of "overbooking" is permitted. The principle of overbooking is that connections request traffic parameters, such as the peak cell rate, which are greater than what is used the majority of the time. When a large number of these connections share a common resource, it is unlikely that they will all use the resource at their peak demand level. Therefore, more connections may be admitted than the traffic parameters would indicate could be supported to still achieve the specified Quality of Service (QoS). However, the QoS can still be statistically achieved if the sources have a predictable statistical activity, as covered in the next chapter.

As in UPC with tagging, the use of CAC with "overbooking" must also be used in conjunction with a congestion recovery mechanism, such as dynamic UPC of disconnection.

14.3.4 Call Blocking

Before the network becomes severely congested, Connection Admission Control (CAC) can simply block any new connection requests. A good example of this type of congestion avoidance is that which occurs in the telephone network — if there is blockage in the network, you will get a fast busy signal, effectively blocking your call attempt. This approach avoids severe congestion for connection-oriented services; however, it is not applicable to connectionless services.

14.4 CONGESTION AVOIDANCE BY FLOW CONTROL

One particular class of congestion avoidance algorithms is so large that this section covers it in some detail. First, some motivation and general principles are covered, followed by the definition of three classes of flow control methods. We then define an example that we use to illustrate each of the flow control methods.

Many data communications applications have a desire to utilize as much of the available bandwidth as possible, thus attempting to continuously operate in a mildly congested state. The basic idea is to back off the offered load just before any loss occurs in the network, thus achieving maximum throughput with no loss. Also, user access to the available bandwidth should be fair. In other words, no one user should get all of the available bandwidth of a bottleneck resource if several users are equally contending for it. Additionally, conforming users should be isolated from the effects of overloading type users.

The generic name given to this balancing act is *flow control*. In essence, the objective is to control the flow of offered load just enough to achieve a throughput that is very close to that of the resource's capacity, with very low loss. This requires cooperation between the users and the network. The network notifies users of congestion in a timely fashion, and the user's application reduces the flow accordingly. When there is available capacity, users can access as much as they have contracted for.

This section summarizes three methods of flow control: window-based, rate-based, and credit-based. Window-based flow control limits the amount of source transmit data that can be sent (called the window) and dynamically adjusts the window size based upon feedback. Rate-based flow control dynamically adapts the source transmit rate in response to feedback. Credit-based flow control is a scheme which returns permission to send (called credits) to a source or intermediate node. Each of these methods has a common approach in that each one attempts to control the flow from the sender to avoid severe network congestion. The methods differ somewhat in how feedback or congestion indication is generated and in what response, or action, is taken to avoid congestion.

An example of two sources (workstations or clients) attempting to continuously send to a single output link (to a server) using the same switch is shown in Figure 14.4. Both the switch and the server are collocated. The example for all three methods of flow control uses this configuration. The speeds of the two access source links (workstations or clients) and the single output link (server) are equal in the example, shown as a DS3, as shown in Figure 14.4. The two sources are separated from the switch by a Round-Trip Time (RTT) of 16 cells (about 8 miles at a DS3 rate). This is because the switch buffer threshold is 8, and the round trip would double the amount of cells outstanding on a round trip to 16. The RTT is measured from a source to the switch (and collocated server) and back to the same source. This is a good model of a file transfer application between two clients and a server. Each client is transmitting independent of, but simultaneous to, the other.

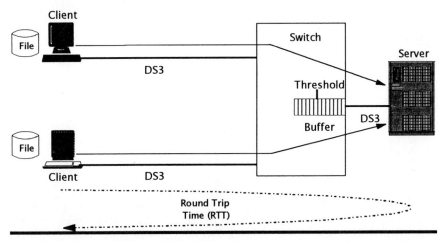

Figure 14.4 Example Flow Control Scenario

The parameters were chosen for each of the examples to result in a throughput that is approximately 75 percent of the maximum throughput. Each example covers 512 cell times, or 32 round-trip times. For each example the following are plotted: the parameter being controlled at the source, window, rate, or credit, along with the transmitted cell stream from the source. The buffer occupancy at the switch is also plotted.

14.4.1 Window-Based Flow Control

An example of window-based flow control is illustrated in Figure 14.5. This example will demonstrate the concepts of window-based flow control. Each source has a transmit window that defines how many cells can be transmitted during each successive RTT. The complete RTT is equal to 16 cells as explained previously. The switch has a shared buffer that has a threshold T, which was chosen to be 8 in this example. Every Round-Trip Time (RTT), the switch provides feedback to each source. If the contents of the buffer are less than the threshold (< 8), then the switch provides feedback so that each of the sources can increase its transmit window linearly by a window increment, which in this example is 1. This can be seen from the linearly increasing window size and the increasing width of the transmit bar graph corresponding to longer and longer burst of cells in Figure 14.5.

If the number of cells in the buffer at the switch is greater than the threshold, then the switch provides feedback for the sources to decrease their transmit window by a multiplicative decrease, which is 0.5 in this example. In other words, when the switch provides feedback indicating congestion, each source reduces its transmit window by one-half, thus decreasing the next cell burst size by one-half. This can be seen from Figure 14.5. As the transmit window size reaches 12, the feedback is received from the switch indicating that the threshold has been crossed, and the window size is thus reduced to 6. While a spreadsheet was used to calculate these plots, observe that the source is only sampled at the RTT value occurring at the ticks on the graph, which occurs at the minimum of the buffer curve. Thus, the buffer content graph peaks out at a window size of 12, even though the threshold was crossed at a window size of 11, the feedback is delayed by the round-trip time so the sender overshoots the window while the feedback is returned. The same effect occurs in rate-based flow control covered later in this section.

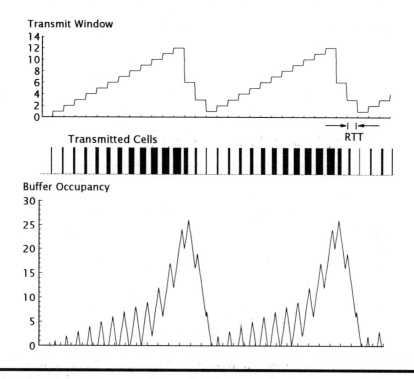

Figure 14.5 Illustration of Window-Based Flow Control

The delay that is encountered by the traffic is directly related to the number of cells contained in the switch buffers. The number of cells in the switch buffer is shown plotted on the right-hand side of Figure 14.5. Once the threshold in the shared buffer (8 in this example) is crossed, there are already a large number of cells in transit from the sources, and since feedback is not returned except at round-trip increments, the buffer can become quite full — in this example with 26 cells. The fact that each source sends at the peak rate is indicated by the very jagged appearance of the plot of buffer contents versus time. Each source transmitting at the peak rate causes the shared buffer to fill rapidly, but then both sources stop (at the RTT) and the buffer empties. This process is repeated each round-trip time until the buffer threshold is exceeded. Doing some simple math allows utilization to be calculated. Each client transmits 179 cell times per source, or 358 cell times for both sources. This equates to about 70 percent of the maximum throughput of 512 cell times.

Window-based flow control is very simple. It was the first type of flow control implemented in data communications. It has been refined further in the Internet Transmission Control Protocol (TCP) to increase throughput and be more adaptive.

14.4.2 Rate-Based Flow Control

The next examples are very similar to the first example, so they are covered in somewhat less detail, pointing out the differences. An example of rate-based flow control is illustrated in Figure 14.6. In this case the transmit *rate* of the sources is controlled instead of the window size, as shown plotted in the upper left-hand corner. The rate is expressed in units of cells per Round-Trip Time (RTT). The transmit rate is initially zero. With every RTT the switch provides feedback on whether the rate can be increased or decreased. The switch has a threshold in the shared buffer (equal to 8 in this example). If the contents of the buffer are below the threshold, then feedback is provided to each source allowing it to *increase* its transmit rate by an additive increase, which is 1 cell per RTT in this example. This is seen from the linear increase in the transmit rate plot shown in the figure. If the contents of the buffer are above the threshold, then feedback is provided by the switch to the sources indicating that they should decrease their transmit rate by a multiplicative decrease, which is 0.5 in this example. This can also be seen from the figure.

The net result of this flow control method is that the cell transmission is more evenly spaced, as shown in Figure 14.6 by the nearly continuous transmission of cells in the bar graph of transmitted cells versus time. The effect upon buffering is shown in the upper right-hand corner of this same figure. As would be expected, a more regular transmission of cells by the sources results in a plot of buffer contents versus time that is smoother than that of window control, and has a maximum value (of 21) that is about 25% less than that of window control. The higher throughput of cells is the key advantage of rate-based control over window-based control. A disadvantage is that the sources must now be more complex and be able to control their transmit rate using some form of traffic shaping as described in the previous chapter.

See Ref. 6 for a description of how Backward Congestion Notification (BCN) cells are used to implement rate control in a local area network environment.

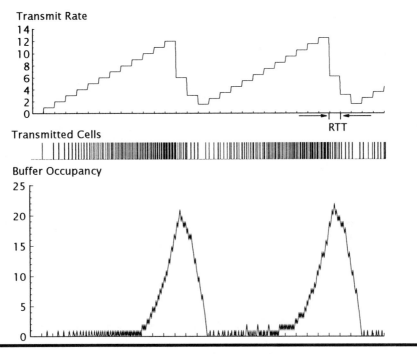

Figure 14.6 Illustration of Rate-Based Flow Control

14.4.3 Credit-Based Flow Control

Figure 14.6 illustrates an example of credit-based flow control. For this method the flow control parameter that is adjusted by the source is the *credit* that is available for transmitting cells. A source may continue to send cells, decrementing its credit counter by 1 for each transmitted cell, as long as the credit counter is greater than zero. The credit counter is initially zero. Each Round-Trip Time (RTT) the switch sends a feedback message indicating the credit counter value to each source (which can vary). In the example, the switch has a buffer dedicated to each virtual connection (either VPC or VCC). The credit is computed as the number of remaining cells in the buffer for each virtual connection. In this example the buffer is 8 cells for each source, chosen to achieve 75 percent throughput for comparison with the other flow control methods. Because the RTT is 16, the buffer cannot be filled because the credit cannot be greater than 8, and the credit is issued only once with each RTT.

This results in a very bursty but regular transmission of cells as seen in the transmit bar graph of Figure 14.7. Because the credit value is sent every RTT, each source bursts that number of cells, and then stops until the next credit is received. This results in a total buffer contents plot that is very jagged as shown in the figure. However, the total buffer contents (two times 8) are never more than half full. Credit-based flow control inherently operates in a region that keeps the buffers relatively full. For further details and references on credit-based flow control see Ref. 9.

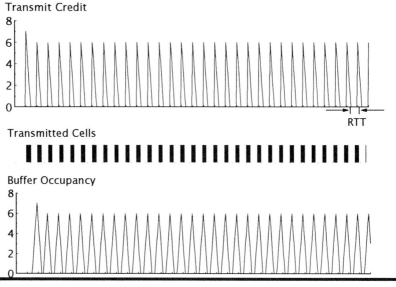

Figure 14.7 Credit-Based Flow Control

In general, the buffer space is allocated in each node proportional to the link delay and maximum VC bandwidth product, in which case nearly maximum throughput can be achieved. This method isolates all virtual connections from each other and causes congestion to back up through the network, on a per-virtual-connection basis. This scheme can be used in conjunction with a priority queueing, or a scheduling, scheme in order to achieve some other type of resource sharing and fairness. Disadvantages of this method are the complexity in the switch and source of implementing the credit control logic, the relatively large amount of storage that is required for longer propagation delays, and the usage of link bandwidth for per-VC credit messages.

14.5   CONGESTION RECOVERY

Congestion recovery is the response to entering (or remaining too long) in the severely congested region. As a real-life example, if one is caught in a particularly bad traffic jam, one will take an alternate route or pull off the road and do something else. Another example is that to obtain a particular service if we arrive in an attempt, and find that the line is too long, we may leave and decide to try again later. The following examples of congestion recovery are covered in detail:

- ⊗ Selective Cell Discard
- ⊗ Dynamic UPC
- ⊗ Loss Feedback
- ⊗ Disconnection
- ⊗ Operations Procedures

14.5.1   Selective Cell Discard

Selective cell discard is defined by the standards as the state where the network may discard CLP=1 cell flow while meeting Quality of Service (QoS) on both the CLP=0 and CLP=1 flows. Recall that the Cell Loss Priority (CLP) bit in the ATM cell header indicates whether a cell is of high priority (CLP=0) or low priority (CLP=1). An objective of selective cell discard is to give preferential treatment to CLP=0 cells over CLP=1 cells during periods of congestion.

Selective cell discard in ATM is a key, standardized network equipment function for performing congestion control. It can be used to avoid congestion or even recover from congestion. The network can use selective cell discard to ensure that connections that request a guaranteed QoS for certain traffic parameters for the CLP=0 cell flow do achieve this performance. If the network is not congested, then the application can achieve higher throughput, but never less than the requested amount for the CLP=0 flow, if the network's UPC implements CLP tagging as described earlier.

The user may also tag cells as CLP=1 if it considers them to be of a lower priority, in which case there is an ambiguity because there is no way to discern whether the user set the CLP, or the UPC set it by tagging, once the cells are at intermediate nodes within the network. If the user sets the CLP bit, and the network does tagging, then it may not be possible to guarantee a cell-loss ratio for the CLP=1 cell flow. This is similar to a user setting the Discard Eligible (DE) bit in frame relay. There is no way to tell if the user of the network set the bit, and if the user is simply being a good citizen during periods of network congestion.

An example of the selective cell discard mechanism is illustrated in Figure 14.8. A buffer of B cell positions is filled with arriving cells on the left. The buffer is emptied by the switch port from the right. Since arrivals may occur simultaneously from multiple inputs, the buffer can become congested. There is a threshold above which CLP=1 cells are immediately discarded. CLP=0 cells may occupy any buffer position. Therefore, by controlling the buffer threshold, CLP=1 loss performance can be controlled, with a side effect on the CLP=0 delay performance. The delay and loss performance for this method of selective cell discard under load will be analyzed in the next chapter. Again, this is similar to the method of allocating buffer preferentially to non-DE frame relay traffic.

Another refinement could support CLP=1 cells being flushed out of the buffer when another threshold is passed [5]. The concept of thresholded buffering can be extended to multiple thresholds, for example, by the use of an internal switch tag that indicates multiple levels of loss priority on a per VPI/VCI basis.

The Cell Loss Priority (CLP) bit in the ATM cell header and its setting by UPC tagging are directly analogous to Frame Relay's Discard Eligible (DE) bit and tagging done when the frame transfer rate exceeds the Committed Information Rate (CIR).

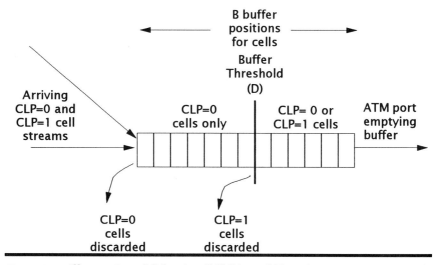

Figure 14.8 Illustration of Selective Cell Discard Function

We give an example of how hierarchical video coding could make use of selective cell discard in Figure 14.9. Hierarchical video coding encodes the critical information required to construct the major parts of the video image sequence as the higher priority CLP=0 marked cells, and the remaining, detailed, or minor change information as lower priority CLP=1 marked cells. Thus, when there is a scene change, there is an increased transmission of CLP=0 cells. The video detail and minor adjustments within the scene are sent as CLP=1 cells. When several such video sources are multiplexed together in a switch which utilizes selective cell discard as shown in Figure 14.9, this congestion recovery scheme ensures that the critical scene change information will get through, even if some of the detail is momentarily lost.

14.5.2 Dynamic Usage Parameter Control (UPC)

Another way to recover from congestion is to dynamically reconfigure the UPC parameters. This could be done by renegotiation with the user [10], or unilaterally by the network for certain types of connections. Ideally, this could be user-definable and user-controlled.

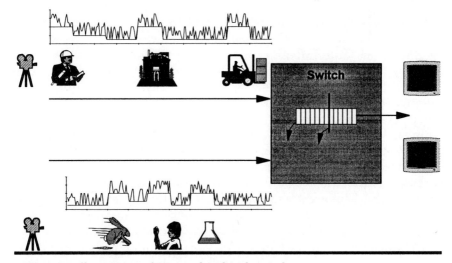

Figure 14.9 Illustration of Hierarchical Video Coding

14.5.3 Loss Feedback

Another method of congestion control is for a higher layer protocol (such as TCP) at the end system to infer that congestion has occurred, and throttle back the end user's application throughput at the end system itself. This approach has the advantage of the end system reducing the offered load, thereby recovering from congestion. There are several possible disadvantages: the response time to congestion is at least the round-trip time through any intervening networks plus any time spent in the end systems, and the possibility that all end systems do not infer congestion and back off in the same manner, creating unfairness.

One noteworthy implicit congestion control scheme widely documented in the technical literature is that of the Internet Transmission Control Protocol (TCP) [11]. The following describes how two TCP sources infer congestion and back off their throughput when they are contending for a common egress line in a cross-country network as illustrated in Figure 14.10. TCP keeps track of a congestion window, which indicates how many unacknowledged segments may be sent. The TCP window size is increased for each acknowledgment received within a time-out period, which is called "slow start," but actually results in a geometric increase because the window is increased by 1 (doubled) for the first acknowledgment, and then increased by 2 for the next two acknowledgments, and so on. This window size is increased in this manner until the size reaches half the previous maximum window size, at which point the window is increased only linearly at a much slower rate. When an acknowledgment time-out occurs, due to either loss of a segment or increased delay, the window is decreased by half, resulting in a sharp drop in the window size. Once an acknowledgment is received, the above process repeats.

In the example the two TCP sources execute this protocol until the buffer on the egress line overflows, at which point they both decrease their windows, only to increase them again. The net effect of this protocol is that each TCP session gets about half of the egress port bandwidth. This sawtooth type of windowing pattern is typical for TCP on other types of networks, such as those with routers containing large buffers. The effect of TCP throughput when DS3 ATM lines for ATM switches with various buffer sizes are used is reported in Ref. 12 and is illustrated in the upper right hand corner of Figure 14.10. In these tests the IP packet size was 4096 bytes, or about 90 cells. When the buffer size of the ATM switch is less than the packet size, then when two packets arrive nearly simultaneously, one or more cells from both packets are lost and both must be retransmitted, resulting in very low throughput as shown in the figure. When the buffer size is increased so that two packets can arrive simultaneously, throughput nearly doubled. As the buffer size increases further, the throughput increases, but not nearly as dramatically. Similar results occur in a local area for switches with buffers that are small with respect to the packet size as reported in Ref. 13.

14.5.4 Disconnection

Another rather drastic response that provides recovery from congestion is to disconnect some connections if and when severe congestion persists. Some connections may be identified as being preemptible. For example, there are certain requirements on carriers to support some national defense traffic or local community emergency services at the highest priority, such that all other traffic is subject to disconnect if capacity is needed for these priority connections.

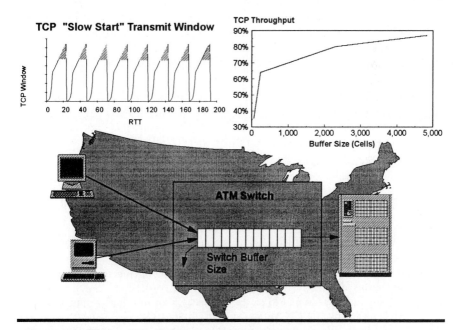

Figure 14.10 Illustration of TCP Protocol Performance

14.5.5 Operational Procedures

If all of the automatic methods fail, then human operators can intervene and manually disconnect certain connections, reroute traffic, or patch in additional resources. Network management actions for controlling the automated reroutes are not standardized, and are therefore a proprietary network implementation. These procedures must be carefully coordinated, especially if multiple networks are involved.

14.6 REVIEW

This chapter defined the notion of congestion as demand in excess of resource capacity. The degree of congestion impacts contention for resources, which can reduce throughput and increase delay, as occurs in vehicular traffic jams. Congestion can occur at multiple levels in time and space. In time, congestion can occur at the cell level, the burst level, or the call level. In space, congestion can occur at a single node or multiple nodes. Congestion control schemes were categorized in terms of the time level and their general philosophy. Congestion management attempts to ensure that congestion never occurs, which may be done at the expense of reduced efficiency, were also covered. Congestion avoidance schemes attempt to operate in a region of mild congestion in order to obtain higher efficiency at nearly optimal performance, while congestion recovery is tasked with moving the network out of a severely congested state in the event that the previous two philosophies fail, sometimes using rather drastic measures.

14.7 REFERENCES

[1]   J. Hui, "Resource Allocation for Broadband Networks," *IEEE JSAC*, December 1988
[2]   G. Awater, F. Schoute, "Optimal Queueing Policies for Fast Packet Switching of Mixed Traffic," *IEEE JSAC*, April 1991
[3]   M. Wernik, O. Aboul-Magd, H. Gilbert, "Traffic Management for B-ISDN Services," *IEEE Network*, September 1992
[4]   A. Eckberg, B. Doshi, R. Zoccolillo, "Controlling Congestion in B-ISDN/ATM: Issues and Strategies," *IEEE Communications Magazine*, September 1991

[5]   H. Kröner, G. Hébuterne, P. Boyer, A. Gravey, "Priority Management in ATM Switching Nodes," *IEEE JSAC*, April 1991

[6]   P. Newman, "Backward Explicit Congestion Notification for ATM Local Area Networks," *IEEE Globecom*, December 1993

[7]   R. Jain, "Congestion Control in Computer Networks: Issues and Trends," *IEEE Network Magazine*, May 1990

[8]   D. Hong, T. Suda, "Congestion Control and Prevention in ATM Networks," *IEEE Network Magazine*, July 1991

[9]   H. T. Kung, R. Morris, T. Charuhas, D. Lin, "Use of Link-by-Link Flow Control in Maximizing ATM Networks Performance: Simulation Results," *Proc. IEEE Hot Interconnects Symposium '93*, August 1993

[10]   A. Eckberg, "B-ISDN/ATM Traffic and Congestion Control," *IEEE Network*, September 1992

[11]   U. Black, *TCP/IP and Related Protocols*, McGraw-Hill, 1992

[12]   L. Boxer, "A Carrier Perspective on ATM," ATM Year 2 Conference, April 1994.

[13]   A. Romanow, "TCP over ATM Simulation Results," FTP Report, playground@sun.com, pub/tcp_atm, July 1993

15

Traffic Engineering

This chapter deals with the modeling of traffic sources and switch performance. It covers some useful approximations for estimating performance. It begins with a discussion of traffic modeling philosophies as an introduction to an application-oriented approach. Several models are introduced for the source traffic, each chosen for simplicity to illustrate the key concepts. The performance of the major buffering schemes used in switches, as described in Chapter 10, is analyzed. Next, the performance of Continuous Bit-Rate (CBR) and Variable Bit-Rate (VBR) traffic source is analyzed. The definition of statistical multiplexing gain and the traffic characteristics for which it is attractive are described. Finally, an analysis of the priority queueing traffic control's performance is then presented. All formulas used in these examples are simple enough for spreadsheet computation so that the reader can use them to evaluate a particular switching machine or network configuration. Throughout this chapter references to more sophisticated models are given for the interested reader.

15.1 PHILOSOPHY

This section discusses several dimensions of traffic engineering philosophy: source model traffic parameter characteristics, performance specification and measurement, and modeling accuracy.

15.1.1 Source Model Traffic Parameter Characteristics

There are two basic philosophies for characterizing source traffic parameters: deterministic and random. There are tradeoffs between obtaining accuracy and simplicity, and a balance between clarity and ambiguity, in both approaches.

Deterministic parameters are based upon the traffic contract outlined in Chapter 12, with conformance verifiable on a cell-by-cell basis using the leaky bucket algorithm. The fraction of cells that conform to the traffic contract can be unambiguously measured at different points to verify conformance to the contract, either by the user or the network. Any changes can be clearly negotiated between the user and the network. In summary, the agreement as to the traffic throughput that achieves a given performance is unambiguously stated.

The other philosophy for modeling source traffic parameters utilizes probabilistic (also called stochastic) models for traffic parameters. Usually these are measurable only over a very long-term average. Since the method and interval for computing the average can differ, conformance testing should define the details of the measurement method. In addition to specifying the parameters, specification of the statistical model is also required. With these additional assumptions the user and network can agree on performance for a certain level of traffic throughput. While these statistical methods are not standardized, they are useful approximations to the deterministic traffic contract behavior. These methods are very useful in analysis if a simple statistical model is chosen, as shown later in this section.

15.1.2 Performance Specification and Measurement

Quality of Service (QoS) performance is either specified and measured for each individual Virtual Path Connection (VPC) or Virtual Circuit Connection (VCC), or measured over the aggregate of many VCCs (or even VPCs). This choice between performance and measurement on an individual or aggregate basis has several implications.

QoS is specified and measured for each connection in the individual case. QoS is measured by inserting Operations, Administration, and Maintenance (OAM) cells on each connection, increasing cell traffic and introducing additional complexity (and cost) for processing these OAM cells. With individual specification and measurement, a situation analogous to a classical digital private line exists, where performance on every line is measured. For example, the error statistics on a DS1 line are estimated from the Extended SuperFrame (ESF) Cyclic Redundancy Check (CRC) for each DS1.

In the aggregate case, QoS is averaged over a large number of connections, which is a more natural interpretation for virtual circuits. Measurement on the aggregate assumes that the performance of all VCCs on a common VPC is identical, which significantly reduces the number of measurement cells that must be transmitted and processed.

Typically, the cost and complexity of individual measurement is justified when it is critical to ensure that the performance of an individual virtual connection is being achieved. Normally, measurement on the aggregate is adequate to ensure that the QoS of a group of virtual connections is being met, and hence the QoS of the individual virtual connection is being met on a statistical basis.

15.1.3 Modeling Accuracy

There is also an aspect of philosophy related to how accurate the traffic engineering model should be. As would be expected, the more complicated the model, the more difficult the results are to understand and calculate. The accuracy of the switch and network model should be comparable to the accuracy of the source model traffic. If you only know approximate, high-level information about the source, then an approximate, simple switch and network model is appropriate. If you know a great deal of accurate information about the source traffic, then an investment in an accurate switch and network model, such as a detailed simulation, is appropriate.

While theoretically optimal, detailed source modeling can be very complex and usually requires computer-based simulation. Often this level of detail is not available for the source traffic. Using source traffic details and an accurate switch and network model will result in the most realistic results.

When either traffic or switch and network details are not available, approximations are the only avenue that remains. Approximate modeling is usually simpler, and can often be done using only analytical methods. One advantage of the analytical method is that insight into relative behavior and tradeoffs can be much clearer. Analytical is the approach used in this book. One word of caution remains, however: these simplified models may yield overly optimistic or pessimistic results, depending upon the relationship of the simplifying assumptions to reality. Modeling should be an ongoing process. As more information about the source characteristics, device performance and quality expectations are obtained, these should be fed back into the modeling effort. For this reason, modeling has a close relationship to the performance measurement aspects of network management.

15.2 SOURCE MODELS

The use of a particular source model, and its accurate representation of real traffic, is a hotly debated topic and the subject of intense on-going research and publication. This chapter utilizes some basic probability and queueing theory, with parameters chosen to model some reasonable traffic assumptions, and comment on how this applies to extreme situations.

15.2.1 General Source Model Parameters

This section defines some general source model parameters used throughout the remainder of this chapter and in many papers and publications which are listed at the end of the chapter.

Burstiness is a commonly used measure of how infrequently a source sends traffic. A source that infrequently sends traffic is said to be very bursty, while a source that always sends at the same rate is said to be nonbursty. The formula that defines burstiness in terms of the peak cell rate and the average cell rate is defined as:

$$\text{Burstiness} = \frac{\text{Peak Rate}}{\text{Average Rate}}$$

The *source activity probability* is a measure of how frequently the source sends, defined by the probability that a source is bursting:

$$\text{Source Activity Probability} = \frac{1}{\text{Burstiness}}$$

Utilization is a commonly used measure of the fraction of a transmission link's capacity that is used by a source, theoretically measured over an infinite period of time; however, in practice measured over a long time interval. The definition of utilization is given in terms of the peak cell rate and transmission link rate as follows:

$$\text{Utilization} = \frac{\text{Peak Rate}}{\text{Link Rate}}$$

15.2.2 Poisson Arrivals and Markov Processes

Random arrival processes are described in general, and the Poisson (or Markov) process in particular, with reference to Figure 15.1. Poisson arrivals occur such that for each increment of time (T), no matter how large or small, the probability of arrivals is independent of any previous history. These events may be either individual cells, a burst of cells, cell or packet service completions, or other, arbitrary events in models.

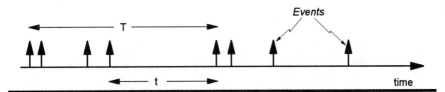

Figure 15.1 Illustration of an Arrival Process

The probability that the interarrival time between events t, as shown in Figure 15.1, has a certain value is called the *interarrival time probability density*. The following formula gives the resulting probability that the interarrival time t is equal to some value x when the average arrival rate is λ events per second:

$$\text{Prob}(t = x) = \lambda \ e^{-\lambda x}$$

This is called a *memoryless process*, because the probability that the interarrival time will be x seconds is independent of the *memory* of how much time has already expired. This fact greatly simplifies the analysis of random processes since no past history, or memory, must be kept. These types of processes are commonly known as

Markov processes, named after the Russian mathematician of the nineteenth century.

The probability that n independent arrivals occur in T seconds is given by the famous *Poisson distribution:*

$$\text{Prob}(n,T) = \frac{(\lambda T)^n}{n!} \, e^{-\lambda T}$$

We combine these two thoughts in a commonly used model called the Markov Modulated Poisson Process (MMPP). There are two basic types of this process: the *discrete* [which corresponds to ATM cells) and the *continuous* [which corresponds better to higher-layer Protocol Data Units (PDUs) which generate bursts of cells]. The next two figures give an equivalent example for the discrete and continuous models.

The labels on the arrows of Figure 15.2 show the probability that the source transitions between active and inactive bursting states, or else remains in the same state for each cell time. In other words, during each cell time the source makes a state transition, either to the other state, or back to itself, with the probability for either action indicated by the arrows in the diagram.

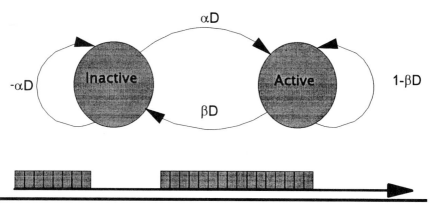

Figure 15.2 Discrete Time Markov Process Model

The burstiness, or peak-to-average ratio, of the *discrete* source model is given by the following formula:

$$b = \frac{\alpha + \beta}{\beta}$$

where α is the average number of bursts arriving per second, and β is the average rate of burst completions. Often we think in terms of β^{-1}, which has units of the average number of seconds per burst. We define D as the cell quantization time having units of seconds per cell. Therefore, αD defines the probability that a burst begins in a particular cell time, and βD defines the probability that a burst ends in a particular cell time. The average burst duration d (in cells) is then computed from the standard geometric series as follows:

$$d \; = \; \frac{1}{\beta D}$$

The second, *continuous* time case is illustrated in Figure 15.3. The time elapsed from the beginning of the burst to the end of the burst is modeled instead of modeling the individual cells. Some accuracy is lost in that the quantization inherent in segmentation and reassembly is not considered; however, we utilize in this book the simplicity in modeling it provides. The diagram is called a *state transition rate diagram* since the variables associated with the arrows refer to the rate exponent in the negative exponential distribution introduced earlier in this chapter. Both the discrete and continuous Markov models yield equivalent results except for the cell quantization factor D.

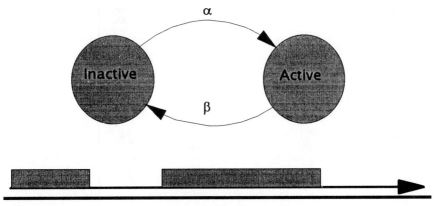

Figure 15.3 Continuous Time Markov Process Model

The corresponding burstiness b for the continuous process is:

$$b \; = \; \frac{\alpha \; + \; \beta}{\beta}$$

and the average burst duration (in seconds) is given by the following formula:

$$d = \frac{1}{\beta}$$

Note how these formulae are identical to the discrete case except for the absence of the discrete cell time D in the denominator of the equation for the average burst duration of the continuous model.

Another distribution that is sometimes used to model extremely bursty traffic is that of the hyperexponential, which is effectively the weighted sum of a number of negative exponential arrivals. This turns out to be a more pessimistic model than Poisson traffic because bursts and burst arrivals are more closely clumped together. For further information on this distribution see Ref. 3.

Recent work based upon actual LAN traffic measurements indicate that these traditional traffic models may be overly optimistic. For further information on this work see Refs. 1 and 2. These results show that the LAN traffic measured at Bellcore is *self-similar,* which means that the traffic has similar properties regardless of the time scale on which it is observed. This is in sharp contrast to the Poisson and Markovian models, where the traffic tends to become smoother, and more predictable, as longer and longer time averages are considered.

15.2.3 Queueing System Models

The usual categorization of queueing systems is given by the notation shown in Figure 15.4. We will use this terminology in this chapter, and it is widely used in the technical literature.

We look at two particular examples of queueing systems, namely, the M/D/1 and M/M/1 systems. From this notation each of these queueing systems has Markovian arrivals (negative exponential or memoryless burst arrivals) at a rate of λ bursts per second. The M/M/1 system has random length bursts with a negative exponential distribution (Markov), while the M/D/1 system has constant length bursts. The parameter μ^{-1} defines how may seconds (on average) are required for the transmission link to send each burst. For the M/M/1 system this is an exponentially-distributed random number with this average length, while in the M/D/1 system this is the constant or fixed length of every burst. Both systems also have a single server (i.e., physical transmission link) and an infinite population (number of potential bursts) and infinite waiting room (buffer space). The units of the buffer in the M/D/1 model are cells, while in the M/M/1 case the units of the buffer are bursts. Figure 15.5 illustrates these physical queueing systems and their specific relationship to ATM.

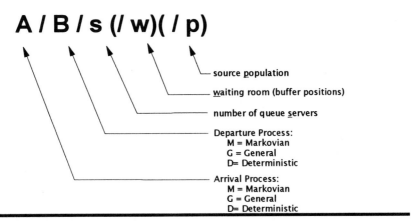

Figure 15.4 Queueing System Notation

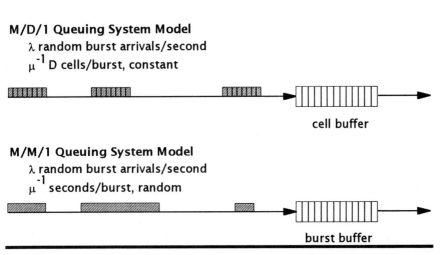

Figure 15.5 Application of M/D/1 and M/M/1 Queueing Systems to ATM

This is a good example of the tradeoffs encountered in modeling. The M/D/1 system accurately represents the fact that the buffers in the switch are in units of cells; however, the bursts are all of fixed length. The M/M/1 system does not model the switch buffers accurately since it is in units of bursts and not cells; however, the modeling of random burst lengths is more appropriate to many traffic sources. The M/M/1 model is also very simple to analyze, and therefore it will be used extensively to illustrate specific tendencies in ATM systems. In general, if the traffic is more deterministic than the M/M/1 model (for example, more like the M/D/1 model), then the M/M/1 model will be pessimistic (there will actually be less queueing and less delay in the modeled network). If the traffic is more bursty than the M/M/1 model, then the M/M/1 results will be optimistic (there will actually be more queueing and more delay in the modeled network).

In many of the following results the system delay and loss performance will be presented in terms of the offered load ρ, given by the following formula:

$$\rho = \frac{\lambda}{\mu}$$

Recalling that λ is the average number of arriving bursts per second, and that μ^{-1} is the average number of seconds per burst, it is seen that the offered load ρ is unitless. Thus the offered load has the interpretation of the average fraction of the resource capacity that is in use.

The service rate μ is computed as follows for a burst of B bytes at a line rate of R bits per second:

$$\mu = \frac{8 \; B}{R} \left(\frac{bursts}{\sec ond} \right)$$

The probability that there are n bursts waiting in the M/M/1 queue is given by the following formula:

$$\text{Prob[n burst in } M/M/1 \text{ queue]} = \rho^{n} \; (1-\rho)$$

The average queueing delay (i.e., waiting time) in the M/M/1 system is given by the following formula:

$$\text{Avg[}M/M/1 \text{ queuing delay]} = \frac{\rho/\mu}{1-\rho}$$

M/D/1 queueing predicts better performance than M/M/1. Indeed the average delay of M/D/1 queueing is exactly one-half of the M/M/1 delay. The probability for the number of cells in the M/D/1 queue is much more complicated, which is one reason the M/M/1 model will be used in many of the following examples.

15.2.4 Bernoulli Processes and Gaussian Approximation

A Bernoulli process is essentially the result of N independent coin flips (or Bernoulli trials) of an "unfair coin." An unfair coin is one where the probabilities of heads and tails are unequal, with p being the probability of that "heads" occurs as the result of a coin flip and (1-p) being the probability that "tails" occurs. The probability that k heads occur [and hence (N minus k) tails] as a result of N repeated Bernoulli trials ("coin flips") is called the *binomial distribution*, as given by:

$$\text{Pr}[k \text{ "heads" in } N \text{ "flips"}] \quad = \quad \binom{N}{k} \ p^k \ (1-p)^{N-k}$$

$$\text{where} \quad \binom{N}{k} \quad \equiv \quad \frac{N!}{(N-k)!\,k!} \, .$$

The Gaussian, or Normal, distribution is a continuous approximation to the binomial distribution when Np is a large number. Figure 15.6 compares the binomial and Gaussian distributions for an example where N=100 and p=0.1. The distributions have basically the same shape, and for large values of Np, in the Np(1-p) region about Np, the Gaussian distribution is a reasonable approximation to the binomial distribution.

This is helpful in analyzing relative performance in that the probability area under the tail of the Gaussian, or normal, distribution is widely tabulated and implemented in many spreadsheets and mathematical programming systems. We approximate the tail of the binomial distribution by the cumulative distribution of the normal density, Q(α). We will use the following approximation several times to estimate loss probability or statistical multiplex gain:

$$\text{Prob}[k > x] \quad \approx \quad Q\left(\frac{x-\mu}{\sigma}\right) \quad = \quad Q(\alpha) \quad \approx \quad \frac{1}{2} \ e^{-\alpha^2/2}$$

where $Q(\alpha) \equiv \dfrac{1}{\sqrt{2\pi}} \displaystyle\int_{\alpha}^{\infty} e^{-x^2/2} \, dx$

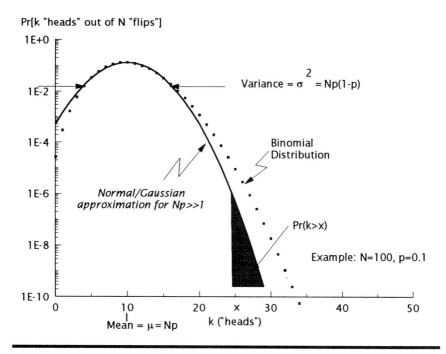

Figure 15.6 Normal Approximation to Binomial Distribution

15.3 PERFORMANCE OF BUFFERING METHODS

This section analyzes several simple models of switch delay and loss performance as impacted by various aspects of the switch buffer architecture. For simplicity, Poisson arrivals and negative exponential service times are assumed. Output queueing delay performance then behaves as a classical M/M/1 system. Input queueing incurs a problem known as Head Of Line (HOL) blocking. HOL blocking occurs when the cell at the head of the input queue cannot enter the switch matrix because the cell at the head of another queue is traversing the matrix.

For uniformly distributed traffic with random message lengths, the maximum supportable offered load for input queueing is limited to 50% (see Ref. 7), while fixed message lengths increase the supportable offered load to only about 58% (see Ref. 6). On the other hand, output queueing is not limited by utilization as in input queueing. Figure 15.7 illustrates this result by plotting average delay versus throughput for input and output queueing. For a more detailed analysis of input versus output queueing see Ref. 8 which shows that these simple types of models are valid for switches with a large number of ports.

Average Waiting Time (Bursts)

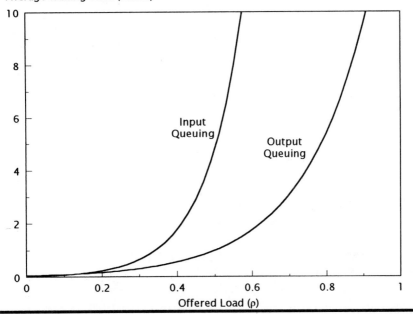

Figure 15.7 Delay versus Load Performance for Input and Output Queueing

The consequence of this result is that almost all ATM switches have some form of output buffering. If input buffering is used on a switch that you are considering, check to make sure that some means to address Head Of Line (HOL) blocking is implemented. Examples of methods to address HOL blocking are: a switch fabric that operates much faster than the cumulative input port rates, schemes where an HOL-blocked cell can be bypassed by other cells, or the use of priority queueing on the input.

The next example gives a simple, useful approximation for the output buffer overflow probability. For simplicity, an M/M/1 queueing system, which has an infinite buffer, is assumed, instead of a M/M/1/B system, which would have a finite buffer. The overflow probability for a buffer of size B cells is approximately the probability that there are B/P bursts in the infinite queue system. Comparison with simulation results and exact analysis has shown that this is a reasonable approximation [9]. When the average higher layer Protocol Data Unit (PDU) burst size is P cells, the approximate buffer overflow probability is given by the formula:

$$\text{Prob[Overflow]} \approx \rho^{B/P+1}$$

Figure 15.8 plots the approximate buffer overflow probability versus buffer size for various levels of throughput ρ assuming a Protocol Data Unit (PDU) size of P=1 cells. The performance for other burst sizes can be read from this chart by multiplying the x axis by the PDU burst size P.

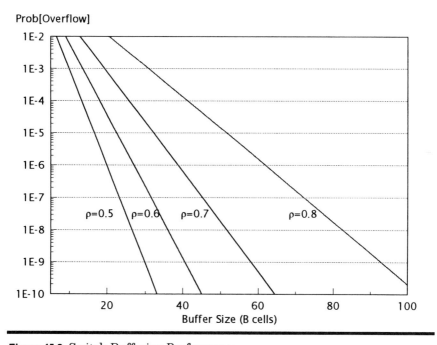

Figure 15.8 Switch Buffering Performance

Note that for a specific overflow probability objective and a fixed buffer size the load must be limited to a maximum value. We illustrate this concept by solving the above equation for the overflow probability in terms of the required buffer size B to achieve an objective Cell Loss Ratio (CLR). The result is the following:

$$B \approx P \; \frac{\log(\text{CLR})}{\log(\rho)}$$

Figure 15.9 shows a plot of the required number of cells in the buffer to achieve an objective Cell Loss Ratio (CLR) for a PDU size P=1 from this equation. Note that as the offered load increases, the required number of buffers increases nonlinearly.

Required Number of Cell Buffers (B)

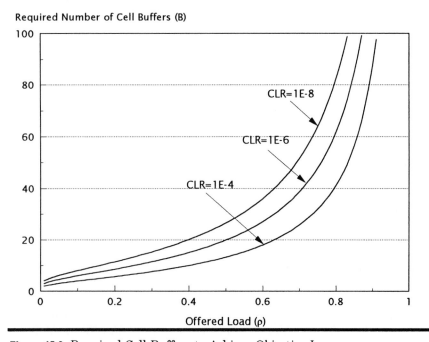

Figure 15.9 Required Cell Buffers to Achieve Objective Loss

Figure 15.10 illustrates the impact of higher layer PDU size (P) on buffer overflow performance for various output buffer sizes (B cells). As expected, the buffer overflow probability increases as the higher layer PDU size increases. When the PDU size approaches the buffer size, the loss rate is almost 100%.

The shared output buffer scheme has a marked improvement on buffer overflow performance because of sharing a single, larger buffer between many ports. Since it is unlikely that all ports are congested at the same time, the loss will be substantially less than an equivalent number of individual output buffer positions dedicated to each port.

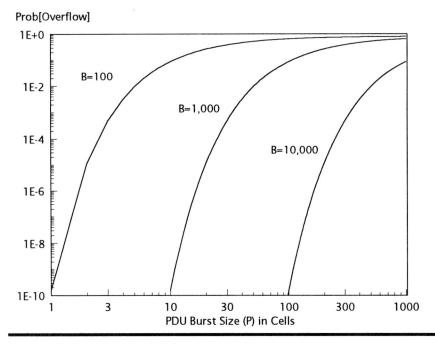

Figure 15.10 Overflow Probability versus PDU Burst Size

Since the exact analysis of shared buffer performance is somewhat complicated [8], we present a simple approximation based on the normal distribution. In the shared-buffer architecture, N switch ports share the common buffer, each with approximately the M/M/1 probability distribution requirement on buffer space. The sum of the individual port demands determines the shared-buffer probability distribution. The normal distribution approximates a sum of such random variables for larger values of N. The mean and variance of the normal approximation are then given by the following:

$$\text{Mean} \ = \ \frac{N\rho}{1-\rho} \qquad \text{Variance} \ = \ \frac{2N\rho^2}{(1-\rho)^2}$$

Figure 15.11 shows a plot of the overflow probability versus the equivalent buffer size per port for shared buffers on switches of increasing port size (N), along with the dedicated output buffer performance for large N from Figure 15.8 for comparison purposes. The offered load is $\rho=0.8$ or 80% load. The total buffer capacity on a shared buffer switch is N times the buffer capacity on the x axis. Note that as N increases, the capacity required per port approaches a constant value. This illustrates the theoretical efficiency of shared buffering. Of course, a practical implementation has limits in terms of the shared buffer access speed.

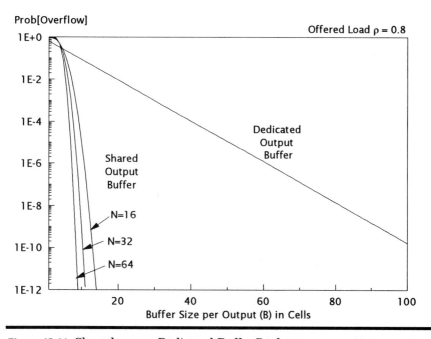

Figure 15.11 Shared versus Dedicated Buffer Performance

15.4 DETERMINISTIC CBR PERFORMANCE

The accurate loss performance measure of another very important traffic type, Continuous Bit Rate (CBR), turns out to be very easy to calculate. Figure 15.12 illustrates the basic traffic source model. N identical sources emitting a cell once every T seconds, each beginning transmission at some random phase in the interval (0,T), define the traffic source model.

The cell loss rate for such a randomly phased CBR traffic input is well approximated by the following formula [10]:

$$P_L = \text{Prob[Cell Loss]} \approx \exp\left[-2B^2/n - 2B(1-\rho)\right]$$

where n is the number of CBR connections
B is the buffer capacity (in cells)
$\rho = nT$ is the offered load

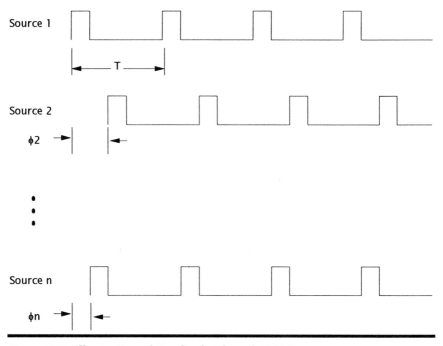

Figure 15.12 Illustration of Randomly Phased CBR Sources

A closed form solution can be easily derived for the number of buffers required to achieve a certain loss probability from the above formula, with the following result (see Ref. 11):

$$B \approx \frac{\sqrt{[n(1-\rho)]^2 - 2n \ \ln(P_L)} - n(1-\rho)}{2}$$

Figure 15.13 illustrates the results of this calculation by plotting the required buffers B versus the number of CBR connections n for various levels of overall throughput. When the switch implements priority queueing, this performance measure can be applied independently of other performance measures.

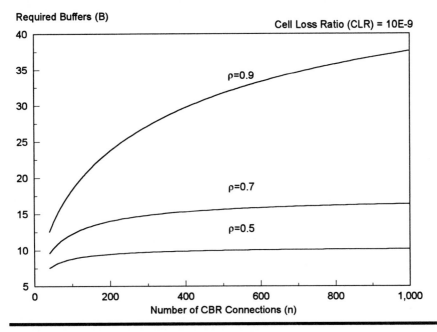

Figure 15.13 Required Buffer Capacity for Deterministic CBR Traffic

15.5 RANDOM VBR PERFORMANCE

The Variable Bit Rate (VBR) traffic distribution can be modeled using a normal approximation to the binomial distribution for V VBR sources with an average activity per source of p, using the following parameters:

Mean = Vp   Variance = $Vp(1-p)$

From these parameters the expected loss can be estimated and used to solve for the required number of buffers as follows:

$$B \approx Vp + \alpha\sqrt{Vp(1-p)}$$

where the parameter α is determined by equating the Cell Loss Ratio (CLR) to the tail of the normal distribution, namely, $Q(\alpha)=CLR$.

Figure 15.14 illustrates the required number of buffers to achieve a specified cell loss objective for various values of source activity p.

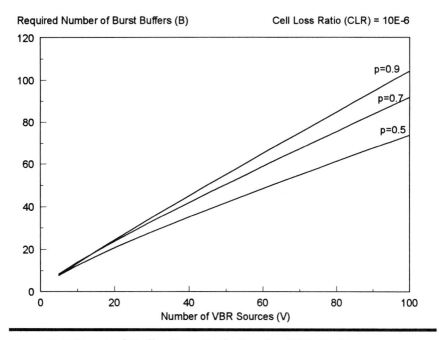

Figure 15.14 Required Buffer Capacity for Random VBR Traffic

Note that the curves become nearly linear for a greater number of sources, V.

15.6   STATISTICAL MULTIPLEXING GAIN

One key capability that ATM enables is that of statistical multiplexing, which attempts to exploit the on-off, bursty nature of many source types as illustrated in Figure 15.15. On the left-hand side there are several sources and sinks of ATM cell bursts: a video display, a server, a camera, and a monitor. The lower trace in the figure shows a plot of the sum of these cell bursts. In this particular simple example, only two channels are required at any point in time. As more and more sources are multiplexed together, the statistics of this composite sum become increasingly more predictable.

Each Source uses an entire channel at the peak rate only a fraction of the time

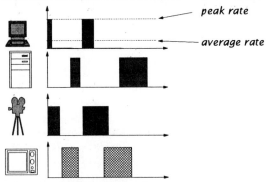

The source outputs can be combined to statistically share channels

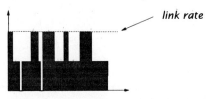

Figure 15.15 Illustration of Statistical Multiplex Gain

The statistical multiplexing gain G is defined as the following ratio:

$$G = \frac{\text{Number of Sources Supported}}{\text{Required Number of Channels}}$$

The statistical multiplex gain G can be computed from the binomial distribution [12], or estimated from the Normal distribution with the following parameters:

Mean = N/b   Variance = N/b(1-1/b)

where   N is the number of sources
          b is the burstiness (peak/average rate)

The required number of channels, C (in units of the number of peak rate sources), to achieve an objective Cell Loss Ratio (CLR) of $Q(\alpha)$ is given by:

$$C \approx N/b + \alpha\sqrt{N(b-1)}/b$$

The parameter η defines the peak source-rate-to-link-rate ratio, which means that the link capacity is 1/η. Therefore the statistical multiplex gain reduces to G=N/C=Nη. Setting C in the above equation equal to this link capacity 1/η and solving for N using the quadratic formula yields the result:

$$G \approx \frac{\eta \left(\sqrt{\alpha^2(b-1) + 4b/\eta} - \alpha\sqrt{b-1}\right)^2}{4}$$

Figure 15.16 plots the achievable statistical multiplex gain G versus the peak-to-link rate ratio η with burstiness b as a parameter for a cell loss ratio of 10^{-6}. This figure illustrates the classical wisdom of statistical multiplexing: the rate of any individual source should be low with respect to the link rate η, and the burstiness of the sources b must be high in order to achieve a high statistical multiplex gain G.

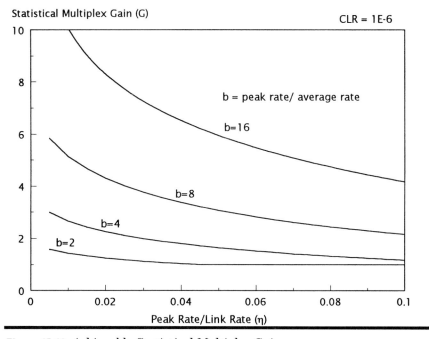

Figure 15.16 Achievable Statistical Multiplex Gain

Figure 15.17 illustrates how many sources must be multiplexed together in order to achieve the statistical multiplex gain predicted in the previous example. This confirms the applicability of the key statistical multiplex gain assumption that a large number N of sources, of low average rate 1/b and modest peak rate, with respect to the link rate η, must be multiplexed together in order to achieve a high statistical multiplex gain G.

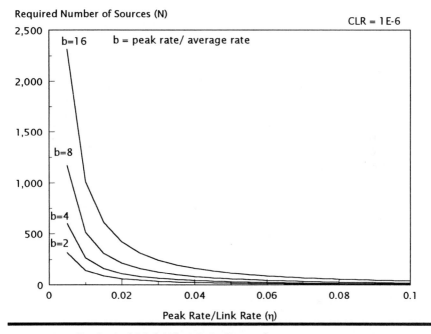

Figure 15.17 Statistical Multiplex Gain Example

15.7   PRIORITY QUEUEING PERFORMANCE

This section illustrates the capability of priority queueing to provide different delay performance for two classes of traffic. Figure 15.18 illustrates cell traffic originating from two virtual connections multiplexed onto the same transmission path, the higher priority traffic being numbered 1, and the lower priority traffic being numbered 2. Priority 1 cells are serviced first, and priority 2 cells are serviced only when they are not contending with priority 1 cells for link resources. In the event that more cells arrive in a servicing interval than can be accommodated, the priority 2 cells are either delayed or discarded as illustrated at the top of the figure.

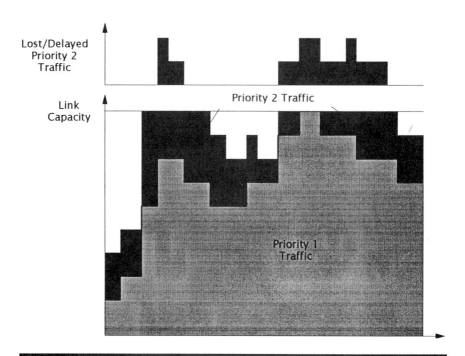

Figure 15.18 Illustration of Priority Queueing

We assume that priority 1 and 2 cells have Poisson arrivals and negative exponential service, priority 2 cells are delayed in an infinite buffer when priority 1 cells are being served (such that no loss can occur), and that 25% of the load is priority 1. The key to priority queueing is that the priority 1 traffic will observe a delay as if the priority 2 traffic did not even exist. On the other hand, the priority 2 traffic sees delay as if the transmission capacity were reduced by the average utilization taken by the priority 1 traffic. The formulas for the average priority 1 and priority 2 queueing delays are given by the following (see Ref. 13):

$$\text{Avg[Queuing Delay for Priority 1]} = \frac{\rho/\mu}{(1 - \rho_1)}$$

$$\text{Avg[Queuing Delay for Priority 2]} = \frac{\rho/\mu}{(1 - \rho)(1 - \rho_1)}$$

where ρ_1 and ρ_2 are the offered loads for priority 1 and 2 traffic, respectively, and $\rho = \rho_1 + \rho_2$ is the total offered load.

Figure 15.19 illustrates the effect of priority queueing by showing the average delay that would be seen by a single priority system according to the M/M/1 model of Section 15.2.3, the priority 1 cell delay, and the priority 2 cell delay. Observe that the priority 1 performance is markedly better than the single priority system, while priority 2 performance degrades only slightly.

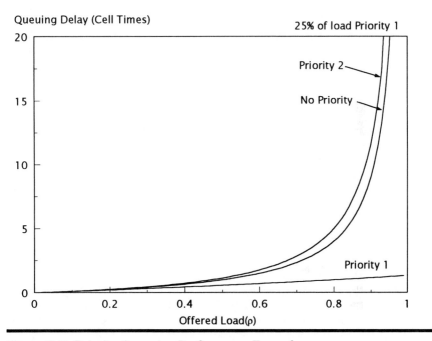

Figure 15.19 Priority Queueing Performance Example

15.8 REVIEW

This chapter first discussed several key aspects of traffic engineering philosophy: source modeling, performance measurement, and switch performance modeling. These affect the accuracy and complexity of the traffic engineering calculations. In general, the more accurate the model, the more complicated the calculation. This book opts for simplicity in modeling and introduces only a few of the popular, simple source models. Next, key aspects of modeling switch performance were covered: a comparison of buffering methods, Continuous Bit-Rate (CBR) loss performance, Variable Bit-Rate (VBR) loss performance, statistical multiplexing, and priority queueing. The superiority of output buffering over input buffering and the increased efficiency of shared output buffering over dedicated buffering were shown. Simple formulas to evaluate loss formulas for CBR and VBR traffic were presented, all of which can be entered into a simple spreadsheet program. A simple statistical multiplex gain model was described to demonstrate the classic wisdom, that in order to achieve gain, there must be a large number of bursty sources, each with a peak rate much less than the link rate. Finally, it was shown how priority queueing can yield markedly improved delay performance for the high-priority traffic over that of a single priority system.

15.9 REFERENCES

[1]   Fowler, Leland, "Local Area Network Traffic Characteristics, with Implications for Broadband Network Congestion Management," *IEEE JSAC*, September 1991

[2]   Leland, Willinger, Taqqu, Wilson, "On the Self-Similar Nature of Ethernet Traffic," *ACM Sigcomm '93*, Septemeber 1993

[3]   L. Kleinrock, *Queueing Systems Volume I: Theory*, Wiley, 1975

[4]   Fowler, Leland, "Local Area Network Traffic Characteristics, with Implications for Broadband Network Congestion Management," *IEEE JSAC*, September 1991

[5]   Leland, Willinger, Taqqu, Wilson, "On the Self-Similar Nature of Ethernet Traffic," *ACM Sigcomm '93*, September 1993

[6]   Hui, Arthurs, "A Broadband Packet Switch for Integrated Transport," *IEEE JSAC*, December 1988

[7]   D. McDysan, "Performance Analysis of Queueing System Models for Resource Allocation in Distributed Computer Networks," D.Sc. Dissertation, George Washington University, 1989

[8]   M. Hluchyj, M. Karol, "Queueing in High-Performance Packet Switching," *IEEE JSAC*, December 1988

[9]   M. Schwartz, *Computer Communication Design and Analysis*, Addison Wesley, 1977

[10]   L. Dron, Ramamurthy, Sengupta, "Delay Analysis of Continuous Bit Rate Traffic over an ATM Network," *IEEE JSAC*, April 1991

[11]   M. Wernik, Aboul-Magd, Gilbert, "Traffic Management for B-ISDN Services," *IEEE Network*, September 1992

[12]   Rasmusen, Sorenson, Kvols, Jacobsen, "Source-Independent Call Acceptance Procedures in ATM Networks," *IEEE JSAC*, April 1991

[13]   Gross, Harris, *Fundamentals of Queueing Theory*, Wiley, 1985

16

Design Considerations

This chapter explores important design considerations in network or future switching applications. First, the way in which delay and loss impact an application's design and performance is considered. Next, the tradeoff between efficiency and features is examined, listing considerations influencing the decision of what is best for your application. Some practical implications in deciding whether to build a dedicated private network, a shared public network, or a hybrid private/public network are then discussed. The factors involved in the decision between a fixed or usage-based service are discussed for a shared public network. The chapter closes by describing several open issues relevant to any network design. The accumulation of delay variation in multiple hop networks is important to delay-variation-sensitive applications, such as video, audio, and real-time interactive application traffic. One should also consider the reaction of the network in response to overloads or significant failures. Finally, practical applications of source modeling and traffic engineering are identified.

16.1 APPLICATION IMPACTS OF DELAY AND LOSS

This section reviews the impact of delay and loss on applications. Applications may be either bandwidth or latency limited. Loss impacts the usable throughput for most network and transport layer protocols.

16.1.1 Impact of Delay on Applications

Two situations occur when a source sends a burst of data at a certain transmission rate across a network with a certain delay, or latency: we call these bandwidth limited and latency limited. A *bandwidth-limited application* occurs when the receiver begins receiving data before the transmitter has completed transmission of the burst. A *latency-limited application* occurs when the transmitter finishes sending the burst of data before the receiver begins receiving any data.

Figure 16.1 illustrates the consequence of sending a burst of length b equal to 100,000 bits (100 kb) at a peak rate of R Mbps across the domestic United States with a propagation delay τ of 30 ms. It takes 30 ms for the bit stream to propagate from the originating station to the receiving station across approximately 4,000 miles of fiber since the speed of light in fiber is less than that in free space, and fiber is usually not routed along the most direct path. When the peak rate between originator and destination is 1 Mbps, and after 30 ms, only about one-third of the burst is in the transmission media, and the remainder is still buffered in the transmitting terminal. This is called a bandwidth-limited application because the lack of bandwidth to hold the transmission is limiting the transmitter from releasing the entire message immediately.

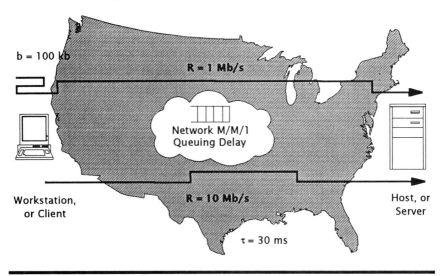

Figure 16.1 Propagation Delay, Burst Length, and Peak Rate

Now let's look at the case where the transmitter has sent the *entire* transmission before the receiver has received any data. When the peak rate is increased to 10 Mbps, the situation changes significantly — the entire burst is sent by the workstation before it can even reach the destination. Indeed, only about one-third of the bits propagating through the fiber transmission system are occupied by the burst! If the sending terminal must receive a response before the next burst is sent, then we see that a significant reduction in throughput will result. This type of situation is called *latency limited*, because the latency of the response from the receiver limits additional transmission of information.

Now let's apply the basic M/M/1 queueing theory from Chapter 15 as an additional element of end-to-end delay that increases non-linearly with increasing load, and thus is of key concern to an application. The average M/M/1 queueing plus transmission delay in the network is $b/R/(1-\rho)$, where ρ is the average trunk utilization in the network. The point where the time to transfer the burst (i.e., the transmission plus queueing time) exactly equals the propagation delay is called the latency/bandwidth crossover point as illustrated in Figure 16.2.

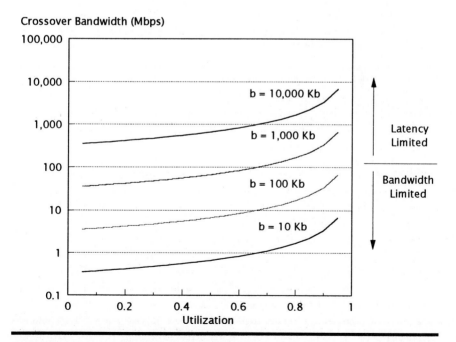

Figure 16.2 Latency/Bandwidth Crossover Point

In the previous example, for a file size of b=100 kb the crossover point is 3.33 Mbps for zero utilization, and increases to 10 Mbps for 66% utilization. A more detailed discussion on this subject can be found in Ref. 1.

16.1.2 Impact of Loss on Applications

Loss can be another enemy of applications. For many applications, the loss of a single cell results in the loss of an entire packet (or an entire protocol data unit) because the SAR sublayer in the AAL will fail in its attempt at reassembly. Loss (or even excessive delay) can result in a time-out or negative acknowledgment in a higher layer protocol, such as at the transport layer. If the round-trip time is long with respect to the application window size, then the achievable throughput can be markedly reduced. This is a basic aspect of all flow control methods, as we saw in Chapter 14 on congestion control. The amount of buffering required in the network is proportional to the delay, bandwidth product. In our example of Figure 16.1, the delay-bandwidth product is 300 bits for a 1-Mbps link (30 ms X 1 Mbps) and 3000 bits for a 10-Mbps link.

This situation is analogous to what occurred in early data communications over satellites, where the data rates were low, but the propagation delay was very high. The delay-bandwidth product is high in satellite communications because the propagation delay is high, while in B-ISDN and ATM communications the delay-bandwidth product becomes large because the transmission speeds are high.

Higher layer protocols recover from detected errors, or time-outs, by one of two basic methods: either all information that was sent after the detected error or time-out is retransmitted, or only the information that was actually in error or timed out is selectively retransmitted. Resending all of the information means that if N packets were sent after the detected error or time-out, then N packets are retransmitted, reducing the usable throughput. This scheme is often called a *Go-Back-N* retransmission strategy. The second method is where the packet that has a detected error, or causes a time-out, is explicitly identified by the higher layer protocol; then only that packet need be retransmitted. This scheme is often called a selective reject retransmission strategy. The usable throughput is increased because only the errored or timed-out information is retransmitted. However, this type of protocol is more complex to implement.

A simple model of the performance of these two retransmission strategies is presented to illustrate the impact of cell loss on higher layer protocols.

The number of cells in the retransmission window W is determined by the transmission rate R, the packet size p (in bytes), and the propagation delay τ as follows:

$$W = \left\lceil \frac{2\tau R}{8p} \right\rceil$$

The probability that an individual packet is lost due to a random cell loss probability π derived from the Cell Loss Ratio (CLR) is approximately the following:

$$\pi \approx \left\lceil \frac{p}{48} \right\rceil CLR$$

In the Go-Back-N strategy, if a single packet is in error of a window of W packets, then the entire window of W packets must be retransmitted. For the Go-Back-N retransmission strategy, the usable throughput η(Go-Back-N) is approximately the inverse of the average number of times the entire window must be sent, which is approximately [2], [3]:

$$\eta(\text{Go Back N}) \approx \frac{1-\pi}{1+\pi W}$$

In the selective reject strategy, if a single packet is in error, then only that packet is retransmitted. For the selective reject retransmission strategy, the usable throughput η(Selective Reject) is approximately the inverse of the average number of times any individual packet must be sent, which is [2], [3]:

$$\eta(\text{Selective Reject}) \approx (1-\pi)$$

This formula is valid for the case in which only one packet needs to be transmitted within the round-trip delay window. It also applies to a more sophisticated protocol that can retransmit multiple packets, such as SSCOP.

Figure 16.3 plots the usable throughput (or "goodput") for Go-Back-N and selective reject retransmission strategies for a DS3 cell rate R of 40 Mbps, a packet size p of 200 bytes, and a propagation delay of 30 ms. The resultant window size W is 1500 packets. The retransmission protocols have nearly 100% usable throughput up to a Cell Loss Ratio (CLR) of 10^{-6}. As the CLR increases, the usable throughput of the Go-Back-N protocol decreases markedly because the probability that an individual window (of 7500 cells) will be received error free decreases markedly. As the CLR increases toward 10^{-2}, the probability of an individual packet having a lost cell starts to become significant, and even the selective reject protocol's usable throughput begins to degrade.

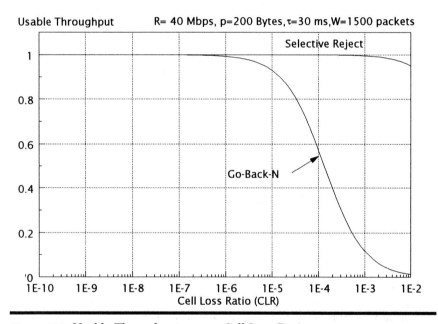

Figure 16.3 Usable Throughput versus Cell Loss Ratio

These examples illustrate the importance of selecting a QoS class with loss performance that meets the application requirements. For example, the Internet Transmission Control Protocol (TCP) uses a Go-Back-N type of protocol and hence works best with low loss rates.

16.2   PRIVATE VERSUS PUBLIC NETWORKING

Is a dedicated private network or a shared public network better for your suite of protocols and applications? Or would a combination of private and public network be better? In this section we give some objective criteria for assistance in making this sometimes difficult decision.

One key element to be considered is the overall cost. Overall costs should include planning, design, implementation, support, service, maintenance, and ongoing enhancements. These can require the dedication of significant resources for a private network, but are often included in public network services.

The current network will likely consist of mainframes, minicomputers, and/or LANs interconnected by bridges and routers. Almost every existing network contains some form of legacy SNA protocols. Their interconnections may use private lines, or an existing private or public data service. Identify if there are some applications, or concentration of traffic, at certain locations that require connectivity with other locations. Try to construct a traffic matrix of what throughput is required between major locations. You can then build a trial network design for a private network to achieve the delay and loss performance that your application requires. In a private network, it is important to estimate the capital and ongoing costs accurately. It is a common mistake to overlook, or underestimate, the planning, support, and upgrade costs of a private network. For a more detailed discussion on the topics of traffic estimation, network design, and other practical considerations, see Chapters 15 and 16 of Ref. 4.

If you have the site locations and estimated traffic between sites, a carrier will often be able to respond with a fixed cost and recurring cost proposal. These proposals often offer both fixed and usage pricing options. The performance of the public service will be guaranteed by the service provider, while in a private network this can be controlled to some extent by the network designer. In a public network switches and trunks can be shared across multiple customers, reducing cost and achieving economies of scale which are difficult to achieve in private networks. Carriers will often implement a shared trunk speed higher than any access line speed, and consequently can achieve lower delay and loss than in a private network due to the economy of scale inherent in the large numbers required for the statistical multiplexing gain we demonstrated in Chapter 15. This decreases costs for the individual user with performance that is suitable for most applications. If your application has unique performance requirements that are not met by any of the carrier's Quality of Service (QoS) offerings, then you may be forced to build a private network, regardless of cost. Be very sure that your current and future application requirements will justify this decision.

For comparable performance, the decision then becomes a matter of choosing which costs less, and assessing any risk. Of course, the least costly (and least risky) design might be a hybrid private and public network. In the very general sense, large volumes of high-utilization point-to-point traffic may be better handled via a private network or circuit switching, while diverse, time-varying connectivity is usually supported better in public packet-based networks. If you are uncertain about your traffic patterns, then a usage-based billing public packet network is often the most economical choice.

If a public network service is chosen, then a key question is whether fixed or usage-based billing should be selected. Not all carriers offer these billing options, so if your application can be more economically supported by choice of billing option, then this should be an important factor in your selection process. A fixed-rate option would charge a fixed monthly price for a certain traffic contract, which specifies a guaranteed throughput and QoS. Examples would be frame relay where the CIR frame delivery rate is guaranteed, or in ATM where the Quality of Service (QoS) is expressed in terms of cell loss and delay as defined in Chapter 12. A usage-based option would specify the guaranteed throughput and QoS, but the actual bill should be based upon the units of data that were successfully received by the destination station. Usage billing is usually subject to a minimum and maximum charging amount, usually a 100%+ percentage of the equivalent fixed rate.

Most users who are unfamiliar with their traffic patterns will choose a fixed-rate option with a guaranteed QoS level. As the users become more familiar with their traffic patterns and volumes, they naturally migrate to usage rates, understanding that they still have the minimums and maximums to protect them from unplanned outages or times of excess traffic. Accurate traffic forecasts are difficult, and due to the accelerating bandwidth principle we believe that the difficulty in future traffic forecasting is increasing. Check to see what the minimum charge is for usage-based billing, and if there is a cap on the usage-based option in relation to the fixed-rate option. If the minimum is low and the cap is reasonable, then usage-based billing may be your best choice.

If the public service provides traffic measurement capabilities, you can use them to obtain more accurate traffic data and make a better informed decision in the future. Examples include the network management offerings from MCI Communications called HyperScope^{SM}, and the network management product from Southwest Network Services called ProWatch^{SM}. These products provide the user with detailed daily, weekly, and monthly traffic reports on a virtual circuit basis — ideal for sizing your network. Be sure and check how frequently you can change the public service traffic contract and billing options when selecting a public carrier. It is also very important to understand how frequently the access options can be changed and what charges are associated with such changes.

A simple example is now given to illustrate the above concepts. The actual distance, economic data, and structure are not intended to be accurate, but are representative of what is available for private and public networks. Figure 16.4 illustrates a comparison of a full mesh private line network to a dual homed public network with diverse, redundant access. The mileage between the four cities that are to be connected: New York (NYC), Los Angeles (LA), Chicago, and Dallas are also shown. The figure shows the traffic matrix between these cities with traffic measured in units of Mbps during the busy hour. For example, users in Los Angeles transmit to Chicago, on the average over a single busy hour period, 16 Megabits of data per second. The busy hour is assumed to be 10% of the entire daily traffic, and that there are 20 working days per month. Therefore, the total daily usage is 10 times the busy hour traffic. Since there are 20 (days/month) times 10 (busy hour's of usage/day) this means that there are 200 hours of the total usage per month.

The private line network is composed of DS3s since peak data requirements exceed several DS1s, and there is a need for redundancy. The assumed private line access charges are $5,000 per month, and the assumed long distance charges are $50 per mile per month.

The public ATM service incurs the same private line access line charge of $5,000 per month, and an additional charge of $2,000 per month per ATM DS3 port. The example ATM public data service has two pricing plans: fixed and usage-based. The fixed-rate plan is $1.00 per Mbps per month per mile, with a granularity of 5 Mbps. The usage-based plan is 0.6¢ per Mbps for one hour (450 Mbytes) per mile, or about 1.3¢ per Gigabyte of data transferred. For 200 hours of monthly traffic the usage-based plan costs $1.20, as compared to the fixed-rate plan cost of $1.00; however, there is no granularity. Note that a rate of 0.5¢ per Mbps would yield even greater savings with usage-based pricing.

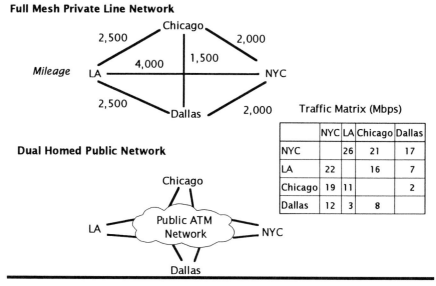

Figure 16.4 Example Private and Public ATM Networks

Table 16.1 shows the resulting costs for the private, fixed-rate public and usage-based public network alternative under the assumed traffic, mileage, and cost elements defined above. It is also broken down into access and network costs. Access costs include the DS3 access line charges for both private and public networks, and include the ATM port charge for the public network options. Network costs are the distance-based DS3 charges for the private line network, and the fixed or usage-based charges are based upon the traffic matrix for the public services.

Table 16.1 Comparison of Example Private and Public Network Costs

Item	Private Network	Public Network Fixed-Rate	Public Network Usage-Based
Access Costs	$60,000	$56,000	$56,000
Network Costs	$725,000	$527,500	$525,000
Total Cost	$785,000	$583,500	$581,000

The access charges are slightly less for the public services because fewer access lines are required, but each line costs more since there are additional charges for the ATM ports. The real differentiator is the network cost element. The network costs are significantly less because the public ATM network charges only for the actual bandwidth used. Even though the usage-based charges would be

20% higher than the fixed rate under our assumptions, the 5-Mbps granularity of the fixed-rate option makes the usage-based option slightly less expensive. There is typically a crossover point where usage rates becomes less expensive (given a known traffic volume) than fixed rates. This crossover point will vary between public services, access rates, and distance (to name a few). Some providers tie the usage rate to the fixed rate, effectively making them the same cost at the crossover point. There are also typically minimums and maximums applied to the usage rates for consumer (maximums) and provider (minimums) protection.

16.3   VARIATIONS IN DELAY

When designing a network of multiple switches connected by trunks, there are several things to consider. Some of these considerations are common to private and public networks, while some are specific to each. This section focuses on those aspects that are common to private and public networks, and point out some unique considerations that occur in each.

16.3.1 Impact of Cell Delay Variation on Applications

Recall the Cell Delay Variation (CDV) QoS parameter, which measures the clumping that can occur when cells are multiplexed together from multiple sources, either in the end system or at any switch, multiplexer, or intermediate system in the network. The resulting delay effect can accumulate when traversing multiple switches in a network. This becomes critical when transporting delay-sensitive applications such as video, audio, and interactive data over ATM. In order to support these applications, there is a need to provide a buffer to absorb the jitter introduced in an end-to-end network.

Jitter refers to the rate of clumping or dispersion that occurs to cells that were nominally spaced prior to transfer across an ATM network. The accumulated jitter must be accommodated by a playback buffer as will be described with reference to Figures 16.5 and 16.6. The playback buffer must be sized to allow the underrun or overrun events to occur infrequently enough, according to the CDV (clumping) and cell dispersion that accrues across a network. In our examples the nominal cell spacing is 4 cell times, and the playback buffer is 4 cells in length and is nominally centered (i.e., holding two cells).

In the overrun scenario of Figure 16.5 cells arrive too closely clumped together, until finally a cell arrives and there is no space in the playback buffer; thus cells are lost. This can be a serious event for a video coded signal because an entire frame may be lost due to the loss of one overrun cell.

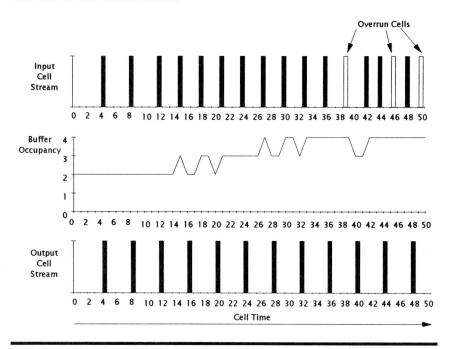

Figure 16.5 Illustration of Playback Buffer Overrun Scenario

In the under-run scenario of Figure 16.6 the cells arrive too dispersed in time, such that when the time arrives for the next cell to leave from the playback buffer, the buffer is empty. This too can have a negative consequence on a video application because the continuity of motion or even the timing may be disrupted.

16.3.2 Estimating Cell Delay Variation in a Network

The accumulation of CDV across multiple switching nodes can be approximated by the square root rule from the ATM Forum B-ICI specification. This states that the end-to-end CDV is approximately equal to the CDV for an individual switch times the square root of the number of switching nodes in the end-to-end connection. The reason that we don't just add the variation per node is that the extremes in variation are unlikely to occur simultaneously, and tend to cancel each other out somewhat. For example, while the variation is high in one node, it may be low in another node.

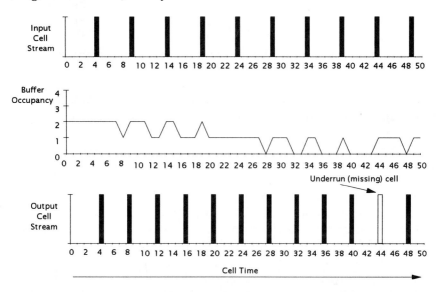

Figure 16.6 Illustration of Playback Buffer Underrun Scenario

Figure 16.7 illustrates this concept. We show in the probability distribution that the delay assumes a certain value at various points in the network. The assumed delay distribution has a fixed delay of 25 µs per ATM switch. The ATM switch also adds a normally distributed delay with mean and standard deviation also equal to 25 µs. Therefore, the average delay added per node is 50 µs. Starting from the left-hand side at node A, the traffic has no variation. The first ATM switch adds a fixed delay plus a random delay resulting in a modified distribution. The next ATM switch adds the same constant delay and an independent random delay. The next two ATM switches add the same constant and random delay characteristics independently. The resulting delay distribution after traversing four nodes is markedly different as can be seen from the plots — not four times worse but only approximately twice as bad.

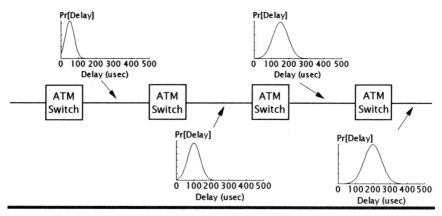

Figure 16.7 Illustration of Cell Delay Variation in a Network

Figure 16.8 plots the probability that the delay exceeds a certain value x after traversing N nodes. The random delays are additive at each node; however, in the normal distribution the standard deviation of the sum of normal variables only grows in proportion to the square root of the sum. This is the basis for the square root rule in the ATM Forum B-ICI specification. The sum of the fixed and average delays is 200 μs, and therefore the additional variation is due to the random delay introduced at each node.

The equation for the plots of Figure 16.8 for an average plus fixed delay given by α (50 μs in this example) and a standard deviation σ (25 μs in this example) for the probability that the delay is greater than x after passing through N ATM switches is given by:

$$\text{Prob[Delay} \geq \text{x]} \approx Q\left(\frac{x - N\alpha}{\sqrt{N}\sigma}\right)$$

Chapter 15 defined the expression for $Q(x)$ for the normal distribution.

This means that the end equipment must be able to absorb the jitter that the intermediate network(s) introduce. One option that exists to balance this tradeoff is the capability to reduce CDV within switches in the network. Chapter 13 showed how to accomplish this by shaping, spacing, priority queueing, or framing.

The calculation of average delay is simple — add up all of the components to delay for each switch and transmission link. Be sure to ask carriers about average delays through their services, as well as variations in delay.

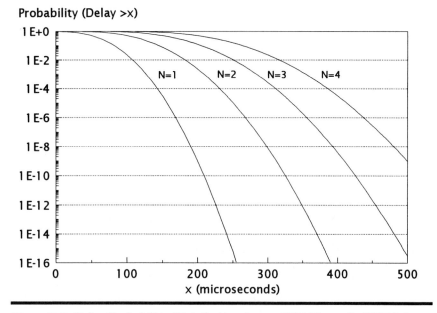

Figure 16.8 Delay Probability Distribution for an ATM Network of N Nodes

Loss is also relatively easy to estimate. Assuming that loss is independent on each switch and link, then the end-to-end loss is 1 minus the product of successful delivery for each element. In practice this is closely approximated when the loss probabilities are small by simply adding up the individual loss probabilities. The error ratio is also computed in an analogous manner.

16.4   NETWORK PERFORMANCE MODELING

The concepts of modeling traffic sources and buffering strategies at the single node level were introduced in Chapter 15. This is a good way to begin approaching the problem; however, most users are concerned with modeling the performance of a network. As discussed in Chapter 15, there are two basic modeling approaches: simulation and analysis. A simulation is usually much more accurate, but can become a formidable computational task when trying to simulate the performance of a large network. Analysis can be less computationally intensive, but is often inaccurate. What is the best approach? The answer is similar to nodal modeling — it depends on how much information you have and how accurate an answer you require.

Simulation models are very useful in investigating the detailed operation of an ATM system, which can lead to key insights into equipment, network, or application design. Simulations generally take too long to execute to be used as an effective network design tool.

A good way to bootstrap the analytical method is to simulate the detailed ATM node's performance under the expected mix of traffic inputs. Often an analytical approximation to the empirical simulation results can be developed as input to an analytical tool. An assumption often made in network modeling is that the nodes operate independently, and that the traffic mixes and splits independently and randomly. If better information is not available, then this is a reasonable assumption, and simulation may be the only recourse. Analytical models become very complex without the assumption that nodes are independent of each other.

The inputs and outputs of a network model are similar for any packet switched network design problem. The inputs are the topology, traffic, and routing. The network topology must be defined, usually as a graph with nodes and links. The characteristics of each node and link relevant to the simulation or analytical model must be described. Next the pattern of traffic offered between the nodes must be defined. For point-to-point traffic this is commonly done via a traffic matrix. The routing, or set of links that traffic follows from source to destination, must be defined.

The principal outputs are measures of performance and cost. The principal performance measures of a model are loss and delay statistics. A model will often produce an economic cost to allow the network designer to select an effective price-performance tradeoff (see Ref. 4).

16.5 REACTION TO EXTREME SITUATIONS

Anther important consideration in network design is the desired behavior of the network under extreme situations. We will consider the extreme situations of significant failure, traffic overload, and unexpected peak traffic patterns.

It is very important to consider what will happen during a significant failure, such as one or more critical trunk failures, or failure of an entire switch. The general guideline is that any element of the network should not become isolated by a single trunk or switch failure. If you can survive with some loss of connectivity, costs can be significantly reduced in a private network; however, public networks are usually designed to survive such single failures. The exception and most vulnerable point is usually access to the public network, in both single-point-of-failure CPE or access circuits. This can be solved with both equipment and circuit diversity. You may have different performance objectives under failure situations than under normal circumstances. Also, you may desire that some traffic be preempted during a failure scenario so that support for mission-critical traffic is maintained. A failure effectively reduces either a bandwidth or switching resource, and hence can be a cause of congestion, which we will cover next.

Traffic overloads and unexpected traffic parameters can also cause congestion. For example, offered traffic in excess of the contract may create congestion. Under normal circumstances the network marks this excessive traffic for selective cell discard, and those cells in excess of the contract will be discarded first. But if the network is overbooked with contracted traffic parameters, the selective cell discard may not be enough. As pointed out earlier, congestion can drive a switch or multiple points in a network into overload, reducing overall throughput significantly if congestion collapse occurs. If you expect this situation to occur, then a mechanism to detect congestion, correlate it with its cause, and provide some feedback in order to isolate various traffic sources and achieve the required measure of fairness is desirable. If you are uncertain as to how long such overloads will persist, then some slow-reacting feedback controls may actually reduce throughput because the reaction may occur after congestion has already abated (such as we saw with higher layer protocols such as TCP).

In order to support time-varying traffic patterns there are two fundamentally different approaches: one based upon the telecommunications concept of reserving bandwidth and the communications concept of fairly shared resources. A Switched Virtual Channel (SVC) capability allows an application to request the network to reserve bandwidth for its exclusive use, very similar to circuit switching. In order to use the current SVC protocols, you will need to estimate your bandwidth accurately since it can only be changed by tearing down the call and setting it up again. Future SVC protocols will allow the bandwidth to be dynamically negotiated without the call being taken down. Both of these scenarios will require the user to estimate the bandwidth requirement and communicate this to the network.

Another possibility would be that of dynamic flow control, using one of the methods presented in Chapter 14, where feedback from the network regarding congestion is used to throttle back sources in a fair manner which also isolates applications which are "good citizens" from those that aren't following the flow control rules. The choice of approach will depend upon the characteristics of your applications, and which approach is adopted by the ATM vendor and service providers.

16.6 TRAFFIC ENGINEERING COMPLEXITY

Realistic source and switch traffic models are not currently amenable to direct analysis, with the results presented in this book providing only approximations under certain circumstances. Such approximate methods may have large inaccuracies, which can only be ascertained by performing detailed simulations or actual tests. Simulations are time consuming, and cannot effectively model low cell loss rates since an inordinate number of cells must be simulated. For example, in order to simulate a cell loss rate of one in a billion, at least 100 billion to 1 trillion cells must be simulated. Even with today's computers this is a lot of computations to obtain a single point on a loss curve. Where possible, methods for extrapolating loss rates and estimating the occurrence of unlikely events should be used instead. Also, constantly changing source, switch, and network characteristics creates a moving target for such traffic engineering models.

Traffic engineering has a real-world impact on the Connection Admission Control (CAC) function's complexity. Recall that CAC only admits a connection request if the QoS of existing connections would still be met. Implicit in this function is the ability to calculate (or at least estimate) the QoS parameters for a large number of sources with different traffic parameters in multiple QoS classes. Add to this the need to execute CAC in real time to support switching of SVC requests, and it becomes apparent that accurate and complete traffic engineering becomes a key step in the design process.

16.7 REVIEW

This chapter covered several key considerations in the design of networks and switches. It began by reviewing the impact of delay and loss on applications and gave guidelines on choosing link speed, burst duration, and loss ratio. Next, some of the factors in whether to choose a private network, a public network, or hybrid of the two were considered. Methods for calculation of delay loss and error rate in a multiple switch node network were described, along with some suggestions and alternatives for managing delay variation. The tradeoffs and critical factors include cost, risk, and performance. Part of the network design should consider what will happen under extremes, such as failures, overload, and unexpected peak traffic patterns, and each subject was explored in detail. Finally, a commentary on the still open issues in traffic engineering was made, along with a discussion of how this relates to the real-time, critical function of Connection Admission Control (CAC).

16.8 REFERENCES

[1]   L. Kleinrock, "The Latency/Bandwidth Tradeoff in Gigabit Networks," *IEEE Communications Magazine*, April 1992.

[2]   D. McDysan, "Critical Review of ARPANET, IBM SNA and DEC DNA," George Washington University, April 1985.

[3]   M. Moussavi, "Performance of Link Level Protocols over Satellite," *MELECON '89*, Lisbon, Portugal, April, 1989.

[4]   D. Spohn, *Data Network Design*, McGraw-Hill, 1993.

6

ATM-Based Protocol Interworking and Public Service Offerings

This part provides the reader with an in-depth study of ATM interworking with higher layer protocols as well as current and planned public service offerings based on ATM technology. Chapter 17 begins with a discussion of protocol interworking and interoperability using ATM. First the frame-relay-based ATM Data eXchange Interface (DXI) which offers low-speed ATM access is described. The discussion then covers how frame relay, SMDS, and IP services interwork using ATM. The chapter closes with a discussion of how the Broadband InterCarrier Interface (B-ICI) is used to interconnect public service ATM networks. Chapter 18 begins with a road map for public ATM service offerings. A summary of announced ATM services is provided. The first is an ATM "bit pipe" or Cell Relay Service (CRS). The potential services of the future and issues involved with them are then covered. The chapter ends with the desirable attributes of public ATM network services and the selection process and criteria for each.

17

ATM-Based
Protocol Interworking

This chapter first covers some general principles of protocol interworking using ATM, and goes on to summarize the current state of standardization. Next, there is a discussion of the Data eXchange Interface (DXI), which offers low-speed ATM access. Frame relay, SMDS, and IP protocol interworking over an ATM network and interworking with ATM end systems are then described. The use of the ATM Forum's Broadband InterCarrier Interface (B-ICI) for connecting public service ATM networks together is also covered in detail.

17.1 INTERWORKING PRINCIPLES

This section covers the general subject of interworking Specific Service (SS) higher layer protocols using ATM. Specific Services (SSs) are either interworked over ATM or to an ATM-based end system using network interworking. The higher layer network interworking protocols that have been defined to date include frame relay, multiprotocol (including IP), and SMDS. This chapter details each of these network interworking protocols. This section concludes with a discussion of the concept of service interworking and the likely direction of protocol standardization and specification.

17.1.1 Network Interworking

The concept of network interworking is formally defined in ITU-T Recommendation I.555 which defines Frame Relay Service (FRS) interworking with B-ISDN. This concept is generalized to any Specific Service (SS) employing network interworking in one of two scenarios: *interworking over an ATM network*, and *interworking with an ATM end system*. In I.555 these are called network interworking scenarios 1 and 2, respectively. It is our belief that our terminology for network interworking is easier to understand and remember, and hence it will be used in place of the I.555 terminology.

Figure 17.1 illustrates network interworking of a Specific Service (SS) *over* an ATM network. The physical configuration is shown in the diagram at the top of the figure, and the logical protocol stack interfaces are shown at the bottom of the figure. Starting at the left-hand side of the physical configuration, a service-specific end system is shown interconnected to an InterWorking Function (IWF) by a service-specific User Network Interface (UNI). The IWF is then connected to the ATM network by an ATM interface, which interconnects to another IWF with identical function that in turn connects to the destination service-specific end system.

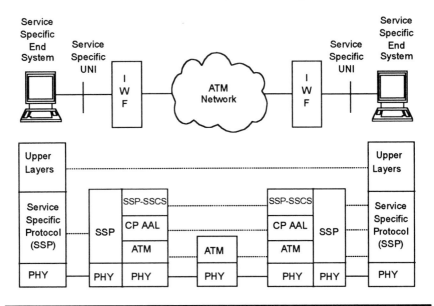

Figure 17.1 Networking Interworking *over* an ATM Network

The protocol stacks at the bottom of the figure illustrate how network interworking transparently conveys the upper layer protocols between the service-specific and ATM end systems. The lower layer protocols are transformed by the InterWorking Function (IWF) as shown in the figure. The Service-Specific Protocol (SSP) is mapped to a Service-Specific Convergence Sublayer (SSCS) AAL that is uniquely defined for the specific service, for example the FR-SSCS defined later. This chapter provides examples of the Frame Relay, SMDS, and multiprotocol encapsulation specific services. The SSP-SSCS AAL is then conveyed over a standard Common Part ATM Adaptation Layer (CP AAL) which is transported over the standard ATM and PHYsical layers as defined in Chapter 6. The net effect of this protocol operation is that the service-specific end systems are unaware that they are interconnected over an ATM network.

Figure 17.2 illustrates network interworking of a Specific Service (SS) *with* an ATM end system. The same style used in the previous figure is employed, and the InterWorking Function (IWF) is identical.

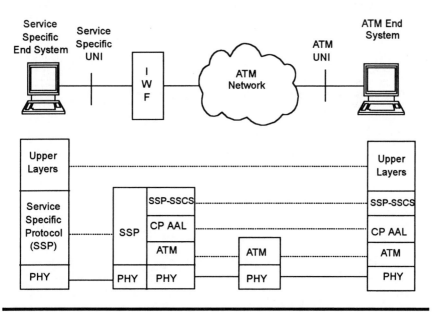

Figure 17.2 Network Interworking *with* an ATM End System

Starting at the left-hand side of the physical configuration, a service-specific end system is shown interconnected to an InterWorking Function (IWF) by a service-specific User Network

Interface (UNI). The IWF is then connected to the ATM network by an ATM interface, which interconnects to an ATM end system via an ATM UNI. The protocol stack at the bottom of the figure is identical to that shown in the previous figure, which means that the InterWorking Function (IWF) need not be concerned with whether it is connected to another IWF or to an ATM end system. The ATM end system must be aware that it is connected to an interworking function on a particular VCC, however, since it must implement the Service-Specific Protocol (SSP) Service-Specific Convergence Sublayer (SSCS) AAL that is uniquely defined for the specific service.

17.1.2 Overview of Network Interworking Protocols

Figure 17.3 depicts a roadmap of the protocol interworking standards covered in this chapter. The protocols that interwork using ATM are listed around the edges of the diagram, with ATM at the center of the diagram. The protocol stack is shown above the line for each interface.

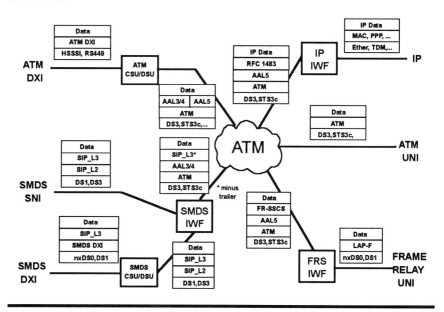

Figure 17.3 Network Interworking Protocol Summary

Each interface is processed by either an InterWorking Function (IWF), or in the case of Data eXchange Interfaces (DXI), by a CSU/DSU. The resulting protocol stack at the output of the IWF or CSU/DSU is shown touching the line connecting to the ATM cloud. The ATM cloud indicates that like interworking functions can be interconnected in interworking over ATM, or connected to an ATM end system via the ATM UNI in interworking with ATM.

This protocol interworking can be used to define translations between the various interfaces based upon the ATM protocol stack. There may be some basic incompatibilities, however. For example, since SMDS is a connectionless service and frame relay and ATM VPs and VCs are connection-oriented, SMDS cannot support status or fault messages as frame relay and ATM VPC/VCCs can.

Functions and data are either mapped or encapsulated in these protocol stack transformations, as will be seen later in the chapter. Higher layer protocol interactions are also required to provide full interworking.

17.1.3 Service Interworking

Figure 17.4 illustrates the concept of service interworking using the same style of a physical configuration and protocol stack model as was used for network interworking.

The physical configuration for service interworking is identical for that of network interworking with an ATM end system. The difference is in the protocol stacks. Service interworking does not require the ATM end system to know whether it is connected to a Service-Specific Protocol (SSP) InterWorking Function (IWF). Note that this interworking function is different than the network IWF. In general, the service IWF should be able to accept a number of service-specific protocols and map these to a single, generic SSCS AAL. Instead, the ATM end system interfaces using a standard Service-specific Convergence Sublayer (SSCS) AAL. The scope of service interworking can also involve interworking between higher layer protocols that may be different between the specific service and ATM end systems via a higher layer interworking function, as shown in the figure. A specific example of higher layer interworking functions will be covered in the review of multiprotocol encapsulation over ATM and frame relay encapsulation interworking.

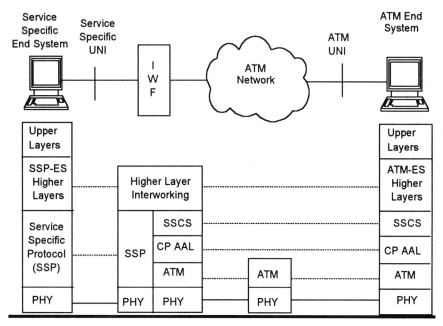

Figure 17.4 Service Interworking Concept

Service interworking is not very well defined at this time; for example, there is only one paragraph in ITU-T Recommendation I.555 on this subject. In general, service interworking is intended to support the interconnection of an ATM terminal using an ATM service with some other service of similar capability and function. The Frame Relay Forum and ATM Forum System Aspects and Applications (SAA) groups have announced that they intend to work together on specifications for service interworking between frame relay and ATM end systems.

17.2   DATA EXCHANGE INTERFACE (DXI)

Many users have asked the following question: What if I want the capabilities of ATM over the WAN, but I can't afford the cost of a DS3 or OC-3 access line?   The answer could be the ATM Forum specified ATM Data eXchange Interface (DXI), which supports either the V.35, RS449, or the HSSI DTE-DCE interface at speeds from several kbps up to and including 50 Mbps. ATM DXI specifies the interface between a DTE, such as a router, and a DCE, usually called an ATM CSU/DSU, which provides the conversion to an ATM UNI, as illustrated in Figure 17.5. Although the ATM DXI is normally thought of as a DTE-DCE interface specification, there is no reason that it cannot be used as a longer distance access protocol over nxDS0, DS1, and nxDS1 access lines. The SMDS DXI is specified to operate in this manner, and it is likely that the ATM DXI will be used in this manner as well. The ATM DXI is an example of relatively simple network interworking with ATM.

The ATM DXI interface is managed by the DTE through a Local Management Interface (LMI), while the ATM UNI Interim Local Management Interface (ILMI) Simple Network Management Protocol (SNMP) messages are passed through to the DTE as shown in Figure 17.5.

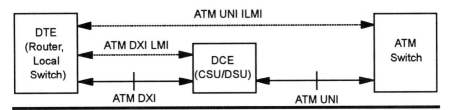

Figure 17.5 ATM Data eXchange Interface (DXI) Configuration

Table 17.1 Summary of ATM DXI Mode Characteristics

Characteristic	Mode 1a	Mode 1b	Mode 2
Maximum number of VCCs	1,023	1,023	16,777,215
AAL5 Support	Yes	Yes	Yes
AAL3/4 Support	No	Yes	Yes
Maximum DTE SDU Length			
AAL5	9,232	9,232	65,535
AAL3/4	N/A	9,224	65,535
Bits in FCS	16	16	32

Let's now take a look at the three major modes of DXI: mode 1a, mode 1b, and mode 2, whose characteristics are summarized in Table 17.1. The maximum number of VCCs supported is determined by the number of addressing bits, 10 for mode 1 and 24 for mode 2. AAL5 is supported in all modes, while AAL3/4 is a configuration option in modes 1b and 2 with maximum DTE Service Data Unit (SDU) length for each AAL, as indicated in Table 17.1. The number of Frame Check Sequence (FCS) bits required usually determines whether ATM DXI support will be a software upgrade for 16 bits in mode 1 or a hardware change for 32 bits in mode 2. The mode 1 option will likely be the choice for existing hardware, while mode 2 will likely be supported by much of the newer hardware.

We now cover the protocols and operation of modes 1a, 1b, and 2.

17.2.1 ATM DXI — Mode 1a and Mode 1b

ATM DXI Mode 1 supports two implementations. Both mode 1a and 1b define DCE support for AAL5 as shown in Figure 17.6. The DTE Service Data Unit (SDU) is encapsulated in the AAL5 CPCS and then segmented into ATM cells using the AAL5 Common Part Convergence Sublayer (CPCS) and Segmentation And Reassembly (SAR) sublayer functions as defined in Chapter 8. The 2-octet DXI header defined later in this section prefixes the DTE SDU. The 2-octet Frame Check Sequence (FCS) is the same as that used in frame relay and HDLC and hence much existing DTE hardware can be modified to support mode 1.

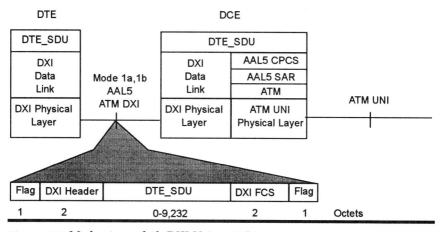

Figure 17.6 Modes 1a and 1b DXI Using AAL5

Mode 1b adds support for the AAL3/4 CPCS and SAR on a per-VCC basis as shown in Figure 17.7. The DTE must know that the DCE is operating in mode 1b AAL3/4 since it must add the 4 octets for both the CPCS PDU header and trailer as indicated in the figure. This decreases the maximum-length DTE SDU by 8 octets. The same 2-octet DXI header used for the AAL5 VCC operation is employed.

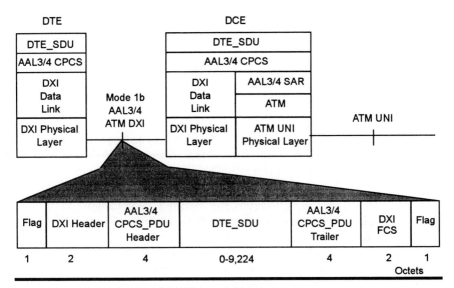

Figure 17.7 Mode 1b ATM DXI Using AAL3/4

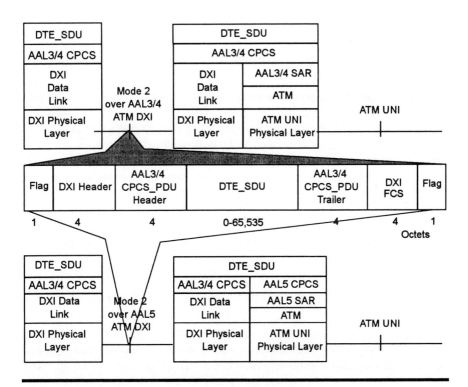

Figure 17.8 Mode 2 ATM DXI Using AAL3/4 or AAL5

17.2.2 ATM DXI — Mode 2

Mode 2 uses the same interface between DTE and DCE regardless of whether the VCC is configured for AAL5 or AAL3/4, as shown in Figure 17.8.

The DTE must place the DTE SDU inside the AAL3/4 CPCS header and trailer, and then the DCE performs the appropriate function depending upon whether the VCC is configured for AAL3/4 or AAL5. The DCE operates the same as in mode 1b for a VCC configured for AAL3/4, performing the AAL3/4 SAR on the AAL3/4 CPCS_PDU as shown in the top part of Figure 17.8. The DCE must first extract the DTE_SDU from the AAL3/4 CPCS_PDU for a VCC configured to operate in AAL5, as shown in the bottom half of the figure. The net effect of these two transformations is that a mode 2 DCE can interoperate with a mode 1 DCE. The mode 2 DXI frame has a 4-octet header and a 4-octet FCS, which will usually require new hardware. Because the FCS is longer, the maximum DTE_SDU length can be larger. The 32-bit FCS used in the DXI is the same as that used for FDDI and AAL5.

17.2.3 ATM DXI Header Formats

Figure 17.9 illustrates the details of the 2- and 4-octet DXI header structure. The DXI Frame Address (DFA) is mapped into the low order bits of the VPI/VCI by the DCE. The Congestion Notification (CN) is mapped from the last ATM cell of the PDU's Payload Type (PT) congestion indication as defined in Chapter 8. The DTE can set the CLP bit so that the DCE will in turn set the CLP bit in the ATM cell header with the same value, thus allowing the user to mark some PDUs as a low loss priority. A great deal of similarity can be found when comparing these formats to the frame relay Service-Specific Convergence Sublayer (SSCS) formats later in this chapter.

17.2.4 Local Management Interface (LMI) Summarized

The DXI Local Management Interface (LMI) defines a protocol for the exchange of SNMP GetRequest, GetNextRequest, SetRequest, Response, and Trap messages between the DTE and the DCE. The LMI allows the DTE to set or query (Get) the mode of the DXI interface as either 1a, 1b, or 2. The LMI also allows the DTE to set or query the AAL assigned on a per-VCC basis as indexed by the DXI Frame Address (DFA). A shortcoming of the current LMI is that the ATM UNI status is not communicated to the DTE by the DCE.

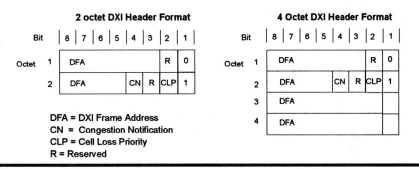

Figure 17.9 Two- and Four-Octet ATM DXI Header Formats

17.3 MULTIPROTOCOL OVER ATM AAL5

IETF standard RFC 1483 defines how a number of commonly used protocols can be routed or bridged over ATM Adaptation Layer 5 (AAL5). RFC 1483 defines two methods: protocol encapsulation and VC multiplexing. *Protocol encapsulation* allows for multiple protocols to be multiplexed over a single ATM "Virtual Circuit (VC)," as defined in RFC 1483, which is the same thing as an ATM Virtual Channel Connection (VCC), as defined in Chapter 8. This is the same term used in most other standards. The *VC multiplexing* method assumes that each protocol is carried over a separate ATM VCC. Both of these methods define the AAL5 Common Part Convergence Sublayer (CPCS) Protocol Data Unit (PDU) format, also called the payload format, that was described in Chapter 8. We first cover the protocol encapsulation and then the VC multiplexing method.

17.3.1 Protocol Encapsulation

Protocol encapsulation operates by prefixing the Protocol Data Unit (PDU) with an IEEE 802.2 Logical Link Control (LLC) header, and hence we call this *LLC encapsulation*. The LLC header identifies the

PDU type. This method is designed for public network or wide area network environments where a premises device would send all protocols over a single VCC, such as in a PVC environment where the pricing structure favors a small number of PVCs.

Figure 17.10a illustrates protocol encapsulation by routers, showing a network of three routers each multiplexing an Ethernet and Token Ring LAN over a single VCC interconnecting them. The Ethernet and Token Ring PDUs are multiplexed onto the same VCC between routers using the encapsulation described below. The drawing in Figure 17.10b illustrates how bridges can multiplex PDUs from an Ethernet and Token Ring interfaces to yield a bridged LAN. The bridges only use a spanning tree of the ATM VCCs at any one point in time.

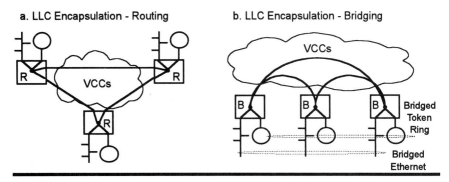

Figure 17.10 Routing and Bridging Use of LLC Encapsulation

17.3.2 LLC Encapsulation For Routed Protocols

The LLC encapsulation for routed protocols is described with reference to Figure 17.11.

The LLC-encapsulated routed ISO protocol payload structure is depicted in Figure 17.11a. The 3 octets of the LLC header contain three 1-octet fields: a Destination Service Access Point (DSAP), a Source Service Access Point (SSAP), and control. The routed ISO protocol is identified by a 1-octet NLPID field which is included in the protocol data. NLPID fields include, but are not limited to, SNAP, ISO CLNP, ISO ES-IS, ISO IS-IS, and Internet IP. The values of these octets are shown in hexadecimal notation as X'zz', where z is a hexadecimal digit representing 4 bits. For example X'1' corresponds to a binary '0001' and X'A' corresponds to a binary '1010'. Figure 17.11b depicts the payload for a routed non-ISO protocol, specifically showing the example for an IP PDU. The 3-octet SNAP header follows the LLC and contains a 3-octet Organizationally Unique Identifier (OUI) and the Protocol IDentifier (PID). Other protocols, such as ISO CLNP and IS-IS, can be encapsulated through the use of different PIDs as defined in RFC1483.

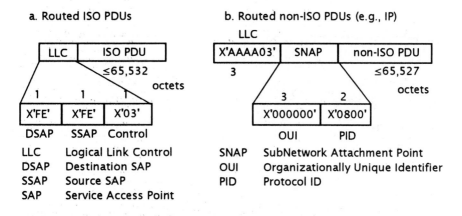

Figure 17.11 LLC Encapsulation Using Routed Protocols

17.3.3 LLC Encapsulation for Bridged Protocols

LAN and MAN protocols are bridged through use of the LLC encapsulation method. The SNAP header is used to identify the type of bridged media and whether the original LAN Frame Check Sequence (FCS) is included with the PDU, as illustrated in Figure 17.12. The LLC header identifies a non-ISO PDU as before. The OUI field identifies an 802.1 identification code. The PID field identifies the protocol. The remaining fields are either padding or the actual LAN PDU. Protocols which can be bridged include 802.3 Ethernet, 802.4 Token Bus, 802.5 Token Ring, Fiber Distributed Data Interface (FDDI), and 802.6 DQDB, as shown in the figure.

	LLC	OUI	PID	PAD			LAN FCS
a. 802.3	X'AAAA03'	X'0080C2'	X'0001', X'0007'	X'0000'	MAC Destination Address	MAC Frame *	If PID X'0001'

	LLC	OUI	PID	PAD				LAN FCS
b. 802.4	X'AAAA03'	X'0080C2'	X'0002', X'0008'	X'000000'	Frame Control	MAC Destination Address	MAC Frame*	If PID X'0002'

	LLC	OUI	PID	PAD				LAN FCS
c. 802.5	X'AAAA03'	X'0080C2'	X'0003', X'0009'	X'0000xx'	Frame Control	MAC Destination Address	MAC Frame*	If PID X'0003'

	LLC	OUI	PID	PAD				LAN FCS
d. FDDI	X'AAAA03'	X'0080C2'	X'0004', X'000A'	X'000000'	Frame Control	MAC Destination Address	MAC Frame*	If PID X'0004'

	LLC	OUI	PID				
e. 802.6	X'AAAA03'	X'0080C2'	X'000B'	Common PDU Header	MAC Destination Address	MAC Frame*	Common PDU Trailer

* Remainder of MAC frame

Figure 17.12 LLC Encapsulation Using Bridged Protocols

17.3.4 VC-Based Multiplexing

The second method of carrying multiple protocols over ATM is through VC-based multiplexing, which supports a single protocol per virtual connection. In other words, the VCs are multiplexed rather than the protocols themselves as done by protocol encapsulation.

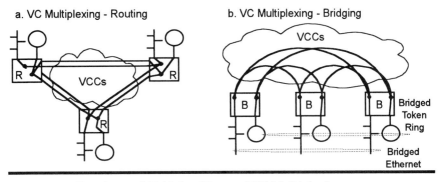

a. VC Multiplexing - Routing b. VC Multiplexing - Bridging

Figure 17.13 Routing and Bridging Usage of VC Multiplexing

Figure 17.13a illustrates this concept for routed protocols, showing a separate VCC connecting the routing point for Ethernet and Token Ring. Figure 17.13b illustrates the same point for bridged protocols, again requiring twice as many VCCs as for protocol encapsulation.

Comparing this to Figure 17.10, observe that the only difference is the use of one VCC for each protocol that is being routed or bridged versus one VCC between each pair of routers or bridges — Ethernet and Token Ring.

The bridged PDU payload is devoid of the LLC and SNAP protocol identifiers used in protocol encapsulation, resulting in less overhead, less processing, and higher overall throughput at the expense of the lost routing function. This method is designed for environments where the user can dynamically create and delete large numbers of ATM VCCs in an economical fashion, which could occur in private ATM networks or ATM SVC networks. Routed protocols can make use of the entire 65,535-octet AAL5 CPCS PDU. Bridged protocols have the same format as shown in Figure 17.12 without the LLC, OUI, or PID fields. The use of LAN FCS in the bridged protocols is implicitly defined by association with the VCC.

17.3.5 Selection of Multiplexing Method

Either of the two types of multiplex methods, encapsulated or VC multiplexing, can be used with PVCs and SVCs. The method is selected by a configuration option for PVCs. SVCs require information elements in the signaling protocol for the two routers to communicate whether protocol encapsulation or VC multiplexing was being used, and when using VC multiplexing, whether the original LAN FCS is being carried.

17.4 ATM AND FRAME RELAY INTERWORKING

Frame relay and ATM interworking are specified in ITU-T Recommendation I.555, the ATM Forum B-ICI specification, and a Frame Relay Forum Implementation Agreement (IA) [1]. Each scenario, along with associated access configuration, protocol, Service-Specific Convergence Sublayer (SSCS), status signaling, traffic and congestion control, mapping, and service interworking considerations, is summarized in this section.

17.4.1 Frame Relay and ATM Interworking Scenarios

Frame Relay (FR) and ATM interworking can operate in one of two scenarios using three access configurations as shown in Figure 17.14. The two scenarios are those described in the section on network interworking at the beginning of this chapter: scenario 1 where FR is interworked *over* ATM, and scenario 2 where FR interworks *with* an ATM end system. The first two access configurations of FR CPE interfacing via a FR UNI to the Interworking Function (IWF) or through a frame relay network apply to both scenarios. The third access configuration of a direct connection of an ATM end system via an ATM UNI to a network that terminates the VCC on the IWF applies to scenario 2 only. The IWF is identical for scenarios 1 and 2.

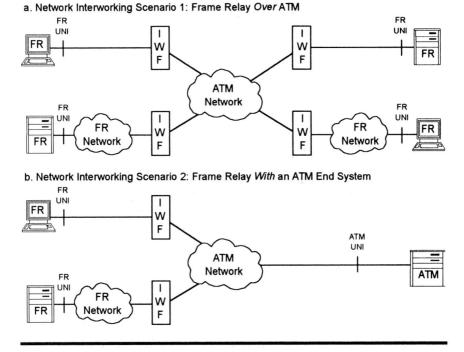

Figure 17.14 FR/ATM Interworking Scenarios and Access Configurations

17.4.2 Frame Relay to ATM Functions and Protocol

The FR to ATM InterWorking Function (IWF) converts between the Q.922 core functions and the FR Service-Specific Convergence Sublayer (FR-SSCS) defined in I.365.1, and the AAL5 Common Convergence Sublayer (CPCS) and Segmentation And Reassembly (SAR) sublayers from I.363, as shown in Figure 17.15a. The IWF must also convert between the Q.933 Annex A PVC status signaling for a single, physical FR UNI port and the VCCs that correspond to the DLCIs. The FR-SSCS Protocol Data Unit (PDU) is the CPCS SDU of the AAL5 Common Part as described in Chapter 8. Figure 7.15b illustrates the FR/ATM interworking protocol of an ATM end system. This function is identical to the right-hand side of the FR/ATM IWF. The ATM end system must support Q.933 Annex A frame relay status signaling.

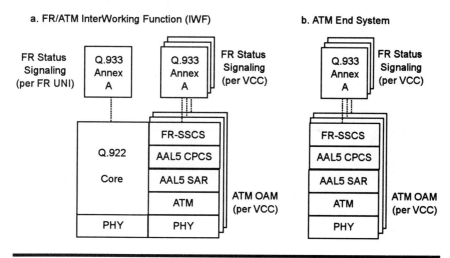

Figure 17.15 FR to ATM Interworking Protocol Stack

Figure 17.16 illustrates the FR-SSCS PDU format (essentially the FR frame summarized in Chapter 5) with inserted zeroes and the trailing CRC both removed. Frame relay supports either 2- , 3- , or 4-octet addressing. The origins of the DXI frame in FR are evident when comparing Figure 17.16 for the FR-SSCS PDU to Figure 17.9 for the DXI header.

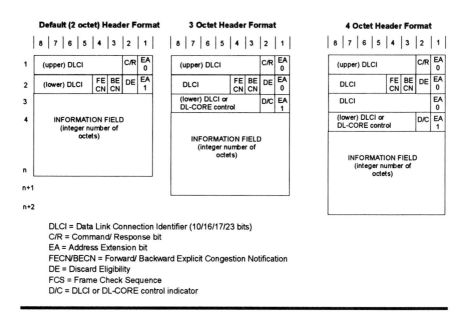

Figure 17.16 FR-SSCS PDU Formats

The FR-SSCS supports multiplexing through the use of the DLCI field, with the ATM layer supporting connection multiplexing using the VPI/VCI. There are two methods of multiplexing FR connections over ATM: many-to-one and one-to-one. Many-to-one multiplexing maps many FR logical connections identified by the Data Link Connection Identifiers (DLCIs) over a single ATM Virtual Channel Connection (VCC). One-to-one multiplexing maps each FR logical connection identified by DLCI to a single ATM VCC via VPI/VCIs at the ATM layer.

17.4.3 Congestion Control and Traffic Parameter Mapping

Control and addressing functions and data are either mapped or encapsulated by the interworking function. For example, in the FR to ATM direction the DLCI, DE, FECN, and BECN fields are encapsulated in the FR-SSCS PDU. The FR DE bit is also mapped to the ATM CLP bit, and the FR FECN bit is mapped to the ATM EFCI bit. FR BECN is encapsulated in the FR-SSCS. BECN may also be mapped if the EFCI was set in the last cell of a frame reassembled from the ATM-to-frame relay direction. The frame relay FCS is mapped to (that is, replaced by) the AAL5 CRC function.

In the ATM-to-frame relay direction the CLP bit may be logically ORed with the DE bit as a configuration option on a per-DLCI basis. The AAL5 CRC is checked, and the FR FCS is recomputed for delivery to a frame relay UNI.   The FR-SSCS PDU carries the encapsulated FECN, BECN, and DE bits intact.

The FR traffic parameters include access line rate (Ra), committed burst size (Bc), excess burst size (Be), and measurement interval, which define a Committed Information Rate (CIR) and an Excess Information Rate (EIR) in T1.617. Appendix A of the ATM Forum B-ICI specification maps these FR traffic parameters to the ATM traffic parameter in terms of Peak Cell Rate (PCR), Sustainable Cell Rate (SCR), and Maximum Burst Size (MBS).

The ATM Quality of Service (QoS) class for the VCC must also be selected.   Usually QoS class 3 as defined by the ATM Forum (see Chapter 12) for connection-oriented data would be used.

17.4.4 Frame Relay and ATM Service Interworking Considerations

The multiprotocol encapsulation for frame relay of RFC 1490 differs from the multiprotocol encapsulation being defined for ATM. RFC 1483 specifies that for direct interworking with a frame relay end system support for RFC 1490 NLPID encapsulation should be provided. This means that an ATM end or intermediate system must also support NLPID if interworking with FR is required.

The issue of status signaling interworking with ATM is currently unresolved.   The ATM ILMI and ATM Layer Operations Administration and Maintenance (OAM) functions do not support all of the frame relay status signaling functions. The ATM Forum plans to address this issue.

17.5 SMDS ACCESS INTERWORKING OVER ATM

The ATM Forum B-ICI document defines how SMDS can be transported between carriers.   The ATM Forum, SMDS Interest Group (SIG), and European SIG (E-SIG) are jointly defining how SMDS can be accessed across an ATM UNI [2]. Access configuration, functions performed, and expected protocol specification based upon the B-ICI specification are provided in this section.

Figure 17.17 depicts the access configuration and logical placement of function for accessing SMDS features over an ATM User-Network Interface (UNI). An ATM end system accessing SMDS over ATM must format an AAL3/4 CPCS PDU containing the SMDS Interface Protocol Layer 3 (SIP_L3) PDU, as shown on the left-hand side of the figure. The ATM network performs Usage Parameter Control (UPC) to emulate the SMDS access class as will be described later. The ATM network relays the cells to an SMDS InterWorking Function (IWF), which may be implemented in a centralized, regionalized, or distributed manner. The SMDS/ATM IWF converts the AAL stream into the SMDS protocol stack and passes this to an SMDS network which implements the SMDS service features, including access class enforcement. The SMDS network can interface to a subscriber using the SMDS Subscriber Network Interface (SNI) [3] or the SMDS Data eXchange Interface (DXI) [4] as shown in the figure.

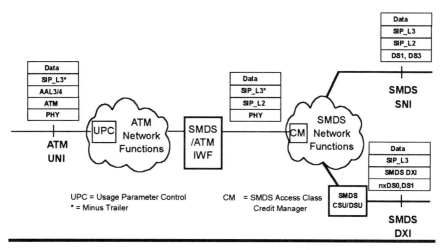

Figure 17.17 Logical Configuration for SMDS Access over ATM

Several of the functions which the InterWorking Function (IWF) performs include:

☞ Conversion between SIP_L3 and the AAL3/4 CPCS
☞ Conversion between 802.6 Layer 2 PDUs (slots) and ATM AAL3/4 SAR
☞ Multiplexing of 802.6 Multiplex IDs (MIDs) into a single ATM VCC

Figure 17.18 illustrates how the SMDS Interface Protocol (SIP) is carried over ATM using the InterWorking Function (IWF). The 4-

octet header and 4-octet trailer of the SMDS SIP_L3 PDU are removed prior to being inserted in the AAL3/4 CPCS PDU (described in Chapter 8). Padding of the truncated SIP_L3 PDU must be performed to ensure that 4-octet alignment is achieved in the AAL3/4 CPCS PDU. The operation of AAL3/4 Segmentation And Reassembly (SAR) using Beginning, Continuation, and End Of Message (BOM, COM, and EOM) mapping to ATM cells is also illustrated in the figure.

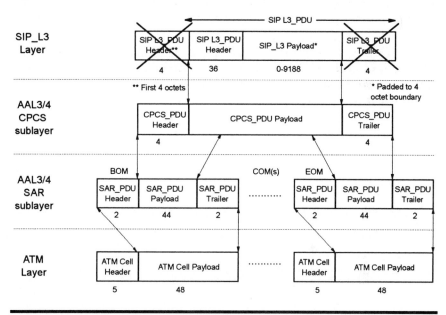

Figure 17.18 SMDS over ATM Protocol Interworking Detail

The mapping between the SMDS access class credit manager parameters and the ATM traffic parameters is a key function. This may be performed as described in section B.2 of Appendix B in the ATM Forum version 3.0 UNI specification using the peak and sustainable cell rates, or using the peak rate only as described in appendix A of the SIG/E-SIG/ATM Forum specification [2].

17.6 MULTISERVICE BROADBAND INTERCARRIER INTERFACE (B-ICI)

The ATM Forum has defined a multiservice Broadband Inter-Carrier Interface (B-ICI) specification. The B-ICI facilitates ATM service connections between two network providers or carriers. The physical and ATM layer is the Network Node Interface (NNI) as defined in Chapter 8.

A number of services may be defined over this interface. These services can be provided between the Local Exchange Carrier (LEC), IntereXchange Carrier (IXC), Independent Local Exchange Carrier (ILEC), and other public ATM network providers. Depending upon regulatory and business arrangements, any one of these carriers may assume any of the roles illustrated in Figure 17.19.

The B-ICI version 1.0 from the ATM Forum currently defines the following services interconnected between carrier networks as illustrated in Figure 17.19:

> ⊳ Cell Relay Service (CRS)
> ⊳ Circuit Emulation Service (CES)
> ⊳ Switched Multimegabit Data Service (SMDS)
> ⊳ Frame Relay Service (FRS).

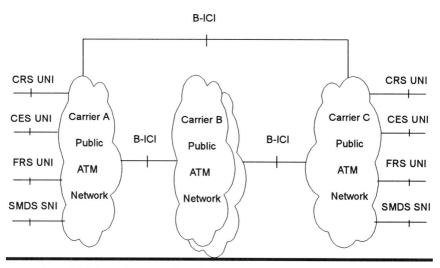

Figure 17.19 Multiservice Broadband InterCarrier Interface (B-ICI)

The current B-ICI specification defines support for both Permanent Virtual Connection (PVC) and SMDS services. The B-ICI specifications for support of ATM Switched Virtual Connections (SVCs) are under development. The areas of the B-ISDN protocol cube (introduced in Chapter 6) that the B-ICI covers are indicated by the shaded user plane and management plane areas in Figure 17.20. The highlights of the B-ICI specification are presented in the following paragraphs.

Physical layer specifications at the B-ICI are the same as defined in Chapter 8 for DS3, STS-3c, and STS12c. Additional operational functions required between carriers to manage the transmission paths are defined, usually requiring some degree of bilateral agreement between carriers (i.e., both carriers must agree).

The ATM layer specification is the ATM NNI as defined in Chapter 8. Only Peak Cell Rate (PCR) traffic management is required at the B-ICI, with other traffic control subject to bilateral agreement. The traffic contract as defined in Chapter 12 is now between two carriers instead of between a user and a network. The Quality of Service (QoS) is now a commitment from one carrier to another, and the B-ICI defines reference traffic loads and assumptions as guidance to carriers in resolving this part of the traffic contract. ATM layer management is covered in great detail in the B-ICI specification.

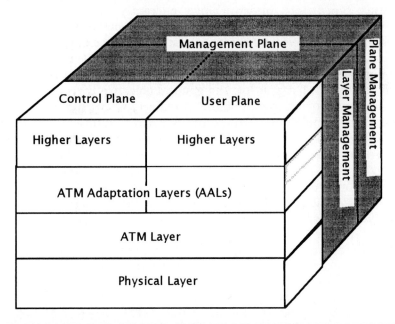

Figure 17.20 ATM Forum B-ICI Specification Coverage

The B-ICI specification not only defines what is required at the B-ICI interface, but also defines the higher layer functions required in the user plane for InterWorking Functions (IWF) required to support frame relay, SMDS, and circuit emulation, since these specifications were not complete. The Frame Relay Forum's FR/ATM network interworking specification is aligned with the B-ICI specification. Several SMDS specific functions in support of intercarrier SMDS are defined in the ATM B-ICI specification, including carrier selection, group addressing, and routing.

The general principles for provisioning, network management, and accounting are stated in the B-ICI specification. Much more work is required to translate these into more detailed guidelines, interfaces, and protocols.

17.7 REVIEW

This chapter introduced the concept of protocol interworking based on an ATM access and network infrastructure. There are two generic types of interworking currently defined: network interworking and service interworking. Network interworking has two scenarios: the first where a specific protocol is carried *over* ATM, and a second where the specific protocol is interworked *with* an ATM-based end system. Service interworking is still in the process of definition. But it still has the general objective that an ATM end system need not be aware of the specific type of protocol that the other user is employing. This is unlike network interworking, where the ATM end system must implement some aspects of the specific protocol as they are mapped and encapsulated to ATM. We then looked at the simple frame-based network interworking capability defined for early support of ATM using existing end system hardware — the ATM Data eXchange Interface (DXI). An example of network interworking over ATM involving the IETF RFC 1483 specification for multiple protocol encapsulation over ATM VCCs was provided. Next, frame relay interworking with ATM was covered, and some service interworking issues between the IETF RFC 1490 for multiple protocol encapsulation for frame relay and RFC 1483 were introduced. We also covered the topic of access to SMDS services over ATM. Finally, the ATM Forum Broadband InterCarrier Interface (B-ICI) specification was introduced, which defined further higher layer details and management principles for how Frame Relay Service (FRS), Circuit Emulation Service (CES),

SMDS, and Cell Relay Service (CRS) can be provided across multiple-carrier ATM networks.

17.8 REFERENCES

[1]   Frame   Relay   Forum,   "Frame   Relay/ATM   Interworking Implementation Agreement - Draft," January 19, 1994.

[2]   SIG, E-SIG, "Protocol Interface Specification for Implementation of SMDS over an ATM-based Public UNI Version 0.08," January 28, 1994.

[3]   Bellcore, "TR-TSV-000772 Issue 1, Generic System Requirements in Support of Switched Multi-Megabit Data Service," May 1991.

[4]   SIG, "SIG-TS-001/1991, SMDS Data Exchange Interface Protocol," Revision 3.2, October 1991.

Public ATM Network Services and Selection

There are many factors to consider when choosing a public network ATM-based service: the type and speed of access, the CPE support, the Quality of Service (QoS) classes offered, interworking support, network management support, and billing options. Chapter 8 defined ATM access speeds, while Chapters 10 and 11 detailed ATM hardware and software available at the time of publication. Chapter 12 defined Quality of Service (QoS) classes and the traffic contract. The last chapter covered standards for interworking with ATM DXI, frame relay, SMDS, and IP — each of which may be available in public network ATM services. This chapter looks at the decision factors for choosing an ATM-based service. First, public ATM network architectures are examined, and then public network ATM services are reviewed. This is followed by discussions of pricing and billing for public ATM network services, and the decision criteria for choice of a public ATM network provider and the questions to ask when making that decision.

18.1 PUBLIC ATM NETWORK ARCHITECTURES

As ATM technology proliferates, it provides an evolutionary path from the traditional networks based on a TDM architecture to an ATM-based architecture. An architecture based upon ATM technology utilizes ATM switches in both the Central Office (CO) and on the Customer Premises Equipment (CPE). ATM can also be employed in cross-connects, routers, gateways, workstations, and InterWorking Functions (IWFs) as shown in Figure 18.1. Also, most public ATM-based networks employ a SONET or SDH backbone. Figure 18.1 shows an example network where users either interface to the ATM network directly via an ATM UNI, an ATM DXI interface, or a frame relay interface, achieve access to an SMDS connectionless server, or interface through multiple other protocols through an MultiProtocol (MP) interworking function.

18.2 PUBLIC ATM NETWORK SERVICES

This section first categorizes possible ATM services, then summarizes the suite of possible ATM services, and follows with a description of services that have been announced to date.

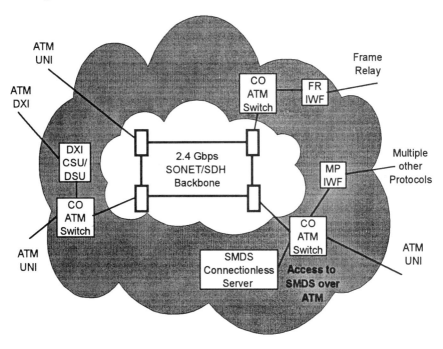

Figure 18.1 ATM-Based Network Services Using a SONET Backbone

18.2.1 B-ISDN Service Classification

ITU-T Recommendation I.211 defines B-ISDN services and the network capabilities required to support them. There are two major classes of service: interactive and distribution.

Interactive (bidirectional transmission) services are defined by three classes of service: conversational (e.g., data, sound, video and sound imaging), messaging (e.g., document and video multimedia electronic mail), and retrieval (e.g., text, data, sound, still-images and full-motion pictures and video).

Distribution services are divided into services with individual presentation control [e.g., data, text, graphics, still images, video with sound such as multimedia (TV) distribution] and without user individual presentation control (e.g., text, graphics, sound, still images such as remote news retrieval and advertising). The services *not* requiring user control are typically broadcast from a central point much like cable TV. The services requiring user control provide users with the capability to interact with the broadcast service, for example, touching the TV screen to select a video-marketed product. The standard also provides an in-depth look at video encoding techniques and methods for B-ISDN usage. Table 18.1 summarizes the B-ISDN classes of service.

Table 18.1 B-ISDN Classes Of Service

Service	Classes of Service
Interactive	Conversational Services
Interactive	Messaging Services
Interactive	Retrieval Services
Distribution	Without user individual presentation control
Distribution	With user individual presentation control

Recommendation I.211 goes on to further categorize other attributes of B-ISDN services, such as:

☺   Quality of Service (QoS)
☺   Continuous or variable bit rate
☺   Timing and synchronization
☺   Connection-oriented versus connectionless
☺   Continuous or variable video coding rates
☺   Signaling differences for interactive and distributive services

18.2.2 ATM Service Suite

The suite of possible ATM services includes the following:

- ❈ ATM Cell Relay Service (CRS)
- ❈ Virtual Path or Virtual Channel Connections (VPC or VCC)
- ❈ Point-to-point or point-to-multipoint connections
- ❈ Permanent Virtual Connection (PVC)
- ❈ Switched Virtual Connection (SVC)
- ❈ Bandwidth reservation
- ❈ Frame relay interworking with ATM
- ❈ SMDS access over ATM
- ❈ Multiprotocol (e.g., IP) interworking over ATM
- ❈ X.25 interworking over ATM

18.2.3 Public ATM Services and Providers

This section summarizes the state of public ATM service offerings announced by major providers at time of publication in the form of two tables. Table 18.2 summarizes commercial ATM services that have been announced by Local Exchange Carriers (LECs), Competitive Access Providers (CAPs), and Independent Service Providers (ISPs) as of the second quarter of 1994 [1]. Table 18.3 lists the commercial ATM services that are available from the IntereXchange Carriers (IXCs), value-added service providers, and international Postal Telephone and Telegraph (PTT) operators as of the second quarter 1994 [1].

PacBell was the first LEC to announce ATM service, a DS3 ATM service through their Bay Area Gigabit Network (BAGNet), and plans to offer ATM tariffed service. US West offers COMPASS, an ATM trial network linking computers, suppliers, and Minnesota-based customers.

BT has recently begun ATM trials. One example is the design of SuperJANET, a network designed to connect universities via ATM and to better understand user requirements and uses of ATM. The initial backbone is comprised of DS3s, with plans to later upgrade to OC-3 with E1 and E3 access speeds. The stated purpose of their network is to understand user requirements and uses of ATM. MCI has taken a similar strategy in understanding user requirements before offering service, and will likely be providing the backbone for the very high-speed Backbone Network Service (vBNS). Sprint has announced CBR and VBR service at DS3 rates, along with PVC Circuit Emulation. WilTel has announced a different strategy, offering Channel Networking Service (CNS), or channel extension at DS1 and DS3 rates. One example of European ATM service is the Societa Italiana per l'Esercizio (SIP) Italian public network operator's choice of the Ericsson FATME as their ATM service node. GTE is also pursuing ATM Cerritos trials in California. Time Warner plans to use the AT&T GCNS-2000 as video servers for "ATM TV." This venture is planned for local service only.

Table 18.2   ATM Services Provided by LECs, CAPs, and ISPs

LEC, CAP, or ISP	Service	Switch Vendor(s)
Ameritech	Planned	AT&T
Bell Atlantic	Trials	Siemens
BellSouth	PVC now, SVC(95)	Fujitsu FETEX 150
GTE	Trials	SPANet, NEC NEAX 61E
MFS Datanet	Service	Newbridge 36150
NYNEX	Commercial Service	Fujitsu FETEX 150
PacBell	DS3 service 1994	Newbridge 36150
SWBell	Testing	Newbridge
US West	Trials, Commercial 2H95	Newbridge, Fujitsu FETEX 150, Siemens

Table 18.3 ATM Services Provided by IXCs and PTTs

IXC Provider	Service	Switch Vendor
AT&T	PVC Circuit Emulation 3Q94	GCNS-2000
BT	Trials	GCNS-2000
MCI	1994/95	Unknown
MFS DataNet	Service	Newbridge
Sprint	Today	TRW BAS 2010
Telecom Finland	34 Mbps service	Netcomm DV2
Time Warner	Video-Servers	AT&T GCNS-2000
WilTel	CNS today	NEC NEAX61

International regional broadband trials include BT, Telia, Deutsche Telekom, France Telecom, Stet/Iritel of Italy, Telefonica, Belgacom, Potugal's TLP, Telecom Finland, Swiss PTT Telecom, Norwegian Telecom, PTT Telecom Netherlands, Tele Danmark, Telecom Eireann, Telecom Portugal, the Austrian PTT, and the National Association of Private Telephone Companies in Finland. See References 2, 3, 4, 5, 6, and 7 for more details.

18.3   PUBLIC ATM SERVICES PRICING AND BILLING

Services such as frame relay, SMDS, and ATM can either be classified as enhanced services, and thus would not be under tariff as is typical with the IXCs, or offered under a tariff, as is the case with the LECs. The pricing and billing policies of these enhanced data services are based on many factors, including:

> Port access speed
> Total virtual connection bandwidth per port
> Traffic contract parameters
   - Committed Information Rate (CIR) for frame relay
   - Access class for SMDS
   - Peak and sustainable cell rates for ATM
> Quality of Service (QoS) class
> Local access line charges based on speed and distance
> Fixed-rate or usage-based billing option
> Different rates based upon priority
   Discard Eligible (DE) for frame relay
   Cell Loss Priority (CLP) for ATM
> Bulk discounts
> Time-of-day discounts
> Points-of-Presence (PoPs) available for customer access
> Access circuit charges based upon distance to serving PoP

Usage-based billing rates may be measured in cells delivered over a time period rounded off to some accuracy. For example, the charge may be based on millions of cells delivered per month.

Fixed charges include a port charge, access charges (for the circuit from the customer premises to the service provider switch), installation charges, and other fees. A fixed-rate billing option will also include a charge based upon the virtual connection bandwidth, priority, and QoS class.

In some cases, the service provider offers an option for fixed or usage pricing. Usually, a better understanding of the traffic patterns allows a user to take advantage of usage-based pricing. This is due to the fact that many service providers offer usage-based rates which favor the intelligent user.

Private line costs often limit users to low-speed dedicated circuits for LAN-to-LAN traffic, creating a wide area network bottleneck. If these circuits are analyzed, it is often seen that they are flooded during the busy hour and not utilized much during off-peak, non-business hours. These applications are ideal candidates for virtual networking as identified by the example in Chapter 16. For many users, the initial deployment of any virtual data service should be limited, allowing the users to learn their traffic patterns. Once the traffic is understood, the decision between usage-based pricing, flat-rate pricing, or a hybrid of both can be intelligently made.

Initial users should look for price caps on a usage-based service so that it can be no greater than a certain percentage above the comparable flat-rate fee or leased line service. Pricing should not penalize users who are just trying out the service and who are not familiar with their traffic patterns. There will still be access charges, but a virtual data network using fewer dedicated access circuits in the local loop will decrease bottom-line access costs. All in all, frame relay, SMDS, and ATM services must be priced to make the risks of these new services minimal and the rewards tempting and cost-effective.

Some providers, such as MCI, have historically offered both usage-based and flat-rate pricing structures across all data services. For frame relay service, MCI has supported unidirectional assignment of virtual circuit configuration parameters for greater traffic billing flexibility and planning, strong customer network management enterprise systems, and the capability for sustained bursting over assigned committed transfer rates (e.g., CIR). MCI has also offered substantial discounts for lower priority traffic, for example, frame relay with zero CIR, carried as Discard Eligible (DE).

Network management reports, often SNMP standards-based, are also an important part of any public data service. In fact, these daily, weekly, or monthly reports should closely track the usage-based billing invoices to provide effective user control of costs.

18.4   CHOICE OF A PUBLIC ATM SERVICE PROVIDER

Users must decide first by analyzing the benefits of accessing a public ATM network or building a private virtual data network. If the decision is made for a public network, the user must decide between IXC, independent service providers, and LEC services. This decision will be based on both intraLATA and interLATA access and pricing. Pricing of virtual data services was covered in the previous section.

Virtual data service offerings have the following advantages over private line networks:

- Little capital investment is required
- Large backbone capacities
- Extensive Points of Presence (PoPs)
- Choice of interface speeds
- Support for multiple services over a single access circuit
- Sophisticated network management support
- Portfolio of services (X.25, FR, SMDS, IP, ATM)
- Smooth migration between services
- Usage-based and flat-rate billing and tariff structures
- Carrier-provided CPE
- Outsourcing packages through joint marketing agreements
- Efficient control of network resources
- Skilled service and equipment support structures
- Inherent reliability, redundancy, and survivability of the redundant, shared backbone
- Public, universal access
- Internetwork connectivity

Another very important decision factor should be the architecture that the service provider employs for the access and backbone network. Whereas the tariff and pricing structure is key in the initial selection, the service provider's architecture is the long-term insurance that costs will remain effective, that service quality will be delivered, and that new features will be offered in a timely manner. With the expansions of international business, as well as the high cost of international private line circuits, many users see virtual data services such as ATM reducing the cost of international data communications. Earlier chapters described additional benefits of public networks as compared to private networks, such as value-added services, economy of scale, network redundancy, and network management.

18.5 QUESTIONS TO ASK ATM SERVICE PROVIDERS

The following are some questions to ask when considering using an ATM-based service (or any data communications public service, for that matter).

18.5.1 Access

What are the access classes and speeds offered? Do they range from nxDS0 through STS-Nc speeds? Do they also offer other public data services, such as SMDS and frame relay, with interworking? What is the access architecture from the PoP to the serving switch?

18.5.2 Network Architecture

What type of equipment does the network use? What is the network architecture? How many network switches does the provider offer, how long is your backhaul to each switch, and what is their end-to-end network delay worse case, and in turn your latency across their network during normal and peak traffic conditions? This figure is heavily dependent on their delay through each switch, which in turn is influenced by the buffer size in each switch. Also ask about their buffering methods and prioritization. What is their network switch redundancy and availability, and are these numbers guaranteed to the user? What is their network Mean Time To Repair (MTTR)?

18.5.3 Features and Functions

Does the service provider offer the features and functions you require? Do they offer PVC service, and when will SVC service be offered? Do they offer truly enhanced services, or just a Cell Relay Service (CRS)? What AALs (1, 2, 3/4, 5) do they support? Is frame relay network interworking supported? Is access to SMDS services over ATM supported? Is access to multiple protocol services over ATM supported? Is point-to-multipoint service supported?

What classes of service are offered? What Quality of Service is guaranteed? What traffic parameters are supported? Peak Cell Rate? Sustainable Cell Rate? Maximum Burst Size? Is there a discount for low-priority service? How are these levels of service monitored and what is the guarantee in case of failure to meet them?

18.5.4 Premises Hardware Support

What Customer Premises Equipment (CPE) is certified to work with the network? Does the provider have a CPE program? What are the CPE lease versus buy tradeoffs? Can the provider always offer the newest revision level of software for their CPE?

18.5.5 Operations and Support

Does the provider have flexible leasing and rental plans and purchase plans? Is the equipment upgradable to future software and hardware releases, or to new technologies? Are they involved in the standards setting bodies? What initial training and ongoing support do they offer?

18.5.6 Pricing and Billing

What are their new order intervals? What are their pricing and billing options? Do they support most or all of that detailed earlier in this chapter?

18.5.7 Network Management

What protocol does the service provider's network management system use? Is it SNMP-compliant, CMIP-compliant, or proprietary? How does their network management system work? How will it work with your network management systems? What is included with the basic service? What has additional fees? What forms of traffic and congestion control can you monitor, or better yet change in near-real time?

The questions could and on, but you get the idea. These are the key areas. You should make your own detailed list, prioritize it, and use it to select the service that best meets your needs.

18.6   REVIEW

This chapter presented the major components of a public ATM network architecture. A standards-based classification of services was then presented, including the class of services offered and characteristics of the service. Next, the chapter reviewed the current state of ATM service offerings from public service providers. Next, a review of the pricing and billing methods in public ATM services was covered. Some considerations in choosing a virtual private network service versus a private line network were presented. Finally, we closed by presenting some criteria for choosing an ATM service provider, including a few key questions to ask before making a final decision.

18.7   REFERENCES

[1]   L. Gasman, "Broadband Networks & Applications Newsletter," Communications Industry Researchers, 1993 through 1994.
[2]   *Telephony*, August 9, 1993.
[3]   *Network World*, 1993.
[4]   *Data Communications*, 1993, 1994.
[5]   *Electronic Business Buyer*, Vol. 19, Issue 9, September 1993.
[6]   *Computer World 1993*, 1994.
[7]   Broadband Networks & Applications, 1994.

7

Operations, Network Management, and ATM Layer Management

Part 7 provides the reader with an overview of operations, network management, and ATM layer management. The philosophy of Operations, Administration, Maintenance, and Provisioning (OAM&P) is first defined to set the stage. A presentation of network management architectures follows as the context for the definition of Management Information Bases (MIBs), ATM layer management, and performance measurement. ATM layer management covers definitions of ATM layer Operations, Administration, and Maintenance (OAM) cell flows and formats. Chapter 20 covers fault management, which is the basic determination of whether the ATM service is operating correctly. Chapter 21 describes the use of performance measurement procedures which ensure that the Quality of Service (QoS) objectives are being met.

19

OAM&P, Network Management Architectures, and Management Information Bases (MIBs)

This chapter first covers the topic of Operations, Administration, Maintenance, and Provisioning (OAM&P) philosophy. We identify generic functions that apply to almost any type of data communication network, and highlight challenges that ATM offers. Next coverage moves to network management architectures by the OSI, ITU, IETF, and ATM Forum. The differences in approach, scope, and use of protocols are covered. A summary of the ATM Forum Interim Local Management Interface (ILMI) and the draft IETF ATM MIB (called the AToMMIB) follows. These summaries give a good flavor for what network management data will be available for ATM interfaces, end systems, switches, and networks.

19.1 OAM&P PHILOSOPHY

A brief definition of each element of Operations, Administration, Maintenance, and Provisioning (OAM&P) and how they interrelate is described below and depicted in the flow diagram of Figure 19.1.

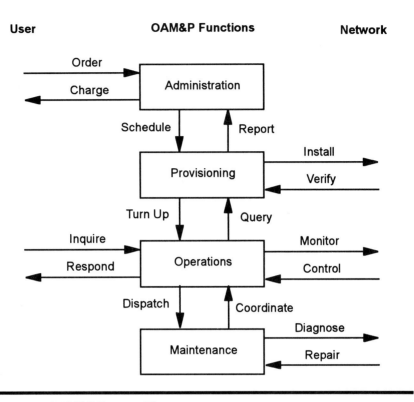

Figure 19.1 OAM&P Process Flow

* Operations involves the day-to-day, and often minute-to-minute, care and feeding of the data network in order to ensure that it is fulfilling its designed purpose.

* Administration involves the set of activities involved with designing the network, processing orders, assigning addresses, tracking usage, and accounting.

* Maintenance involves the inevitable circumstances that arise when everything does not work as planned, or it is necessary to diagnose what went wrong and repair it.

* Provisioning involves installing equipment, setting parameters, verifying that the service is operational, and also deinstallation.

19.1.1 Operations

Some key functions involved in operating an ATM network are described in this section. Monitoring the network involves watching for faults and invoking corrective commands and/or maintenance actions to repair them. It also involves comparing measured performance against objectives and taking corrective action and/or invoking maintenance. Invoking corrective actions involves operators issuing controls to correct a fault or performance problem, or resolving a customer complaint. A key operational function involves assisting users to resolve troubles and effectively utilizing the ATM network capabilities. Operations coordinates actions between administration, maintenance, and provisioning throughout all phases of the ATM connection's life.

19.1.2 Administration

Some key functions involved in administering an ATM network are described in this section. First, a network must be designed, either as a private, public, or hybrid. Once designed, the elements must then be ordered, along with scheduling installation and associated support. An administrative plan for staging the provisioning, operations, and maintenance activities must be developed. This often involves automated system support for order entry, order processing, work order management, and trouble ticketing. Orders must be accepted from users and the provisioning process initiated. Address assignments are made where needed. Once the service is installed, usage data must be collected for traffic analysis and possible accounting use. Based upon forecasts, business requirements, and traffic analysis, changes to the network must be periodically updated. Network planning may also make use of automated tools.

19.1.3 Maintenance

Some key functions involved in maintaining an ATM network are described in this section. Once operations has identified a problem, it must be isolated and diagnosed to determine the cause(s). Fixes to identified problems must be applied in a manner coordinated by operations. This often involves dispatching, parts delivery, or coordination with service suppliers. An important activity is to perform routine maintenance so that faults and performance degradations are less likely to occur.

19.1.4 Provisioning

Some key functions involved in provisioning an ATM network are described in this section. New or upgraded equipment and software must be installed. Hardware requires on-site support, while some software upgrades may be done remotely. Circuits or access line software and hardware installation are also part of provisioning. A key part of processing orders is the setup of ATM-related service parameters. This may be done manually, but the trend should be toward automation. Once the hardware and software parameters are in place, the final step of the provisioning process is to ensure that the service performs according to the objectives prior to releasing service to the end user.

19.1.5 Unique Issues Created by ATM

Standards are still being developed in many of these areas, and in a sense they are never done because of the rapid pace of technological change. Standardization of network management is usually considered later in the technology life cycle because only after you have built the network, determined what can go wrong, and discovered what is needed to make it work can you finalize how to operate, administer, maintain, and provision it. Good planning can provide these OAM&P functions in a much more productive manner soon after the introduction of technology; however, there is no substitute for experience.

In several senses the OAM&P of ATM is similar to that of other private data networking and public data service capabilities. Many of the management functions required for management of an ATM network are identical to a data- or circuit-based network. Network management is complicated by the fact that ATM/B-ISDN is targeted to support older, legacy systems as well as to support significant new functions and applications. Also, ATM changes the underlying, lowest layers of data communications through its new paradigm, which will make many tried and true troubleshooting procedures developed for the circuit-oriented paradigm obsolete, or at least require substantial revision. In short, there is no magic in ATM to solve the problem of network management.

With ATM the multiplicative factor of logical, or virtual, channels that can be defined on each physical interface is orders of magnitude greater than that of many current data services. ATM simultaneously includes all of the complexities involved in LAN, MAN, and WAN services along with currently undefined future extendibility. Also, ATM adds some new types of capabilities, such as point-to-multipoint connections. Furthermore, multiservice interworking aspects may require backward compatibility with all of the existing data communications, and should provide at least the same level of network management, automated provisioning, and billing as done for existing LAN, MAN, and WAN services.

There may be a great advantage for ATM in supporting all of these service types [1]. If done well, ATM-based Network Management Systems (NMS) may provide a seamless level of network management — the goal in the interconnection of LAN, MAN, and WAN user interfaces. Standard Network Management interfaces (e.g., SNMP or CMIP) to Network Elements (NEs) will allow a great deal of visibility and control upon which an NMS could be developed.

The new features of ATM switching and multiplex equipment will require either enhancements to existing network management systems and/or new systems. The additional protocol parameters of AALs and higher layers will also require management. Monitoring every virtual connection using either OAM cells and/or retrieved counter measurements will result in tremendous volumes of network management data. Sampled or hierarchical data reduction and analysis must be done to reduce this data volume.

19.1.6 Centralized versus Distributed Network Management?

When designing your ATM network, it is important to consider how the network management systems impact operational philosophy. A key decision is whether to adopt a centralized or distributed Network Management System (NMS) architecture for managing a network of ATM Network Elements (NEs), as depicted in Figure 19.2.

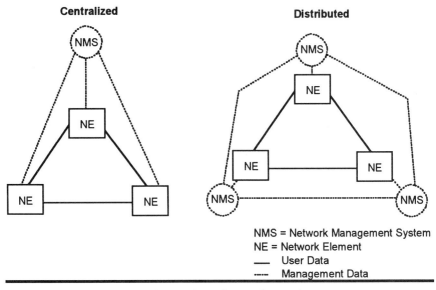

Figure 19.2 Centralized and Distributed OAM&P Architectures

Some will opt for a centralized approach with the expertise concentrated at one, or at most a few, locations with remote site support for only the basic physical actions, such as installing the equipment, making physical connections to interfaces, and replacing cards. In this approach the software updates, configurations, and troubleshooting can be done by experts at the central site. The centralized approach requires that the requisite network management functions defined in the previous section are well developed, highly available, and effective. Functions that need not be performed in real time are often best done in a centralized manner.

Others may only have a few sites, and may be on the leading edge of technology. Hence they may want to have expertise at every site. This approach may be required if the network management system is not sophisticated or the equipment has a number of actions that can only be done at the site. Some lower level, well-defined, automated functions are best performed in a distributed manner. Functions that need to be performed in real-time are often best done in a distributed manner.

There is a performance tradeoff between the centralized and distributed architectures. Transferring large volumes of management data to a central site often requires a large centralized processor. It is all too easy to get in the situation of having too much data and not enough information. On the other hand, defining the interfaces and protocols for a distributed system is complex. One study [2] compares centralized versus distributed Connection Admission Control (CAC), as defined in Chapter 13. CAC depends upon accurate, timely information about the connections in progress in order to admit, or block, connection attempts. If the information is not current, as it would be in a centralized system, then the throughput carried by the network is reduced over that of a totally distributed system.

19.2   NETWORK MANAGEMENT ARCHITECTURE

This section summarizes the OSI, ITU, ATM Forum, and IETF network management architectures.

19.2.1 OSI Network Management Functional Model

OSI has defined the following five generic functional areas for network management, commonly abbreivated as "FCAPS" [6]:

♦   Fault Management
♦   Performance Management
♦   Configuration Management
♦   Accounting Management
♦   Security Management

The current state of ATM OAM-based management covers fault management and performance management only. Some aspects of configuration management, fault management and performance management are covered by the ILMI and AToMMIB. There remains a great deal of work to be done, particularly in the configuration and accounting areas. The work in the performance measurement area is also incomplete. Work in fault management is the most mature, and meets the minimum requirements for initial ATM deployment.

19.2.2 ITU Telecommunications Management Network (TMN)

The ITU-T and ANSI have been utilizing OSI standardized Common Management Information Service Elements (CMISE) and the associated Common Management Information Protocol (CMIP) for the Q3 interface in the Telecommunications Management Network (TMN) architecture [3], [7] depicted in Figure 19.3. The other interfaces are also being specified based upon OSI protocols and management principles. This is the standardized vision of interoperable network management of the future.

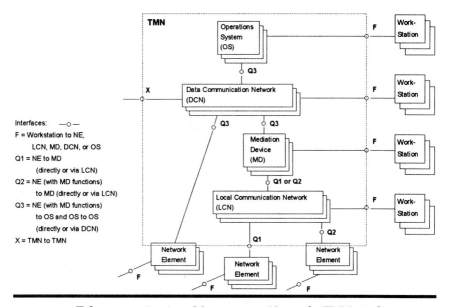

Figure 19.3 Telecommunications Management Network (TMN) Architecture

The software architecture of TMN includes functionally grouped capabilities called *operations systems functions* as follows:

- Business management support of the implementation of policies, strategies, and specific services.

- Service management functions that are necessary to support particular services, such as subscriber administration and accounting management.

- Network management support of configuration, performance management, and maintenance.

- Element management support for management of one or more Network Elements (NEs) certainly concerned with maintenance, but also involved in performance management, configuration, and possibly accounting.

- Network element functions at the individual network element or device level that is the source and sink of all network management observations and actions. This includes traffic control, congestion control, ATM layer management, statistics collection, and other ATM-related functions.

The mapping of these software functions onto the hardware architecture is an implementation decision. For a detailed description of an example implementation in the European RACE ATM technology testbed, see Refs. 4 and 5.

19.2.3 SNMP-Based Network Management Systems (NMS)

The Simple Network Management Protocol (SNMP) defined in IETF standards has five messages types: GET REQUEST (or simply GET), GET NEXT REQUEST (or simply GET NEXT), SET REQUEST (or simply SET), RESPONSE, and TRAP (which is like an alarm). The SET, GET, and GET NEXT messages are all replied to by the RESPONSE message. The TRAP message is very important since it is the notification of an unexpected event, such as a failure or a system restart. SNMP normally operates over the User Datagram Protocol (UDP), which then usually operates over IP in the Internet Protocol (IP) stack, but may operate over some other protocol.

SNMP utilizes a subset of Abstract Syntax Notation 1 (ASN.1) to define a Management Information Base (MIB) as a data structure that can be referenced in SNMP messages. The MIB defines objects in terms of primitives such as strings, integers, and bit maps, and allows a simple form of indexing. Each object has a name, a syntax, and an encoding. The MIB variables have a textual Object IDentifier (OID) which is commonly used to refer to the objects. The MIB objects are defined as a tree structure that allows organizational ownership of subtrees to be defined. The branches of the tree are identified by a dotted decimal notation. For example, the prefix of the subtree registered to the ATM Forum is 1.3.6.1.4.1.353. Each of the other branches is identified by the decimal number assigned to the OID as defined in the ATM Forum UNI specification version 3.0.

The ATM Forum ILMI design provides essential interface management functions to early users of ATM until these functions are standardized in ATM OAM cells. Figure 19.4 illustrates the configuration in which the ILMI operates. Each ATM End System (ES), and every network that implements a Private Network UNI or Public Network UNI, has a UNI Management Entity (UME) which is responsible for maintaining the information and responding to SNMP commands received over the ATM UNI. The information in the ILMI MIB can be actually contained on a separate private or public Network Management System (NMS) or may be accessed over another physical interface. NMSs may also be connected to networks or end systems by other network management interfaces. An example of another Management Information Base (MIB) that could be located at any of the agents depicted in Figure 19.4 is the ATM MIB (AToMMIB) that is being defined in the IETF. Of course, other proprietary MIBs are also supported by this architecture. Indeed, SNMP based MIBs (formally known as MIB-II) have achieved such a high degree of interoperability that it is highly likely that if you obtain an electronic copy of a proprietary MIB, it will run on your SNMP network management system!

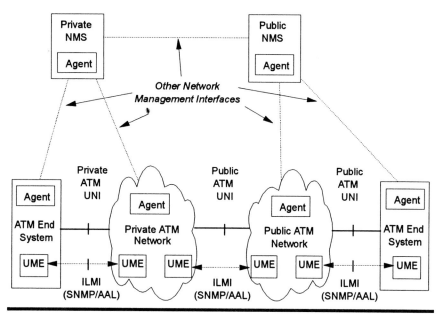

Figure 19.4 ATM Forum ILMI Configuration

A set of upgrades to the Simple Network Management Protocol (SNMP) specifications now define a second, improved version (SNMP version 2) with better security features and improved performance when manipulating large numbers of objects.

The Customer Network Management (CNM) capability is important for carrier-based services. This should include at least physical port status, VPC/VCC status, order parameters, and selected performance metrics. Delivery of detailed performance counts will involve additional complexity and cost. Usage counts from the originating and terminating switch by VPC/VCC may also be used to track performance delivered to customers.

There is general agreement that SNMP and SNMPv2 are best for CPE and private networks, while CMIP/CMISE is appropriate for carrier interconnection. Many early ATM switches implement SNMP since it is simpler, and easier to achieve interoperability with than the CMIP protocols.

It is likely that SNMP-based management will be the de facto standard for the network and higher layers since these exist primarily in end systems and CPE.

19.3 MANAGEMENT INFORMATION BASE (MIB)

In this section we summarize two Management Information Bases (MIBs) as examples of the types of information that can be accessed and manipulated in ATM interfaces, end systems, switches, and networks.

19.3.1 ATM Forum ILMI

The ATM Forum has defined an Interim Local Management Interface (ILMI) based upon Simple Network Management Protocol (SNMP) in advance of a standardized ATM layer management interface. A default value of VPI=0, or VCI=16 for the ILMI, was chosen because CCITT/ITU has reserved VPIs 0 through 15 (i.e., the first 16) for future standardization. Alternatively, another VPI/VCI value can be manually configured identically on each side of the UNI for ILMI use. Use of this method is undesirable since it is not automatic.

The ILMI operates over AAL3/4 or AAL5 as a configuration option, with support for a UDP/IP configurable option. Therefore, in order for UNI Management Entities (UMEs) to interoperate, the AAL (either 3/4 or 5) and the higher layer protocol (either UDP/IP or Null) must be chosen.

Figure 19.5 defines the structure of the ILMI MIB. Each interface has the set of object groups indicated by the branches in the tree. The physical layer group has an additional two groups for common and specific objects (or attributes). The interface index for accessing the data associated with a particular UNI interface is always part of the object index, as shown at the bottom of the figure. The default interface index of zero means that the SNMP command refers to the UNI interface over which the SNMP message was received. Other interfaces can be referenced by supplying non zero interface indices in SNMP requests. As a basic form of security access control, the network implementation should ensure that only interface indices applicable to a particular user UME can be accessed. Some object groups, such as VPC and VCC, have multiple entries per interface and have additional indices, as indicated in the bottom of the figure.

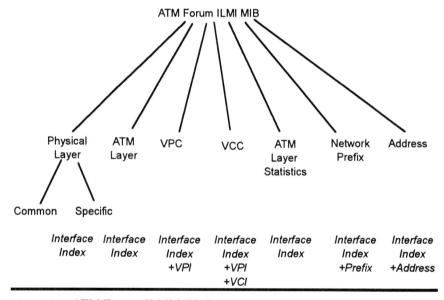

Figure 19.5 ATM Forum ILMI MIB Structure

Two versions of the ILMI MIB have been specified by ATM Forum UNI specification version 2.0 [8] and version 3.0. The version 3.0 MIB is backward compatible with the version 2.0 MIB and is the version described in this section.

The ILMI uses the standard systems group by reference, which supports things such as identification of the system name, and the time that the system has been up. The systems group also provides standard TRAPs, such as when a system is restarted or an interface failure is detected.

Each object in the groups of the ILMI MIB tree is now briefly described. The reader can refer to the ATM Forum UNI specification version 3.0 for further details.

The Physical Layer Group is Required, Read Only, and has an index that is the **Interface Index**, which refers to the following objects:

☞A **Port Type** identification, such as DS3, STS-3, 100 Mbps, or 150 Mbps Fiber is specified from a defined list.

☞A **Media Type** identification, such as coax, single-mode fiber, or multimode fiber is specified from a defined list.

☞The **Operational Status** is specified as a choice from the following: Other, In Service, Out Of Service, or Loopback.

☞**Other Specific Information** is specified to another MIB, such as the standard DS3 MIB.

The ATM Layer Group is Required, Read Only, and has an index that is the **Interface Index**, which refers to the following objects:

☞An integer that specifies the **Maximum Number of VPCs** that can be defined on this interface.

☞An integer that specifies the **Maximum Number of VCCs** that can be defined on this interface.

☞An integer which specifies the **Number of Configured VPCs** that are defined on this interface.

☞An integer that specifies the **Number of Configured VCCs** that are defined on this interface.

☞An integer that defines that **Maximum Number of VPI Bits** that are active on this interface.

☞An integer that defines that **Maximum Number of VCI Bits** that are active on this interface.

The Virtual Path Connection (VPC) Group is Required, Read Only, and has an index that is the **Interface Index** plus the **VPI**, which refers to the following objects:

☞The **VPI** value for this VPC.

☞The **Operational Status** for this VPC, chosen from the following list: Unknown, end2endup, end2enddown, localupend2enddown, and localdown for this VPC.

☞The **Transmit Traffic Descriptor** which contains the traffic parameters from the traffic contract for the transmit direction on this VPC.

☞The **Receive Traffic Descriptor** which contains the traffic parameters from the traffic contract for the receive direction on this VPC.

☞The **Transmit QoS Class** that contains the requested QoS class from the traffic contract for the transmit direction on this VPC.

☞The **Receive QoS Class** that contains the requested QoS class from the traffic contract for the receive direction on this VPC.

The Virtual Channel Connection (VCC) Group is Required, Read Only, and has an index that is the **Interface Index** plus the **VPI** and **VCI**, which refers to the following objects:

☞The **VPI** value for this VCC.

☞The **VCI** value for this VCC.

☞The **Operational Status** for this VPC, chosen from the following list: Unknown, end2endup, end2enddown, localupend2enddown, and localdown for this VCC.

☞The **Transmit Traffic Descriptor** that contains the traffic parameters from the traffic contract for the transmit direction on this VCC.

☞The **Receive Traffic Descriptor** that contains the traffic parameters from the traffic contract for the receive direction on this VCC.

☞The **Transmit QoS Class** that contains the requested QoS class from the traffic contract for the transmit direction on this VCC.

☞The **Receive QoS Class** that contains the requested QoS class from the traffic contract for the receive direction on this VCC.

The ATM Statistics Group is Optional, Read Only, and has an index that is the **Interface Index**, which refers to the following objects:

☞The total number of **Received Cells** on this interface.

☞The total number of **Transmitted Cells** on this interface that contained user data.

☞The total number of **Dropped Received Cells**, which were dropped for one of the following reasons: invalid header, HEC detected header error, or not configured.

The Network Prefix Group is Required at the Private UNI and Optional at the Public UNI, Read/Write, and has an index which is the **Interface Index** plus the **Network Prefix**, which refers to the following objects:

☞A variable length string of between 8 and 13 bytes defining the **Network Prefix** as either the NSAP or E.164 formats in the ATM Forum version 3.0 UNI specification.

☞A **Network Prefix Status** that can have the value of either valid or invalid.

The User Part ATM Address (or ATM Address for short) Group is Required at the Private UNI and Optional at the Public UNI, Read/Write, and has an index that is the **Interface Index** plus the **ATM Address**, which refers to the following objects:

☞An **ATM Address** of 8 bytes defining the low-order bytes of the NSAP signaling address format in the ATM Forum version 3.0 UNI specification.

☞An **ATM Address Status** that can have the value of either valid or invalid.

Figure 19.6 illustrates the SNMP message flows associated with the address registration portion of the ILMI MIB. The Network Prefix Group resides in the ILMI MIB on the User side of the UNI interface, while the Address Group resides on the Network side of the UNI interface. The Address Group is not applicable for E.164 addressing since the 8-byte Network Prefix completely specifies the address. Registration occurs at initialization time (i.e., a cold start trap), or whenever a prefix or address is to be added or deleted. At initialization, the address and prefix tables are initialized to empty. One side registers a prefix or address with the other by first SETting the address value and its status to be valid. The other side sends a RESPONSE with either a *No Error* or *Bad Value* parameter to indicate the success or failure, respectively, of the registration. If a SET message is sent and no RESPONSE is received, then the SET request should be retransmitted. If the SET message is received again, then a RESPONSE indicating no error is returned. If a SET request is received which attempts to change the status of an unregistered prefix or address to invalid, then a RESPONSE with a "NoSuchName" error is returned.

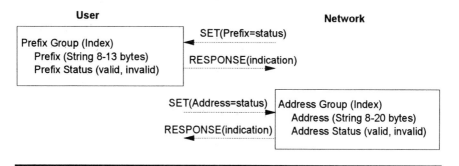

Figure 19.6 Illustration of ILMI MIB Address Registration

19.3.2 IETF AToMMIB

The IETF is in the process of specifying an experimental ATM Management Information Base, called the AToMMIB [9]. The scope of the AToMMIB covers the management of ATM PVC-based interfaces, devices, and services. Managed objects are defined for ATM interfaces, ATM VP/VC virtual links, ATM VP/VC cross-connects, AAL5 entities, and AAL5 connections supported by ATM end systems, ATM switches, and ATM networks.

The AToMMIB uses a grouping structure similar to the ILMI described earlier to collect objects referring to related information and provide indexing. The AToMMIB defines the following groups:

⊙ ATM interface configuration
⊙ ATM interface DS3 PLCP
⊙ ATM interface TC Sublayer
⊙ ATM interface virtual link (VPL/VCL) configuration
⊙ ATM VP/VC cross-connect
⊙ AAL5 connection performance statistics

The ATM interface configuration group contains ATM cell layer information and configuration of local ATM interfaces. This includes information such as the port identifier, interface speed, number of transmitted cells, number of received cells, number of cells with uncorrectable HEC errors, physical transmission type, operational status, administrative status, active VPI/VCI fields, and the maximum number of VPCs/VCCs.

The ATM interface DS3 PLCP and the TC sublayer groups provide the physical layer performance statistics for DS3 or SONET transmission paths. This includes statististics on the bit error rate and errored seconds.

The ATM virtual link and cross-connect groups allow management of ATM VP/VC virtual links (VPL/VCL) and VP/VC cross-connects. The virtual link group is implemented on end systems, switches, and networks, while the cross-connect group is implemented on switches and networks only. This includes the operational status, VPI/VCI value, and the physical port identifier of the other end of the cross-connect.

The AAL5 connection performance statistics group is based upon the standard interface MIB for IP packets. It is defined for an end system, switch, or network that terminates the AAL5 protocol. It defines objects such as the number of received octets, number of transmitted octets, number of octets passed to the AAL5 user, number of octets received from the AAL5 user, and number of errored AAL5 CPCS PDUs.

19.4 REVIEW

This chapter introduced a model of Operations, Administration, Maintenance, and Provisioning (OAM&P) functions and how they interact to meet the overall needs of network users. It then covered specific issues that ATM brings to the area of OAM&P, in particular the factors involved in a choice between centralized and distributed network management. Next network architectures from the OSI, ITU, ATM Forum, and IETF were introduced. The OSI architecture defines the functions of Fault, Configuration, Accounting, Performance, and Security (FCAPS). The ITU defines a physical and logical Telecommunications Management Network (TMN) architecture. The IETF has defined the Simple Network Management Protocol (SNMP) and Management Interface Base (MIB) and has achieved a high degree of interoperability in the industry. We then summarized the ATM Forum's SNMP based Interim Local Management Interface (ILMI) for the ATM UNI and the IETF's ATM Management Information Base (AToMMIB) for management of interfaces, end systems, switches and networks.

19.5 REFERENCES

[1] N. Lippis, "ATM's Biggest Benefit: Better Net Management," *Data Communications*, October 1992.

[2] I. Rubin, T. Cheng, "The Effect of Management Structure on the Performance of Interconnected High-Speed Packet-Switched Networks," *GLOBECOM*, December 1991.

[3] CCITT, "Recommendation M.3010 — Principles for Telecommunications Management Network," 1992.

[4] G. Schapeler, E. Scharf, S. Manthorpe, J. Appleton, E. Garcia Lopez, T. Koussev, J. Weng, "ATD-specific network management functions and TMN architecture," *Electronics & Communication Engineering Journal*, October 1992.

[5] J. Callaghan, G. Williamson, "RACE telecommunications management network architecture," *Electronics & Communication Engineering Journal*, October 1992.

[6] U. Black, *Network Management Standards*, McGraw-Hill, 1992.

[7] B. Hebrawi, *OSI Upper Layer Standards and Practices*, McGraw-Hill, 1992.

[8] ATM Forum, "ATM User-Network Interface (UNI) Specification, Version 2.0," June 1992.

[9] M. Ahmed, K. Tesnik, "ATM Managed Objects (AToMMIB) Version 4.0," IETF, December 23, 1993.

20

ATM Layer Management and Fault Management

This chapter first introduces the integrated physical and ATM layer OAM information flow architecture. The details of the ATM OAM cell formats are defined for the description of fault management in this chapter, and performance management in the next chapter. The desccription then continues with the usage of OAM cells for fault detection and identification using the Alarm Indication Signal (AIS) and Far End Reporting Failure (FERF) ATM OAM cells, as demonstrated in several examples. Finally, the chapter presents example uses of the loopback capability to verify connectivity and diagnose problems that AIS/FERF cannot.

20.1 OAM FLOW REFERENCE ARCHITECTURE

Currently OAM flows are only defined for point-to-point connections. A fundamental part of the infrastructure for network management is that of Operations, Administration, and Maintenance (OAM) information. Figure 20.1 shows the reference architecture that describes how ATM OAM flows relate to SONET/SDH management flows [1]. The F1 flows are for the regenerator section level (called the Section level in SONET), F2 flows are for the digital section level (called the Line level in SONET), and F3 flows are for the transmission path (call the Path level in SONET). ATM adds F4 flows for Virtual Paths (VPs) and F5 flows for Virtual Channels

(VCs), where multiple VCs are completely contained within a single VP. Each flow can be either connected or terminated at an endpoint. Each of the F4/F5 flows may be either end-to-end or segment-oriented. End-to-end flows are received only by the device that terminates the ATM connection. Only network nodes receive segment OAM flows. Indeed, network nodes must remove segment flows before they ever reach devices that terminate an ATM (VP or VC) connection.

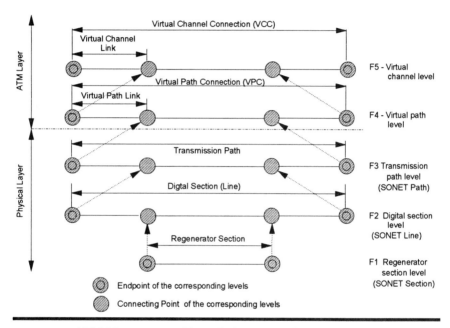

Figure 20.1 ATM Management Plane Reference Architecture

OAM flows may be either segment or end-to-end. An end-to-end flow is from one endpoint at the same level to the other endpoint. A segment flow is from one connection point to another connection point.

Figure 20.2 shows a less abstract, real-world example of these OAM flows for an end-to-end VC connection in terms of end-to-end flows. The example shows an end-to-end Virtual Channel Connection (VCC) connecting two end systems. Starting from the left-hand side, end system 1 is connected to Lightwave Terminal Equipment (LTE) 1 that terminates the digital section OAM flow (F2). The transmission path flow (F3) terminates on the Virtual Path (VP) Cell Relaying Function (CRF). The VP flow (F4) passes through the VP CRF, since it is only a connection point; that is, only the Virtual Path Identifier (VPI) value changes in cells that pass through that specific VP. The Virtual Channel Identifier (VCI) value is not changed. Next, a typical transmission path in the wide area is traversed from LTE2 to LTE3 through a repeater (as indicated by the "bow tie" symbol as shown in Figure 20.2). The regenerator section flow (F1) is used between LTEs 2 and 3 and the repeater (and between repeaters). The OAM flow between LTE2 and LTE3 is called a digital section flow (F2). The transmission path (F3) flow terminates on the VC CRF. The VP flow (F4) also terminates on the VC CRF because in its relaying function it can change the VCI as well as the VPI. A separate digital section OAM flow (F2) then extends from LTE4 to a CPE device (NT-1) as another line flow (F2). The OAM flow to end system 2 from the NT-1 is also a digital section level flow (F2). The transmission path flow (F3) extends from VC CRF to end system 2, as does the VP Flow (F4) since the VPI cannot change in this portion of the connection. Finally, note that the Virtual Channel (VC) flow (F5) is preserved from end system 1 to end system 2.

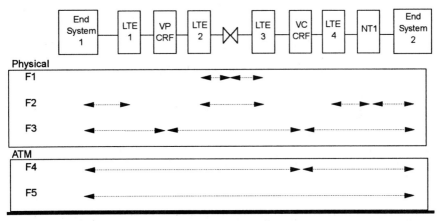

Figure 20.2 Illustrative Example of OAM Flow Layering

20.2 OAM CELL FORMATS

Figure 20.3 shows the ATM OAM cell format [3], [4]. There are Virtual Path (VP) flows (F4) and Virtual Channel (VC) flows (F5) between connection endpoints that are defined as end-to-end OAM flows. There are also F4 and F5 OAM flows that occur across one or more interconnected VC or VP links that are called segment OAM flows. As described in Chapter 8, VP flows (F4) utilize different VCIs to identify whether the flow is either end-to-end (VCI=3) or segment (VCI=4). Recall that the first 16 VCIs are reserved for future standardization. For a VC flow (F5) a specific VCI cannot be used because all VCIs are available to users in the VCC service. Therefore, the Payload Type (PT) differentiates between the end-to-end (PT=100) and segment (PT=101) flows in a VCC.

F4 (VPC) OAM Cell Format

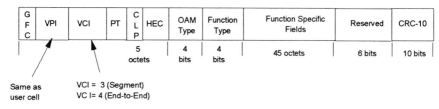

F5 (VCC) OAM Cell Format

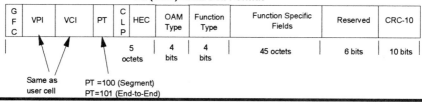

Figure 20.3 ATM OAM Cell Types and Format

Table 20.1 summarizes the OAM type and function type fields in the OAM cells from Figure 20.3 [3], [4]. The three OAM types are fault management, performance management, and activation/deactivation. Each OAM type has further function types with codepoints as identified in Table 20.1. For the fault management OAM type there are Alarm Indication Signal (AIS), Remote Defect Indication (RDI) (also called Far End Reporting Failure (FERF), and continuity check function types. For the performance management OAM type there are forward monitoring and backward reporting types, or a third type that is a combination of these two, called monitoring and reporting. The third OAM type defines activation and deactivation of the other OAM types. Currently, there are activation and deactivation function types for performance management and the continuity check.

Table 20.1 OAM Types and OAM Function Types

OAM Type		Function Type	
Fault Management	0001	AIS	0000
	0001	RDI/FERF	0001
	0001	Continuity Check	0100
	0001	Loopback	1000
Performance Management	0010	Forward Monitoring	0000
	0010	Backward Reporting	0001
	0010	Monitoring & Reporting	0010
Activation/Deactivation	1000	Performance Monitoring	0000
	1000	Continuity Check	0001

Note that there are a significant number of unassigned codepoints in the OAM and function types. The definition of OAM cell formats, functions, and protocols is ongoing and evolving in standards and specification development. For this reason, the ATM Forum UNI specification recommends that these OAM functions be implemented in software. Furthermore, as much as a 1 second response time to an OAM cell is allowed, which would easily allow a software implementation. Further refinements of OAM functionality will be standardized and specified, along with completely new types and associated functions, as more experience is gained with ATM, and additional capabilities are added.

The "function-specific" fields of the ATM OAM cells defined in subsequent sections are based primarily on the ATM Forum B-ICI Specification [3]. The specific formats are the subject of ongoing standardization and specification and are subject to change. We describe the functional elements of each type, and where standardized, the particular values from the B-ICI specification. There are two function-specific fields for the fault management type, one for Alarm Indication Signal (AIS) and Remote Defect Indication (RDI)/Far End Reporting Failure (FERF) and a second for loopback. The continuity check format as defined in the draft ANSI and ITU-TSS work [4] currently has no function-specific fields. There is only one format for the performance management type and the activation/deactivation type.

See the I.610 standard [1], the ANSI working document [4], the ATM Forum documents, [3], [2], or one of several papers [6], [7], [8] for more information on ATM OAM details.

20.3 FAULT MANAGEMENT

Fault management determines when there is a failure, notifying other elements of the connection regarding the failure and providing the means to diagnose and isolate the failure. The ATM OAM cell function types involved with this function are AIS, RDI/FERF, loopback, and continuity check. The following sections describe and illustrate each of these.

We provide a real-life analogy for fault management with vehicular traffic. Recall how the notion of virtual paths and channels was described as vehicles obeying different lane disciplines in Chapter 7. Chapter 14 on congestion control drew the analogy with the busy hour of mild congestion, and that of an accident as severe congestion. The analogy for fault management is one where the road actually fails! For example, imagine a divided highway with separate bridges crossing a river. A flash flood may wash out one, or both, of the bridges. The vehicles crossing one bridge cannot see the other bridge because this is very rugged country. The motorists who just passed over the bridge and saw it collapse will travel to the next police station and report the bridge failure. If both bridges wash out, then the police know to divert traffic away from the bridge in each direction. If the bridge washes out in one direction, then the failure must be reported in one direction, and another vehicle must travel across the remaining bridge in the other direction in order to divert traffic away from the failed bridge.

20.3.1 AIS and RDI/FERF Theory and Operation

Figure 20.4 illustrates the ATM OAM cell AIS and RDI/FERF function-specific fields. The meaning of each field is described below.

- ✒ **Failure Type** is an indication of what type of failure has occurred. Currently no specific values are standardized.

- ✒ **Failure Location** is an indication of where the failure occurred. Currently no specific values are standardized.

Failure Type*	Failure Location*	Unused*
8	9x8	35x8

bits

* Default Coding = '6A'Hex for all octets

Figure 20.4 Function-Specific Fields for AIS and RDI/FERF

Figure 20.5 illustrates the operation and theory of the Alarm Indication Signal (AIS) and Far End Reporting Failure (FERF) [or equivalently Remote Defect Indication (RDI)] ATM OAM cell function types. We cover two examples, (a) where a failure occurs in both directions simultaneously, and (b) where a failure occurs in only one direction. In both examples there is a VP (or VC) connection between node 1 and node 4.

Part (a) illustrates the typical failure of both directions of the physical layer between nodes 2 and 3 that causes the underlying VPs and VCs to simultaneously fail. The failures in each direction are indicated as "Failure-A" and "Failure-B" in the figure so that the resulting AIS and RDI/FERF cells can be traced to the failure location. A node adjacent to the failure generates an AIS signal in the downstream direction to indicate that an upstream failure has occurred, as indicated in the figure. As can be seen from example (a), both ends of the connection (nodes 1 and 4) are aware of the failure because of the AIS alarm that they receive. However, by convention, each generates a RDI/FERF signal.

Example (b) illustrates the purpose of the FERF (or RDI) signal. In most communications applications the connection should be considered failed, even if it fails in only one direction. This is especially true in data communications. Example (b) illustrates the case of a failure that affects only one direction of a full duplex connection between nodes 2 and 3. Node 3, which is downstream from the failure, generates an AIS alarm, which propagates to the connection end (node 4), which in turn generates the RDI/FERF signal. The RDI/FERF signal propagates to the other connection end (node 1), which is now aware that the connection has failed. Without the RDI/FERF signal, node 1 would not be aware that there was a failure in the connection between nodes 2 and 3. This method will also detect any combination of single-direction failures. Note that the node(s) that generate the AIS signals know exactly where the failure is, and could report this to a centralized network management system, or take a distributed rerouting response.

(a) Failure in Both Directions

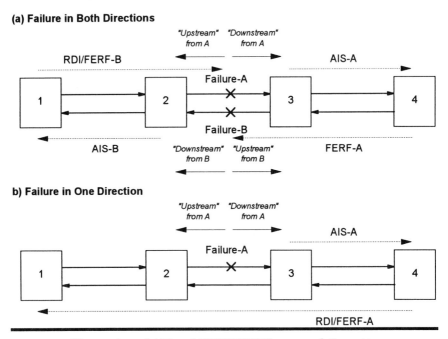

Figure 20.5 Illustration of AIS and RDI/FERF Theory and Operation

Figure 20.6 illustrates the relationship between the physical layer, VP layer and VC layer failures, AIS, and RDI/FERF indications at a VP and VC level connection point. Note that a VC connection point is a VP endpoint. The figure shows an exploded view of a VP/VC switch that has PHYsical (PHY) path, Virtual Path (VP), and Virtual Channel (VC) interfaces entering and leaving physical, VP, and VC connection points indicated by the circles. In the top half of the figure the receive PHY path and the ATM VP and VC links enter from the left, terminate on the connection points, and are switched through to the downstream VC, VP, and PHY links. The bottom half of the figure illustrates the upstream physical link and VP link for the VP endpoint. Logic in equipment associated with the respective connection endpoints process the failure, AIS, and FERF indications as indicated in the figure.

At every ATM device (which is always a transmission path termination) either PHY-AIS or physical layer failure is detected (as shown by the "?" in the diamond-shaped decision symbol), resulting in PHY-RDI/FERF being sent upstream. At a VP connection point a detected PHY-AIS causes VP-AIS to be generated in the downstream direction. For a VP endpoint, if VP-AIS is received then VP-RDI/FERF is sent in the upstream direction. If VP-AIS, a physical failure, or PHY-AIS is received, for the VP endpoint corresponding to the VC connection points, then AIS is sent downstream for those VCs on which the AIS feature is activated.

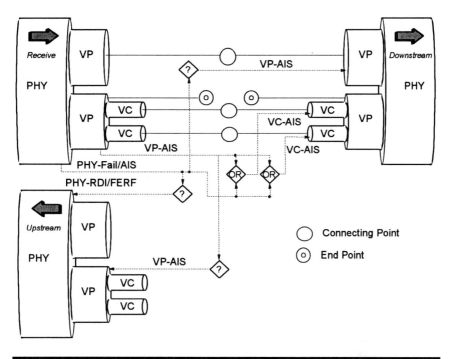

Figure 20.6 Physical/ATM AIS and RDI/FERF Connecting Point Functions

Figure 20.7 illustrates the relationship between the physical layer, VP layer and VC layer failures, AIS, and RDI/FERF indications at a VP and VC endpoint. This would correspond to the actions taken at an ATM end system. Exactly the same as in the switch, if either PHY-AIS or physical layer failure is detected, then PHY-RDI/FERF is sent upstream. For a VP endpoint, if VP-AIS is received, a physical failure is detected, or PHY-AIS is detected, then a VP-RDI/FERF is returned in the upstream direction. A VP endpoint exists for sets of VC endpoints. For those VC endpoints for which fault management is activated, the receipt of VP-AIS, detection of a physical failure, detection of PHY-AIS, or detection of VC-AIS results in VC-RDI/FERF being returned in the upstream direction.

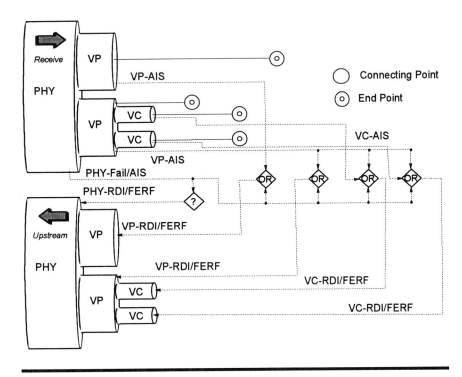

Figure 20.7 Physical/ATM AIS and RDI/FERF Endpoint Functions

If you got the impression that the preceding AIS and RDI/FERF function is complicated, and that a single physical failure can generate many VP and/or VC OAM cells, then you were correct. Once a fault condition is detected, the OAM cell is sent periodically. In order to limit the number of OAM fault management cells, the period for generating OAM cells is on the order of seconds. Furthermore, the ATM Forum UNI does not require VC-AIS and VC-RDI/FERF, while the ATM Forum B-ICI specification specifies that VC-AIS and VC-RDI/FERF is to be used on a selective basis only. These restrictions constrain the amount of OAM cells that are generated (and processed), such that the very useful function of AIS and RDI/FERF is delivered in an efficient manner.

20.3.2 Loopback Operation and Diagnostic Usage

Figure 20.8 illustrates the ATM OAM cell Loopback function-specific fields.

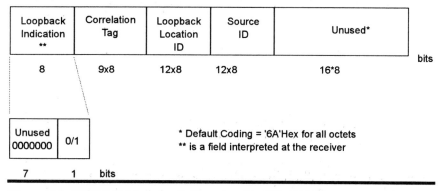

Figure 20.8 ATM OAM Loopback Function-Specific Fields

A summary of the ATM OAM cell loopback function-specific fields is:

↳ **Loopback Indication** is a field that contains "01" when originated, and is decremented by the receiver. It should be extracted by the sender when it is received with a value of "00." This prevents the cell from looping around the network indefinitely.

↳ **Correlation Tag** is a field defined for use by the OAM cell originator since there may be multiple OAM cells in transit on a particular VPC/VCC, and this allows the sender to identify which one of these has been received.

↳ **Loopback Location ID** is a field provided to the sender and receiver for use in segment loopbacks to identify where the loopback should occur. The default value of all 1s indicates that the loopback should occur at the end point.

↳ **Source ID** is a field provided so that the loopback source can be identified in the cell. This can be used by nodes to extract OAM cells that they have inserted for extraction after they have looped back to the source.

As seen in the preceding section, AIS and RDI/FERF are most useful in detecting and identifying to the connection endpoints that a failure has occurred. In some cases there may not be a hard failure, or we may just want to verify continuity. An example of a failure that AIS/FERF does not detect is a misconfiguration of VPI and/or VCI translations in a VPC/VCC service, such that cells do not reach the destination endpoint. The loopback function type meets these requirements.

Figure 20.9 illustrates the two basic loopback functions: segment, and end-to-end. The OAM cell flow type as defined earlier determines whether the loopback is segment or end-to-end.

Part (a) of Figure 20.9 illustrates a segment loopback that would normally be used within a network. The originator of the segment OAM loopback cell (node 1) includes a Loopback Indication of 1, a loopback ID indicating the endpoint (of the network), and a correlation tag of "X" that it can look for in received OAM cells. The loopback destination, endpoint switch (node 3), extracts the loopback cell, decrements the Loopback Indication field, and transmits the loopback cell in the opposite direction as shown in the figure. Note that every node is extracting and processing every OAM cell, which results in node 2 transparently conveying the segment loopback cell between nodes 1 and 3. Eventually, node 1 extracts the loopback cell and matches the correlation tag, thus verifying continuity between the segment of VP (or VC) links from node 1 to node 3.

Part (b) of Figure 20.9 illustrates an end-to-end loopback that could be used by a network switch to verify connectivity with an endpoint, or by an endpoint to verify connectivity with the distant endpoint. In the example, node 1 performs an end-to-end loopback to endpoint 2. Node 1 inserts an end-to-end OAM loopback cell that has a Loopback Indication of 1, a Loopback ID indicating the endpoint and a correlation tag of "X" that it can look for in received OAM cells. The loopback destination, endpoint 2, extracts the loopback cell, decrements the Loopback Indication field, and transmits the loopback cell in the opposite direction as shown in the figure. Note that switch nodes and the endpoints are extracting and processing every OAM cell, which results in nodes 2 and 3 transparently conveying the end-to-end loopback cell between node 1 and end point 2. Eventually, node 1 extracts the loopback cell and matches the correlation tag, thus verifying continuity the of VP (or VC) links from node 1 to endpoint 2.

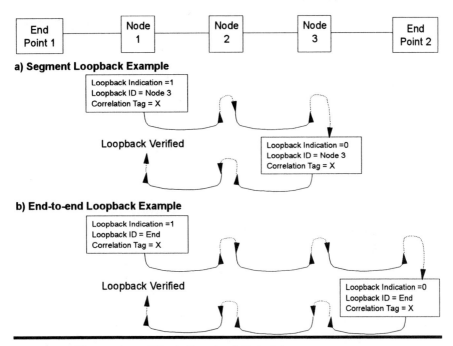

Figure 20.9 Illustration of Basic Loopback Primitives

Figure 20.10 illustrates how these loopback primitives can be used to diagnose a failure that would not be detected by AIS and RDI/FERF at any node. An example of such a failure would be a misconfigured VP or VC cross-connect. The example shows two endpoints and two intervening networks, each with three nodes. Part (a) shows the verification of end-to-end continuity via an end-to-end loopback to endpoint 1. If this were to fail, then network 2 could diagnose the problem as follows. Part (b) shows verification of connectivity between a node in network 2 and endpoint 2 via an end-to-end loopback. If this fails, then the problem is between network 2 and endpoint 2. Part (c) shows verification of connectivity to end point 1 via an end-to-end loopback. If this fails, there is a problem in the link between endpoint 1 and network 1, a problem in network 1, or a problem in the link between network 1 and 2. Part (d) shows verification of connectivity across networks 1 and 2 via a segment loopback. If this succeeds, then the problem is the access line from endpoint1 to network 1. Part (e) shows verification of connectivity from entry to exit in network 1. If this succeeds, then the problem is in network 1. Verification within any of the networks could also be done using the segment loopback.

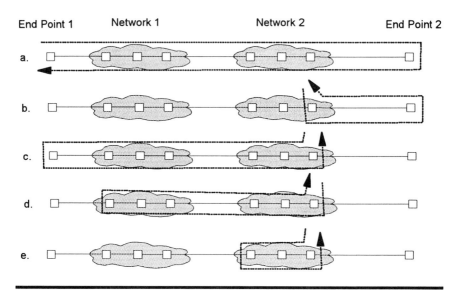

Figure 20.10 Usage of Loopback in Verification/ Problem Diagnosis

20.3.3 Continuity Check

The idea behind continuity checking is that the endpoint sends a cell periodically at some predetermined interval so that the connecting points and the other endpoint can distinguish between a connection that is idle and one that has failed. The current draft ANSI document [4] extends this concept by determining whether any user cell or a continuity check cell was received in a time interval (2 to 20 s) to declare loss of continuity that would result in VP-RDI/FERF being returned in the upstream direction.

The continuity check cell currently has no standardized function-specific fields. This function is being considered for the Virtual Path (VP) only, and may not be active for all Virtual Paths. Continuity checking is activated and deactivated by the procedures that we will describe in the next chapter. The ATM Forum currently does not specify support for the continuity check for interoperability at the ATM User Network Interface.

The continuity check can detect failures that AIS cannot, such as an erroneous VP cross-connect change, as illustrated in Figure 20.11. Part (a) shows a VP connection traversing three VP cross-connect nodes with VPI mappings shown in the figure carrying only Continuity Check (CC) cell traffic. In part (b) an erroneous cross-

connect is made at node 2, interrupting the flow of CC cells. In part
(c) node 3 detects this continuity failure and generates a VP-
RDI/FERF OAM cell in the opposite (upstream) direction.

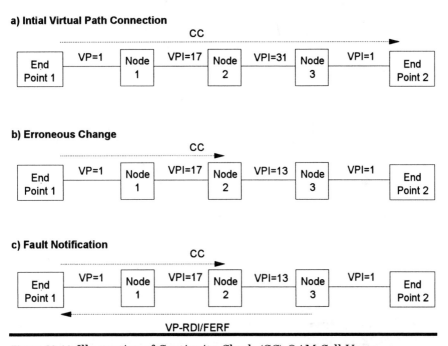

Figure 20.11 Illustration of Continuity Check (CC) OAM Cell Usage

20.4 RESTORATION

The standards currently do not specify what can be done in response
to a fault at the ATM layer. There are SONET and SDH standards,
however, that define physical layer protection switching on a point-
to-point, 1:N redundant basis or a ring configuration. There are also
restoration strategies for partial mesh networks. These same
concepts could also be applied to restore ATM connections. Restoring
Virtual Paths (VPs) that carry a large number of Virtual Channels
(VCs) would be an efficient way to perform ATM-level restoration.
We briefly discuss these three restoration methods with reference to
Figure 20.12.

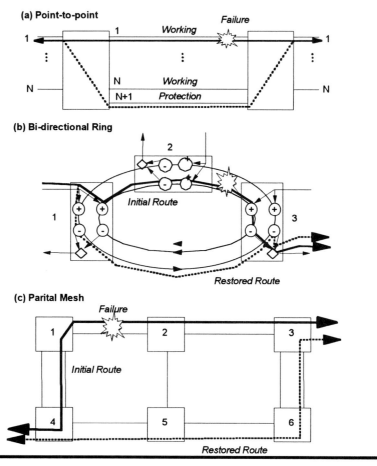

(a) Point-to-point

(b) Bi-directional Ring

(c) Parital Mesh

Figure 20.12 Basic Restoration Method Examples

The term 1:N (read as "one for N") redundancy means that there is one bidirectional protection channel for up to N working bi-directional channels, as illustrated in Figure 20.12(a). If a working channel fails, its endpoints are switched to the protection channel. If a failure occurs and the protection channel is already in use or unavailable, then the failed working channel cannot be restored.

Figure 20.12(b) illustrates a bidirectional ring. Traffic from node 1 to node 3 is sent in both directions around the ring. At each node signals may be added as shown by the plus sign inside the circle, or dropped as shown by the minus sign inside the circle. At the receiver (node 3) only one of the signals is selected for output. Upon detection of a failure the receiver will switch to the other, redundant signal, as shown in the example for a failure between nodes 2 and 3.

The mirror image of this capability is in place for traffic between nodes 3 and 1. Note that the ring architecture achieves 1:1 redundancy, or in other words, only half of the transmission bandwidth is available to traffic. These types of ring architectures are economically attractive for metropolitan areas and can be further optimized for ATM using Virtual Path level Add Drop Multiplexing (ADM) to improve multiplexing efficiency [9].

Figure 20.12(c) illustrates a partial mesh network. In the example, traffic between nodes 3 and 4 initially follows the route shown by the solid line. A failure between nodes 1 and 2 impacts the node 3 to 4 traffic on its current route. This can be detected in a centralized, distributed, or hybrid network management system to find a new route shown by the dashed line in the example. Longer distance networks tend to have the type of lattice structure shown in the example and tend to use this type of restoration. In the example of Figure 20.12(c), 1:2 redundancy is provided for some routes.

20.5 REVIEW

This chapter first defined the reference configuration for ATM Operations Administration, and Maintenance (OAM) flows at the physical layer and the ATM Virtual Path (VP) and Virtual Channel (VC) levels. The reference model defines connecting points and endpoints for VPs and VCs at the ATM layer. Coverage moved to definition of the OAM cell format and how VPs and VCs use it on either an end-to-end or segment basis. The discussion then proceeded to the topic of fault management in detail, giving an automotive analogy of a bridge failure as an example. This includes the definition of the Alarm Indication Signal (AIS) and Far End Reporting Failure (FERF), also called a Remote Defect Indication (RDI) in recent standards work. Next, the chapter described the use of loopback and continuity check functions in determining and diagnosing faults where AIS and RDI/FERF does not.

20.6 REFERENCES

[1]   CCITT, I.610, "OAM Principles for the B-ISDN Access," Geneva, 1991.
[2]   ATM Forum, "ATM User-Network Interface (UNI) Specification, Version 3.0," 1993.

[3]   ATM Forum, "BISDN Inter Carrier Interface (B-ICI) Specification, Version 1.0," August 1993.

[4]   ANSI, "B-ISDN Operations and Maintenance Principles and Functions (T1S1.5/93-004R2)," January 1994.

[5]   Bellcore, TA-NWT-001248, "Generic Requirements for Operations of Broadband Switching Systems Issue 1," October 1992.

[6]   S. Farkouh, "Managing ATM-based Broadband Networks," *IEEE Communications*, May 1993.

[7]   H. Breuer, "ATM-Layer OAM: Principles and Open Issues," *IEEE Communications*, September 1991.

[8]   J. Anderson, M. Nguyen, "ATM-Layer OAM Implementation Issues," *IEEE Communications*, September 1991.

[9]   T. Wu, "Cost Effective Network Evolution," *IEEE Communications*, September 1991.

21

ATM Layer Performance Measurement

This chapter defines reference configurations for specifying and measuring Network Performance (NP) and user Quality of Service (QoS). Quality of Service is user perception, while Network Performance finds use in network management, OAM, and network design. Next OAM cell formats and procedures used to activate and deactivate the performance measurement and continuity check functions are defined. Descriptions and examples illustrate the functions performed by performance measurement OAM cells. Then the chapter gives detailed examples of how the OAM performance measurement cells and procedures estimate each of the QoS parameters from the traffic contract described in Chapter 12. The chapter concludes with a discussion of other topics including open issues in the estimation of NP/QoS using the current methods, other elements of performance that require measurement, and the direction of industry standardization and specification.

21.1 NP/QOS REFERENCE CONFIGURATION

Network Performance (NP) is observed by the network at various points. Quality of Service (QoS) is observed by the user on an end-to-end basis. The starting point for defining this difference in perception is a reference configuration, against which some observations can be objectively measured and compared.

Some phrases that ITU-T Recommendation I.350 uses to define Quality of Service (QoS) are:

☞ "... parameters that can be directly observed and measured ... by the user ..."

☞ "... may be assured to a user at service access points by the network provider ..."

☞ "... are described in network independent terms ..."

By comparison, I.350 defines Network Performance (NP) using the following phrase:

☞ "... parameters ... meaningful to the network provider ... for ... system design, configuration, operation and maintenance."

Figure 21.1 illustrates these concepts of QoS and Network Performance in a reference configuration derived from ITU-T Recommendation I.350.

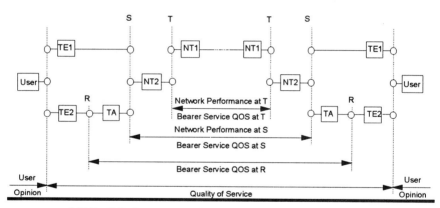

Figure 21.1 NP/QoS Reference Configuration

Figure 21.2 illustrates how Network Performance and QoS apply only to the ATM layer. The NP/QoS for AALs is an item for future study and standardization. This includes common part and service-specific aspects. This will be the QoS that applications actually see.

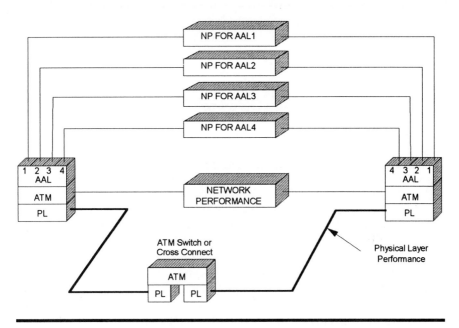

Figure 21.2 QoS/NP for Physical ATM and Adaptation Layers

21.2   ATM NP/QOS MEASUREMENT

QoSmeasureindexThis section describes how the Performance Measurment (PM) procedure is activated and deactivated. We then describe the performance measurement process. A key objetive is to measure or estimate these parameters *in-service*; that is, the customer traffic is not impacted by the measurement process.

21.2.1 Activation/ Deactivation Procedures

Figure 21.3 depicts the ATM OAM cell Activation/Deactivation function-specific fields. The following text defines the meaning of each field.

Message ID	Directions of Action	Correlation Tag	PM Block Sizes A-B**	PM Block Sizes B-A**	Unused Octets *	
6	2	8	4	4	42x8	bits

* Default coding = '6A' Hex
** Default coding = '0000'

Figure 21.3 ATM Activation/Deactivation OAM Cell Function-Specific Fields

⊠ **Message ID** is defined as follows:

'000001'	Activate (Request)
'000010'	Activation Confirmed
'000011'	Activation Request Denied
'000101'	Deactivate(Request)
'000110'	Deactivation Confirmed
'000111'	Deactivation Request Denied

⊠ **Direction of Activation** is specified as A-B ('10') from the activator or B-A ('01') towards the activator.

⊠ **PM Block Size A-B** identifies the Performance Measurement (PM) block size which can be supported from A to B by a bit mask for sizes of 1024, 512, 256, or 128 from most significant bit (msb) to least significant bit (lsb), respectively.

⊠ **PM Block Size B-A** identifies the block sizes that can be supported from B to A using the same convention as above. Note that the block sizes can be different in each direction.

Figure 21.4 illustrates the Activation/ Deactivation procedure for performance monitoring or a continuity check. Connection/segment endpoint A generates a De(Activate) request toward B requesting action on either the A-to-B, B-to-A, or both directions. If B can comply with all of the requests, then a (De)Activation confirmed message is returned. If B cannot comply with the request(s) then a (De)Activation Request Denied message is returned. A denial can result if the endpoint is unable to support the performance management function. If the Deactivation refers to a single function that is currently not operating, then the request is also denied. If the Deactivation request refers to both directions, yet only one direction is operating, then the request is confirmed, with the reference to the nonoperational function ignored.

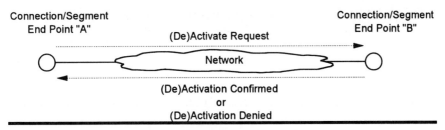

Figure 21.4 Illustration of Activation/Deactivation Flow

Once a performance measurement flow is activated, the procedure described in the following section is performed. Activation and Deactivation allow the performance measurement to be performed on selected VPCs and VCCs. This reduces the total processing load required for performance measurements. Continuous monitoring or sampled monitoring can use these Activation and Deactivation procedures.

21.2.2 Performance Measurement Procedure

Figure 21.5 depicts the ATM OAM cell Performance Management (PM) function-specific fields. The following text defines the meaning of each field.

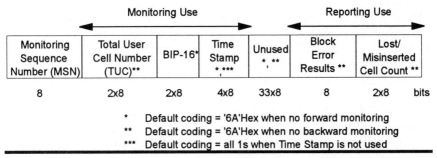

Figure 21.5 Performance Measurement Function-Specific Fields

⊠ **Function Type** indicates the type of function that is being performed: Forward Monitoring ('0000'), Backward Reporting ('0001'), or both ('0010').

⊠ **Monitoring Sequence Number (MSN)** is the PM cell number, modulo 256.

⧫ **Total User Cell (TUC) Number** is the total number of cells containing user data sent since the last PM cell.

⧫ **BIP-16** is a block error detection code computed over all of the user cells since the last PM cell. It is used for error rate estimation.

⧫ **Time Stamp** is a field used in the backward reporting direction to estimate delay. It is specified to have an accuracy of 1 µs or less.

⧫ **Block Error Result** is a count of the number of errored BIP-16 bits in the last block for backward reporting.

⧫ **Lost/Misinserted Cells** is a signed number indicating the number of cells actually received minus the TUC number since the last PM cell.

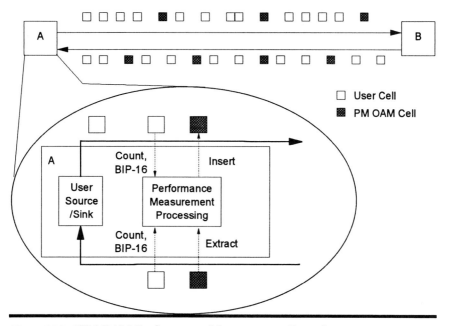

Figure 21.6 ATM OAM Performance Measurement Procedure

Figure 21.6 illustrates the operation of the performance measurement OAM cell insertion and processing. The connection or segment endpoints A and B that are involved are determined by the Activation/Deactivation procedure. In this example there is a PM cell flow in each direction. The example also illustrates the use of PM cell flows with different block sizes in each direction. The functions involved are insertion of OAM cells, counting user cells, and computing the BIP-16 on the transmit side. On the receive side OAM cells are extracted. The same counts are made, and the BIP-16 is recomputed for comparison with the value received in the monitoring cell as determined by the transmitter. Note that the monitoring cell contains the results for the cells in the preceding block.

21.3 NP/QOS PARAMETER ESTIMATION

QoSparmindexThis section defines the Quality of Service (QoS) and Network Performance (NP) parameters in terms of basic cell transfer outcomes. The method to estimate these QoS parameters from the Performance Measurement (PM) OAM cells is also described.

21.3.1 ATM Cell Transfer Outcomes

QoS is measured based upon cell entry events at one end and cell exit events at the other end, as illustrated in Figure 21.7. These two events, defined in ITU-T Recommendation I.356, determine the cell transfer outcomes. These outcomes are then used to define the ATM cell transfer performance parameters.

- A *cell exit event* occurs when the first bit of an ATM cell has completed transmission out of an End User Device to a ATM network element across the source UNI Measurement Point 1.
- A *cell entry event* occurs when the last bit of an ATM cell has completed transmission into an End User Device from an ATM network element across the destination UNI Measurement Point 2.

As illustrated in Figure 21.7, a cell that arrives at the exit point without errors and is not too late (i.e., it arrives in less than or equal to a delay of Tmax seconds) is considered successfully transferred. If a cell arrives within a certain delay (Tmax) but has errors, then it is considered an errored outcome. If a cell arrives too late, or never arrives at all, then it is considered lost. There is also a possibility of a misinserted cell, which is defined as the case when a cell arrives at the exit point for which there was no corresponding input cell at the entry point. This can occur due to undetected cell header errors. Depending upon the higher layer protocol, a misinserted cell can be a very serious problem since a misinserted cell cannot be distinguished from a transmitted cell by the receiver. Fortunately, this is a very unlikely event under normal circumstances.

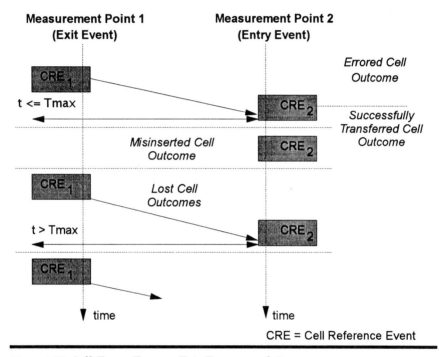

Figure 21.7 Cell Entry Events, Exit Events, and Outcomes

The following possible cell transfer outcomes between measurement points for transmitted cells are defined based on ITU-T Recommendation I.356.

☺ **Successful Cell Transfer Outcome:** The cell is received corresponding to the transmitted cell within a specified time Tmax. The binary content of the received cell conforms exactly to the corresponding cell payload, and the cell is received with a valid header field after header error control procedures are completed.

☺ **Errored Cell Outcome:** The cell is received corresponding to the transmitted cell within a specified time Tmax. The binary content of the received cell payload differs from that of the corresponding transmitted cell, or the cell is received with an invalid header field after header error control procedures are completed.

☺ **Lost Cell Outcome:** No cell is received corresponding to the transmitted cell within a specified time Tmax (examples include "never arrived" or "arrived too late").

☹ **Misinserted Cell Outcome:** A received cell for which there is no corresponding transmitted cell.

☹ **Severely Errored Cell Block Outcome:** When M or more Lost Cell outcomes, Misinserted Cell Outcomes, or Errored Cell outcomes are observed in a received cell block of N cells transmitted consecutively on a given connection.

21.3.2 ATM Performance Parameters

This section summarizes the set of ATM cell transfer performance parameters defined in I.356. The definitions of these performance parameters use the cell transfer outcomes defined above. This set of ATM cell transfer performance parameters correspond to the generic QoS criteria shown in parentheses as follows:

+ Cell Error Ratio	(Accuracy)
+ Severely Errored Cell Block Ratio	(Accuracy)
+ Cell Loss Ratio	(Dependability)
+ Cell Misinsertion Rate	(Accuracy)
+ Cell Transfer Delay	(Speed)
+ Mean Cell Transfer Delay	(Speed)
+ Cell Delay Variation	(Speed)

The QoS definitions in I.356 summarized here apply to cells conforming to the traffic contract only. Extending the QoS definitions to include nonconforming cells is an area of future standardization. This means that nonconforming cells must be excluded from the cell transfer outcomes. Principally, this affects the cell loss ratio calculation.

The draft ANSI standard [1] assigns the cell counts to all cells, conforming and non-conforming. The spirit of I.356 is still preserved in the ANSI work in that it specifies that counts of the number of cells discarded or tagged due to noncompliance be kept if the monitoring is done in the vicinity of the Usage Parameter Control (UPC) function.

21.3.3 Cell Error Ratio

Cell Error Ratio is defined as follows for one or more connection(s):

$$\text{Cell Error Ratio} = \frac{\text{Errored Cells}}{\text{Successfully Transferred Cells} + \text{Errored Cell}}$$

Successfully Transferred Cells and Errored Cells contained in cell blocks counted as Severely Errored Cell Blocks should be excluded from the population used in calculating the Cell Error Ratio.

Errored Cells (ECs) can only be estimated by counting the number of up to M ($2 \le M \le 16$, with a default of 4) parity errors in the BIP-16 code for the block. The successfuly transferred cell count is the Total User Cell number (TUC) from the PM OAM cell.

21.3.4 Severely Errored Cell Block Ratio

The Severely Errored Cell Block Ratio for one or more connection(s) is defined as:

$$\text{Severely Errored Cell Block Ratio} = \frac{\text{Severely Errored Cell Blocks}}{\text{Total Transmitted Cell Blocks}}$$

A cell block is a sequence of N cells transmitted consecutively on a given connection. A severely errored cell block outcome occurs when more than a specified number of errored cells, lost cells, or misinserted cells are observed in a received cell block.

An Errored Cell Block (ECB) contains one or more BIP-16 errors, lost cells, or misinserted cells.

A Severely Errored Cell Block (SECB) is a cell block with more than M ($2 \le M \le 16$, with a default of 4) BIP-16 errors, or more than K ($2 \le K \le M$, with a default of 2) lost or misinserted cells.

21.3.5 Cell Loss Ratio

The Cell Loss Ratio is defined for one or more connection(s) as:

$$\text{Cell Loss Ratio} = \frac{\text{Lost Cells}}{\text{Total Transmitted Cell}}$$

Lost and transmitted cells counted in severely errored cell blocks should be excluded from the cell population in computing cell loss ratio.

The number of lost cells can be estimated as the difference in the past two Total User Counts (TUC) received in the PM OAM cells from the distant end minus the number of cells actually received in a cell block. If this result is negative, then the estimate is that no cells were lost, and cells were misinserted as defined below.

Note that this estimation method would report zero loss and misinsertion if there are an equal number of cell loss and misinsertion outcomes in a cell block.

21.3.6 Cell Misinsertion Rate

The Cell Misinsertion rate for one or more connection(s) is defined as:

$$\text{Cell Misinsertion Rate} = \frac{\text{Misinserted Cell}}{\text{Time Interval}}$$

Severely Errored Cell Blocks should be excluded from the population when calculating the cell misinsertion rate. Cell misinsertion on a particular connection is most often caused by an undetected error in the header of a cell being transmitted on a different connection. This performance parameter is defined as a rate (rather than the ratio) since the mechanism producing misinserted cells is independent of the number of transmitted cells received on the corresponding connection.

The number of misinserted cells can be estimated as the number of cells actually received in a cell block minus the difference in the past two Total User Counts (TUC) received in the PM OAM cells from the distant end. If this result is negative, then cell loss has occurred, and the number of misinserted cells is estimated as zero.

21.3.7 Measuring Cell Transfer Delay

The Cell Transfer Delay is defined as the elapsed time between a cell exit event at the measurement point 1 (e.g., at the source UNI) and a corresponding cell entry event at measurement point 2 (e.g., the destination UNI) for a particular connection. The cell transfer delay between two measurement points is the sum of the total inter-ATM node transmission delay and the total ATM node processing delay between measurement point 1 and measurement point 2.

Of course, only the total delay can be estimated using the Performance Measurement (PM) OAM cells, and not the individual components that were described in Chapter 12. The methods currently proposed are optional and utilize either the time stamp function-specific field or a well-defined test signal.

For the time stamp method, tight requirements for Time-Of-Day (TOD) setting accuracy are essential to the measurement of absolute delay, and are currently not standardized. CDV can be estimated by taking differences in time stamps. Any estimation of delay also assumes that the PM OAM cells are processed exclusively in hardware, and not in software as all other OAM cell types could be. For more information on the time stamp estimation method, see the draft ANSI standard [1].

Figure 21.8 illustrates how absolute delay and differential delay can be measured using the time stamp method. The source and destination have highly accurate time stamp clocks which are set to nearly the same time. The source periodically sends OAM Performance Measurement (PM) cells and inserts its time stamp. These cells enter an ATM network and experience varying delays. As an OAM PM cell leaves the network and arrives at its destination, the time stamp is extracted and several operations are performed on it. First, the absolute delay is calculated as the (non-negative) difference between the local time stamp clock and the time stamp received in the OAM PM cell. Next, the value in a memory is subtracted from the absolute delay to yield a differential delay. Finally, the current absolute delay calculation is stored in the memory for use in calculation of the next differential delay.

The Mean Cell Transfer Delay is the average of a specified number of absolute cell transfer delay estimates for one or more connections. The above 2-point Cell Delay Variation (CDV) defined in I.356 can be estimated from the differential delays. A histogram of the differential delay can be computed, as well as the mean, the variance, or other statistics.

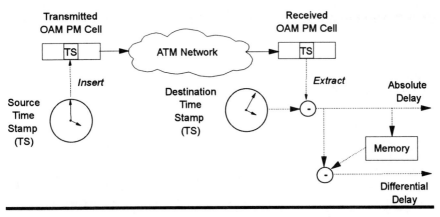

Figure 21.8 Time Stamp Method Delay Estimation

ITU-T Recommendation I.356 also define the 1-point CDV as the variability in the pattern of cell arrival events observed at a single measurement point with reference to the negotiated peak rate 1/T as defined in the traffic contract. Figure 21.9 illustrates a method to implement this measurement. A Continuous Bit-Rate (CBR) source emits a cell once every T seconds (note this implies that T is a multiple of the cell slot time in the TDM transmission convergence sublayer). This cell stream, perfectly spaced, is transmitted across an ATM network that introduces variations in delay that we wish to measure. The receiver knows the spacing interval T, and can compute the interarrival times of successive cells and subtract the time T to result in a 1-point CDV estimate. Positive values of the 1-point CDV estimate correspond to cell clumping, while negative values of the 1-point CDV estimate correspond to gaps, or dispersion, in the cell stream. As shown in Figure 21.9 cell clumping occurs for cells that are closer together, while dispersion occurs for cells spaced too far apart. This is important for determining the likelihood of overrun and underrun for CBR services as described in Chapter 16.

21.4 FUTURE DIRECTIONS

Several key areas still need to be addressed in performance measurement. The need to extend the definitions to include non-conforming cells is an ongoing effort in standards. Standardized means for measuring cell dispersion, in addition to cell clumping, are also being defined. The difficulties in estimating delay using the time stamp method need to be resolved by developing an accurate clock setting method and requiring that OAM PM cells be processed in hardware. Otherwise, other methods for estimating delay must be developed.

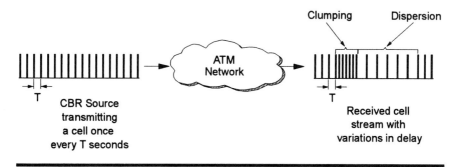

Figure 21.9 CBR Test Source Measurement of CDV

21.5 REVIEW

This chapter defined the reference configuration for Network Performance (NP) and Quality of Service (QoS) measurement. Basically, QoS is what the user perceives, while NP is what the network uses to make design, operational, or capacity decisions. Next the ATM OAM cell activation/deactivation format and procedure, which applies to the continuity check of Chapter 20, as well as Performance Measurement (PM) was defined. Then description and examples explained the ATM cell measurement method and the possible outcomes: successfully transferred cell, a lost cell, an errored cell, or a misinserted cell. These outcomes define the NP/QoS parameters of the traffic contract and determine how the OAM PM cells are used to estimate them. The chapter closed with a look at the future direction of some key issues in performance measurement.

21.6 REFERENCES

[1]   ANSI, "B-ISDN Operations and Maintenance Principles and Functions (T1S1.5/93-004R2)," January 1994.
[2]   Bellcore, TA-NWT-001248, "Generic Requirements for Operations of Broadband Switching Systems Issue 1," October 1992.

[3]   Farkouh, "Managing ATM-based Broadband Networks," *IEEE Communications*, May 1993.

8

Technology Comparison and Future Directions Involving ATM

This part begins with a comparison of ATM with other data communication technologies. Data communications technologies are categorized as the basis for comparison and analysis. The first comparison is between circuit, message, and packet switching. A detailed comparison is made with ATM for the various techniques of performing wide area packet switching: X.25, IP, frame relay, and SMDS. Local area data communication technologies, such as Fast Ethernet and FDDI, are compared to ATM. These comparisons are done based upon objective measures, suitability to application needs, network attributes, and services provided to end users. The final chapter outlines possible scenarios for the future of ATM, along with a discussion of potentially competing or collaborating technologies. We do not forecast that ATM will do everything, or capture the entire data communications market, but we believe that it is likely to leave a significant impact on data communications.

22

Technology Comparison
with ATM

Earlier in this book we studied many of the technologies and services that both compete with and compliment ATM. This chapter compares ATM and ATM-based services to technologies and services such as circuit switching, X.25, IP, frame relay, SMDS, and FDDI. This chapter draws comparisons between these technologies and services, their protocols and interfaces, and details of each switching technique.

22.1   CIRCUIT, MESSAGE, AND PACKET SWITCHING

This section introduces a comparison of technologies with ATM by summarizing all of the data communications methods that have been developed throughout history. Basic motivations for the choice between circuit switching versus packet switching in terms of some simple application examples are then provided. Finally, comments are made on the fate of messaging switching, which began the modern era of communications approximately 150 years ago.

22.1.1 A Taxonomy of Data Communication Methods

There are basically three methods that can be used for communication of data: circuit switching, message switching, or packet switching. All other categories of data communications can be placed into this taxonomy as illustrated in Figure 22.1. The packet switching branch is further broken down into packet, frame, and cell switching. The large number of techniques under the packet switching branch is indicative of the industry focus on this method.

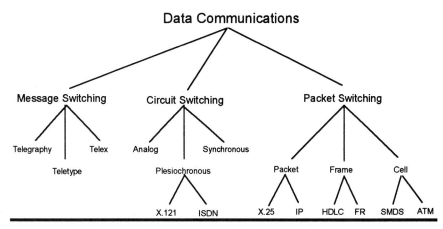

Figure 22.1 A Taxonomy of Data Communication Methods

Message switching started the era of modern data communications in 1847 with the invention of the telegraph by Morse and the system of message relays. Message switching evolved into paper tape teletype relay systems and the modern telex messaging system used for financial transfers and access to remote locations.

The use of circuit switching in data communications started over 50 years ago through modems designed for use on analog transmission systems. This evolved into plesiochronous digital transmission systems over the past 30 to 40 years. The main technologies in use today are digital private lines, X.21 fast circuit switching, and ISDN.

Packet switching was developed in the past 30 years to overcome the expense and poor performance of transmission systems. Three major classes of packet switching have evolved over time. The first packet switching systems of X.25 and IP were designed to operate over very poor transmission networks. A simpler protocol like HDLC could be used for local connections when the quality of the links was better. HDLC in turn has evolved into the wide area with frame relay. The desire to achieve even higher performance and flexibility lead to the development of SMDS, which is further enhanced by the development of ATM. Improvements in transmission technology, electronics, and protocol design are key enablers in this latest generation of data communications technology. These technologies are now examined and compared in more detail.

22.1.2 To Switch or Not to Switch? — An Answer to This Question

This section presents a few generalizations regarding the engineering economics of data communications services in order to make several points. First, there are tradeoffs between dedicated and switched data communications, as illustrated in Figure 22.2a. In general, there is a point when a dedicated facility between two points will be more economical than a switched facility for a certain daily usage. This crossover can be different if the individual usage duration is so small that the required switching capability is too expensive. If the point-to-point usage indicates switching is economically desirable, then there are two type of switching to choose from: either connectionless like the Internet Protocol (IP) or Switched Multimegabit Data Service (SMDS), or connection-oriented like Integrated Services Digital Network (ISDN) or frame relay. As a general rule of thumb, the average transaction duration should be an order of magnitude or greater than the circuit (either physical or logical) setup time, as shown in the setup time region of Figure 22.2b.

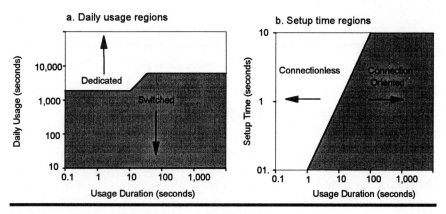

Figure 22.2 Ranges of Data Service Application

Figure 22.3 illustrates this example of usage duration by showing the time required to transfer an object of a certain size for a range of representative PDH circuit transfer rates. There obviously exists a linear relationship. The range of transfer times is divided into three regions based upon applicability of technology: connectionless, connection-oriented, and bulk data transfer (e.g., overnight mail). Note that there is a gap in the transfer time between 1.5 Mbps and 45 Mbps in the PDH and SDH transmission granularity. This is an example of where the flexibility of packet or cell transfer over fixed-rate circuits is evident.

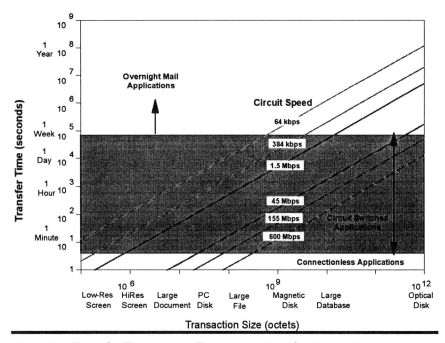

Figure 22.3 Transfer Time versus Transaction Size for Circuit Rates

22.1.3 Oh, Where Has Message Switching Gone?

Although digital data communications began with the telegraph, which initially employed human operators to perform the message switching function, it is not a prevalent data communications technology today. The telegraph was replaced by the "high-speed" 50-bps teletype, which employed human operators to relay punched paper tape messages. These messaging systems have evolved into the current Telex system of today, without which you could not wire money or reach remote locations.

Why hasn't message switching blossomed in the modern era of data communications? The answer is that messaging has become a higher layer protocol, and in many cases, an application that operates over packet switching in modern data communication networks. The ISO X.400 messaging systems, the Simple Mail Transfer Protocol (SMTP) of the Internet, and various proprietary implementations have become the dominant messaging systems applications. Packet switching networks allowed non-real-time message applications to coexist with real-time interactive data applications. It is the experience gained through this integration of applications with dissimilar performance requirements that will enable the networks of tomorrow to carry real-time images, interactive data applications, file transfer, and messaging on the ATM-based networks of the future.

22.2 STM COMPARED WITH ATM

This section compares aspects of Synchronous Transfer Mode (STM) with Asynchronous Transfer Mode (ATM). Recall from Chapter 4 the definition of STM as a special case of Time Division Multiplexing (TDM) where the clock frequency is controlled very accurately. The North American SONET standard and the international Synchronous Digital Hierarchy (SDH) standard described in Chapter 3 are the premier examples of STM today. The terms STM and circuit switching are used interchangeably in this chapter.

22.2.1 ATM versus STM CBR Multiplexing Efficiency

A key advantage of ATM over STM is the ability to flexibly, and efficiently, assign multiple Continuous Bit-Rate (CBR) circuit rates on a SONET transmission facility, as illustrated in Figure 22.4.

The upper left-hand corner of Figure 22.4 illustrates a specific example of STM for the SONET hierarchy carrying STS-1 (51.84 Mbps), VT1.5 (1.792 Mbps), and H0 (0.384 Mbps) within an OC-3 channel. The STS-1 channel must be dedicated. The VT1.5 channels are allocated within an STS-1, and the H0 channels are allocated in another set of VT1.5s as shown in the figure.

The same OC-3 channel is allocated as an STS-3c ATM channel as illustrated in the lower left-hand corner of Figure 22.4. There are approximately 42 ATM cells per time slot (125-μs interval), shown as a series of rectangles. The STS-1 channel is comprised of 16 cells per time slot. The VT1.5 channels require approximately one cell every other time slot. The H0 channels require approximately one cell every eighth time slot. The cells assigned can be assigned in any position or order, as illustrated in the figure. The inefficiency of the 5 bytes of cell overhead and 4 bytes of AAL overhead results in a maximum utilization of 91% for ATM versus the channelized, SONET mapping. This is shown by the line labeled STM 100% STS-1 and the line labeled ATM.

A graphical representation of a calculation where 5% of the traffic is for STS-1, 80% for VT1.5, and 15% for H0 is shown on the right-hand side of Figure 22.4. The carried load is shown plotted versus the offered load. Carried load is proportional to the offered load, minus any blocked load. SONET and ATM are equivalent at lower levels of offered load, up until the offered load exceeds 50% of the trunk capacity. At higher loads, the higher efficiency of ATM becomes evident.

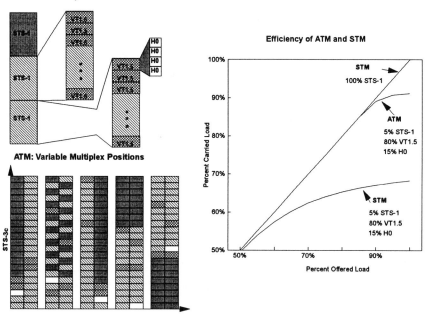

Figure 22.4 ATM versus STM Efficiency

For this particular example, the maximum efficiency of STM is about 65%, compared to 91% for ATM. This represents about a 25%

improvement in efficiency. When the mix of circuit rates becomes even more varied, or unpredictable, the advantage of ATM is even greater.

22.2.2 Statistical Multiplex Gain

Another key advantage of ATM over STM is the opportunity to achieve cost savings by statistical multiplex gain. The basic mathematics of statistical multiplex gain were covered in Chapter 15. The conclusion reached there was that statistical multiplex gain is higher for bursty traffic which has a peak rate less than the trunk line rate. Recall that burstiness is defined as the ratio of the source's peak rate to the average rate. When a source is very bursty, it does not have anything to send most of the time. However, when it does send, it may have a significant amount of data that is to be transmitted in a short time. Approximate values for a number of applications are provided later in this chapter.

22.3 PACKET SWITCHING SERVICE ASPECTS

This section compares the service aspects of packet switching technologies. These include the philosophy of data delivery, congestion control, flow control, protocol functions, and how the service relates to the OSI Reference Model (OSIRM).

22.3.1 Data Delivery Philosophy

A fundamental aspect of a packet switching service is whether the packets are guaranteed to be delivered or not. An assured data service guarantees delivery of the packets, while an unassured data service does not. The assured mode is tolerant of errors. The unassured mode is also called a datagram service.

X.25 assures that packets are delivered reliably on a link-by-link basis. Frame relay and SMDS operate in unassured mode. ATM can offer assured mode service through the Service-Specific Connection-Oriented Protocol (SSCOP) described in Chapter 9.

The packet switching service may guarantee that data arrives in the same sequence as it was transmitted, or may reorder the packets upon delivery. If a packet service has sequence integrity, the order of transmitted packets is preserved end-to-end.

22.3.2 Switching Approach

The packet switched network may operate as either a Connection-Oriented Network Service (CONS) or a ConnectionLess Network Service (CLNS) as defined in Chapter 3. Connection-oriented packet services may be operated as either Permanent Virtual Connections (PVCs) or Switched Virtual Connections (SVCs).

22.3.3 Traffic, Congestion, and Flow Control Approach

A packet service may reserve bandwidth for a connection-oriented service, or it may enforce other limits on access to the service. Chapter 5 summarized the frame relay Committed Information Rate (CIR), while Chapter 12 described the ATM Peak Cell Rate (PCR) and Sustainable Cell Rate (SCR) traffic control functions that are provided on a per virtual connection basis. Chapter 5 summarized the SMDS access class mechanism.

A packet service may implement flow control. The flow control method may be window-based, rate-based, or credit-based as described in Chapter 14. Alternatively a packet service may have no flow control whatsoever. A packet service may detect and react to congestion on either a link-by-link basis, or on an end-to-end basis. The reaction to congestion may be receiver-controlled, transmitter-controlled, both, or neither.

X.25 employs a window-based flow control of the transmitter, while frame relay provides forward and backward notification that can be used to control a transmitter or a receiver, respectively. ATM currently provides only forward congestion indication, which means that only receiver flow control is supported.

22.3.4 Comparison of Protocol Functions

Table 22.1 shows a comparison of public data service protocol functions. The terms used in the table were summarized in the preceding sections.

Table 22.1 Comparison of Protocol Function

Function	X.25	IP	FR	SMDS	ATM
Switching Type	Packet	Packet	Frame	Cell	Cell
Mode	CONS	CLNS	CONS	CLNS	CONS, CLNS
CONS Types	PVC, SVC	N/A	PVC, SVC	N/A	PVC, SVC
Assured Mode	Yes	No	No	No	via SSCOP
Sequence Integrity	Yes	No	Yes	Yes	Yes
Retransmission	Yes	No	No	No	via SSCOP
Traffic Control	None	None	CIR, EIR	Access Class	PCR, SCR
Flow Control	Yes	No	No	No	No
Congestion Control	Transmit	No	Transmit, Receiver	No	Receiver

22.3.5 OSI Functional Mapping

Figure 22.5 depicts an approximate mapping of data communications services, protocols, and common applications to the OSI reference model. Chapter 5 summarized the IP, X.25, SMDS Interface Protocol (SIP), and Frame Relay (FR) protocols. Chapter 8 summarized the ATM layer and common ATM Adaptation Layers (AALs). IP is defined to operate over all of these protocols and services.

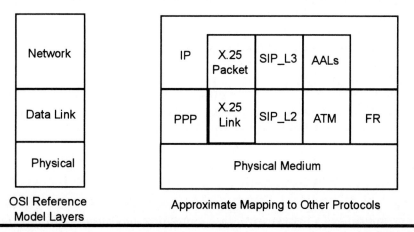

Figure 22.5 Mapping of Services and Protocols to the OSI Reference Model

22.4 GENERIC PACKET SWITCHING NETWORK CHARACTERISTICS

This section covers aspects of addressing, switching, routing, and network design. A tabular presentation summarizes the comparison of the networking aspects of data communication services.

22.4.1 Philosophy of Addressing

Addressing may be geographically oriented like the telephone numbering plan, network-oriented like the Internet, or hardware-oriented. The addressing schemes employed include IP, X.121, E.164, NSAP, and IEEE MAC, as covered in earlier chapters.

The assignment and agreements to interconnect addresses range from monopoly-oriented like the telephone network to cooperative as in the Internet. Agreement on a technical standard for addressing is usually an easy task compared with resolving the political, economic, business, and social issues that arise in addressing.

The IP address is 32 bits and is assigned on a network basis. The E.164 addressing plan is 15 Binary Coded Decimal (BCD) digits and is currently assigned on a geographic basis. The 48-bit IEEE MAC addresses are assigned on a hardware basis. X.25 utilizes a Logical Channel Number (LCN). Switched X.25 connections utilize a 14-digit X.121 address, which is allocated to networks based upon the first four digits, with the network provider specifying use of the remaining digits.

Frame Relay (FR) frames are assigned a Data Link Connection Identifier (DLCI). FR SVCs utilize the E.164 numbering plan.

ATM cells are assigned a Virtual Path Identifier (VPI) and Virtual Channel Identifier (VCI) that can have meaning on an end-to-end basis through translation at intermediate nodes. Switched virtual ATM connections may utilize either E.164 or NSAP addresses. The ISO NSAP addressing plan is assigned on a network administrator basis. Chapter 9 summarized the NSAP numbering plan.

22.4.2 Routing Approach

The routing of packets in data communications services may range from static to dynamic. The routing may be unspecified for the service, or may be standardized. Examples of standard routing protocols are the Open Shortest Path First (OSPF) and the Routing Information Protocol (RIP) of the Internet. The ATM Forum is specifying a routing protocol that encompasses bandwidth reservation in the P-NNI group as summarized in Chapter 6.

22.4.3 Network Access and Topology

A packet service may provide user access via a shared medium or a dedicated medium. The access medium may be dedicated to a single type of service, or may be shared between multiple service types. The service may provide either point-to-point, multicast (that is multipoint-to-multipoint or broadcast), or point-to-multipoint connectivity as defined in Chapter 4.

22.4.4 Protocol-Specific Functions

A key characteristics of packet services is the existence of standard methods to carry other protocols, defined as protocol interworking. A standard is defined for how IP can be carried over each data communication service protocol. Chapter 17 described how multiple protocols are encapsulated and carried over ATM, as defined in IETF RFC 1483.

22.4.5 Summary of Networking Aspects

Table 22.2 compares the networking aspects of public data communication services. The preceding sections defined the terms, concepts, and acronyms listed in this table.

Table 22.2 Comparison of Networking Aspects

Aspect	X.25	IP	FR	SMDS	ATM
Addressing Plan	X.121, LCN	IP	E.164, DLCI	E.164	E.164, NSAP, VPI/VCI
Maximum Packet Length	< 1024 octets	<65,535 octets	<8192 octets	<9188 octets	<65,535 octets
IP Encapsulation	RFC 877	Native	RFC 1490	RFC 1209	RFC 1483
Dynamic Routing	Opt	Rqd	Opt	Opt	Opt
Routing Standard	None	RIP,OSPF, BGP	None	ISSI	P-NNI
Access Medium	Dedicated	Dedicated	Dedicated	Shared or Dedicated	Dedicated
Access Sharing	No	No	Yes	Yes	Yes
Point-to-Point	Yes	Yes	Yes	Yes	Yes
Multicast	No	Yes	Yes	Yes	No
Pt-to-Mpt	No	No	No	No	Yes

22.5 PACKET SWITCHING IN THE LOCAL AREA

This section compares and contrasts aspects of packet switching that are unique to local area environments. The status of FDDI and its follow-on technology FDDI-II is summarized, along with the emerging 100 Mbps Ethernet standard. These are then compared with ATM. See References 1 and 2 more for more details on these subjects.

22.5.1 FDDI and FDDI-II

The Fiber Distributed Data Interface (FDDI) was designed to provide either a high-performance LAN or a campus backbone. Shared FDDI MANs can be connected via OC-3 SONET pipes to form a wider area network, subject to distance constraints. FDDI is also being defined to operate over copper interfaces, in a Copper Distributed Data Interface (CDDI). FDDI does not provide the address screening and security features found in SMDS. FDDI also contains some distance limitations. FDDI-II may eventually correct the lack of FDDI isochronous service capability and allow FDDI to compete more effectively with ATM. But it does seem that FDDI will coexist with ATM for some time. In fact, FDDI and ATM can be viewed as competing technologies. FDDI can also coexist with ATM on a feeder basis. Some projections show that ATM may displace FDDI by late 1995. There are others who believe that the announcement of the death of FDDI may be premature [2].

22.5.2 100-Mbps Ethernet

There is no question that Ethernet is the king of the LAN today. It is extremely cost-effective, with adapter cards typically priced at less than $100. The 10-Mbps limit, or realistically 3 or 4 Mbps when the medium is shared, is not a limit for many clients. Power users and servers can be accommodated by segmenting the LAN to have fewer users per shared medium. A recent trend is to move toward Ethernet switching with only one client or server per Ethernet segment. Servers can be connected via multiple single-user Ethernet segments when more bandwidth is needed. This allows better throughput and extends the life cycle and investment of the existing LAN and wiring structure.

The IEEE has been working to standardize a *Fast Ethernet* operating at 100 Mbps in response to the accelerating need for more bandwidth. Two standards are expected: 802.12 and an addendum to 802.3. The IEEE has also been specifying a 16-Mbps isochronous Ethernet designed to support multimedia, something which the current Ethernet cannot do well.

22.5.3 Comparison with ATM

Table 22.3 shows a comparison matrix of FDDI, FDDI-II, 100-Mbps Ethernet, and ATM.

22.6 BUSINESS ASPECTS OF PACKET SWITCHING

This section compares several business aspects of data communication services. These include the tradeoff between efficiency and features, a quantification of the savings due to integration, an assessment of the market demand, and impacts on user hardware and software.

Table 22.3 Comparison of FDDI, FDDI-II, 100-Mbps Ethernet, and ATM

Attribute	FDDI	FDDI-II	100-Mbps Ethernet	ATM
Throughput	100-Mbps Simplex	100-Mbps Simplex	100-Mbps Simplex	25-Mbps to 600-Mbps Duplex
Evolution Potential	Little	Some	Some	Best
Reserved Bandwidth	No	Yes	Yes	Yes
Isochronous Support	No	Yes	Yes	Yes
Multiple Traffic Classes	No	Yes	Yes	Yes
Projected Cost	High	High	Low	Medium
Use of Existing Wiring	No	No	Yes	Yes
Scalable in Speed	No	No	No	Yes
Scalable to Wide Area	No	No	No	Yes

22.6.1 Efficiency versus Features

One thing that you may want to consider is the tradeoff of efficiency versus features. Increasing the number or quality of features often comes at the expense of decreased efficiency. In this section the efficiency and features of three major data protocols that can be used in private or public networking are compared: frame relay, ATM (using AAL5), and 802.6/SMDS (using AAL3/4). The protocol efficiency, and not the impact of the Physical Layer Convergence Protocol (PLCP) on efficiency, is compared in this section and is shown in Figure 22.6.

Frame relay supports variable-length packets, with an overhead of 5 to 7 bytes per packet (excluding zero insertion). This is the most efficient protocol with respect to overhead of the three protocols considered here. However, frame relay may not support multiple QoS classes, especially if some frames are very long. The longest standardized frame size required is 1600 bytes, while the protocol will support frames up to 8192 bytes long. Its efficiency approaches 100% for very long user data packets (for example, IP packets).

Percentage Efficiency

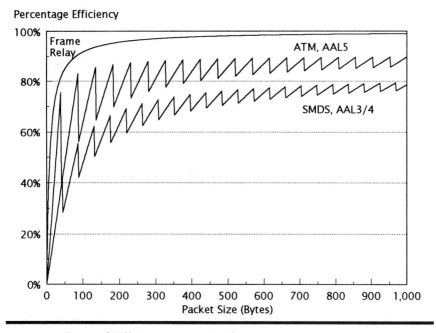

Figure 22.6 Protocol Efficiency versus Packet Size

ATM using AAL5 provides functions very similar to frame relay, and provides the additional flexibility of mixing very long packets with other delay-sensitive traffic. AAL5 also allows support for up to a 64-kbyte packet length, which frame relay and SMDS do not. The 8 bytes in the trailer combined with the 5 bytes of ATM cell header overhead reduce the achievable efficiency by 17%. Because the variable-length packet must be segmented into fixed-length cells, the resultant efficiency decreases markedly when this segmentation results in one or a few bytes of packet data in the last cell. Its efficiency approaches 90% for very large packets.

SMDS currently utilizes the IEEE 802.6 Distributed Queue Dual Bus (DQDB) cell formatting. The per-cell overhead and formatting of the payload field are identical to AAL3/4. AAL3/4 provides an additional level of multiplexing in the MID field and a per-cell CRC, consuming an additional 4 bytes per cell. The first cell in an SMDS packet contains 44 bytes of information, including the source/destination addresses and other fields. Much of the flexibility and many of the features of SMDS derive from this header. Packets may be up to 9188 octets long, slightly longer than the maximum in frame relay. This flexibility reduces efficiency, with the best limited to about 80% maximum efficiency for very large packets. Figure 22.6

plots the resulting efficiency of each protocol versus user packet size. Note how very short packets are handled very inefficiently by ATM Adaptation Layers (AALs) 3/4 and 5.

In one further example, in a study performed on current Internet traffic, a calculation showed that the efficiency of offering this traffic over ATM would be approximately 60% because of the large proportion of very short packets!

Which protocol is best for your application? If you need a feature that only the less efficient protocols support, then the choice is clear; you can't use a more efficient protocol if it doesn't support a critical feature. If your network will require support for multiple QoS classes for different concurrent applications, then ATM is probably a good choice. If you need to connect to other networks via a public service, or require an additional level of multiplexing, then 802.6/SMDS is probably a good choice, either as a standalone service or carried over an ATM-based service. If raw efficiency is key, and support for multimedia applications is not required, then frame relay is a good choice.

22.6.2 CPE Hardware and Software Impacts

X.25 and IP already work on most existing hardware platforms. Software support for both of these protocols is also widely available. Few hardware changes are generally required for frame relay at lower speeds, and software is becoming widely available. Upgrading a data communications network to SMDS generally requires new hardware and software. Hardware includes new interface cards, switches, hubs, and routers. Software includes operating systems, device drivers, and applications. The SMDS DXI is usually packaged as a separate CSU/DSU that interfaces to existing hardware at lower speeds, but it is now finding its way into many internetworking products.

ATM also requires new hardware and software support that is only now finding its way into the marketplace, at rather high initial prices. These prices are coming down rapidly. The ATM DXI interface also promises to be a lower cost, smaller hardware change entry point for ATM similar to the SMDS DXI interface that enables an earlier start for the deployment of ATM.

22.6.3 Integration Savings

X.25, IP, frame relay, and SMDS can pass data traffic only, whereas ATM supports transport of voice, data, and video. The need for multimedia and mixed services over a single integrated access line or shared switch port is a key advantage of ATM. How much savings does this integration achieve? Does it justify the additional cost of new hardware and software?

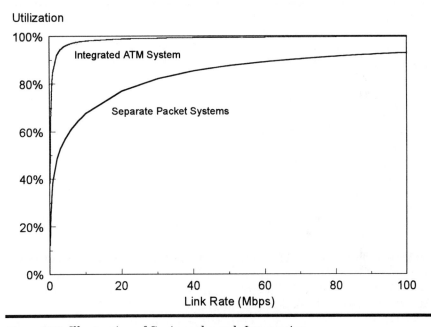

Figure 22.7 Illustration of Savings through Integration

The case of multiple services with different delay requirements sharing a link is examined versus establishing separate links for each service. Costs are assumed proportional to the link speed. The delay performance for an individual M/M/1 system and a priority queueing system was analyzed in Chapter 15. In this simple example, it is assumed that there are two classes of traffic with different average delay requirements: a real-time class with an average delay of no more than 100 μs and a non-real-time class with average delay of 5 ms. The integrated ATM system uses priority queueing as described in Chapters 13 and 15 for the entire link bandwidth. The separate system uses two separate links, each half the capacity of the integrated system. A fair comparison is made by forcing the integrated system real-time load to be equal to the separate system. Figure 22.7 plots total utilization of the two systems versus link bandwidth. The integrated system has higher utilization because it is carrying much more low-priority traffic than separate packet systems for the high- and low-priority with equal capacity.

Of course, this is only the bandwidth saving for two classes of service. The utilization advantage of an integrated ATM system becomes even greater as more traffic classes are considered. Further savings due to integration of network management, provisioning, operations, equipment costs, and the like can also be very significant.

22.6.4 Market Demand

X.25 has been around for many years, and is much more widely available in Europe than in North America. Part of the reason for this is the relatively expensive cost of a private line in Europe as compared to X.25 service. Furthermore, X.25 is slow and has high processing overhead and long delays.

IP networks are experiencing phenomenal growth, which indicates the pent-up demand for ubiquitous data communications access. However, IP currently cannot support multiple classes of service or reserve bandwidth.

Frame relay has also seen substantial interest from large corporations constructing virtual private networks. Frame relay has ineffective congestion control, and the large number of PVCs required make it challenging to manage.

SMDS has seen limited growth so far, primarily because it is more complex than frame relay or basic ATM and because of the lack of carriers' agreement on interconnection. SMDS promises to provide ubiquitous access if these agreements can be reached. SMDS does not support isochronous traffic and has no congestion control mechanisms.

ATM is just being introduced as a PVC service which will likely be used by corporate customers in a manner similar to frame relay to construct virtual private networks. ATM is not yet widely deployed and can be difficult to economically justify in some cases. The true promise of ATM can only be realized through SVCs, which will require assignment of addresses and carrier interconnection. As outlined in Chapter 11, the initial usage of IP over ATM is predicated on an ATM SVC network capability. ATM also is backward compatible with the frame relay and SMDS technologies and can interconnect these different protocols over virtual private networks.

22.6.5 Summary of Business Aspects

Table 22.4 provides a comparison of business aspects of the currently available public data services.

Table 22.4 Comparison of Business Aspects

Aspect	X.25	IP	FR	SMDS	ATM
Efficiency (1-5, best to worst)	2nd	3rd	1st	5th	4th
Multiple Service Support	None	None	Possible	None	Best
CPE Hardware Use	Existing	Existing	Existing	New	New
CPE Software Use	Existing	Existing	New	New	New
Ubiquitous Access	Yes	Yes	No	TBD	TBD
Evolution Potential	Low	Some	Some	Some	High

22.7 MEETING APPLICATION PERFORMANCE NEEDS

This section covers compares how the various services are positioned to meet the performance needs of applications. First, the relative throughput of each service type and underlying technology is summarized. Next, the tradeoffs between burstiness, throughput, and delay are examined. Finally, the approximate performance requirements of various applications are summarized.

The data communications network industry is offering an increasingly wider range of services — providing some very attractive features such as enhanced user control and management, increased maximum bandwidth, improved price-performance ratio, and the support for multiple applications by the same service. This section analyzes how the attributes of user application needs for throughput, burstiness, and delay tolerance are applicable to a range of previously introduced public data services along with ATM.

22.7.1 Throughput

Figure 22.8 shows the history of maximum available throughput for commercially available data services. The time frame is either the historical or projected year in which the maximum data rate is widely offered and is cost-effective. From this perspective we see that there is some differentiation in maximum throughput with ever-increasing maximum bandwidths on the horizon. However, throughput is not the only dimension of performance.

Maximum Data Rate (Mbps)

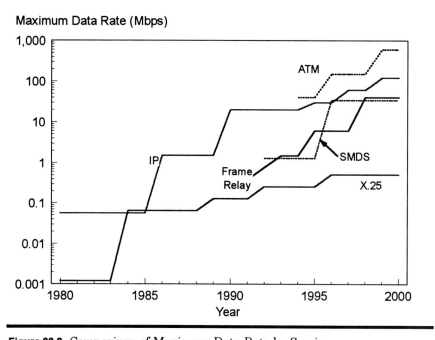

Figure 22.8 Comparison of Maximum Data Rate by Service

22.7.2 Burstiness

Figure 22.9 plots the characteristic of burstiness, defined as the ratio of peak to average rate on the vertical axis versus the supportable peak rate of the service, or throughput, on the horizontal axis. The term "circuits" encompasses both circuit switching and private lines, the choice being based upon the economics summarized earlier in this chapter. The enclosed region for a particular service indicates that it is applicable to that region of burstiness and throughput characteristic. Note that a number of the services overlap in their range of applicability. As described in the previous section, note that the time frame in which peak rate throughput has been available progresses from left to right.

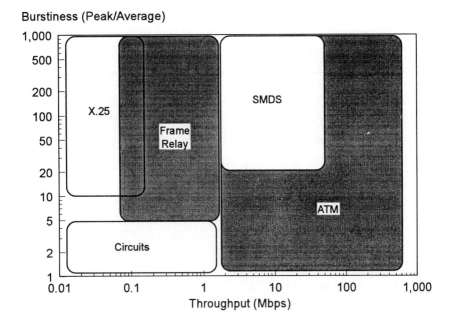

Figure 22.9 Service Applicability to Burstiness and Peak Throughput

22.7.3 Response Time and Delay Tolerance

Figure 22.10 depicts the applicability of services in another dimension, range of nodal delay on the vertical axis versus peak throughput on the horizontal axis. The chart shows that circuit switches have essentially constant nodal delay. Any form of packet switching will introduce variations in delay, and typically, the lower the speed of the packet switching trunks, the more the variation in delay as shown by the general trend of the services to support better (i.e., lower) nodal delays as the peak throughput increases. X.25 and SMDS are designed to support data-only services and thus are not well suited to supporting specific values of delay. Frame relay can support more specific values of delay through the use of prioritization. ATM can support delay close to that of dedicated circuits up to ranges exceeding those of the other services through the implementation of priority queueing in support of multiple Quality of Service (QoS) classes, a topic that Part 5 covered extensively.

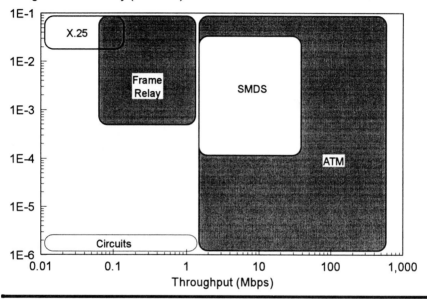

Figure 22.10 Service Applicability to Delay Tolerance and Peak Throughput

Figure 22.11 depicts the applicability of services in the dimension of range of nodal delay on the vertical axis versus burstiness on the horizontal axis. As can be observed from the previous charts, dedicated circuits are best suited to applications that are not bursty and have a strict delay tolerance, such as video, audio, and telemetry data. The X.25 service depicts the classic wisdom of packet switching targeted to bursty applications that can accept significant variations in delay. Frame relay and SMDS support a broader range of application burstiness, while frame relay has the potential to support better delay tolerance through prioritization. ATM supports the broadest range of burstiness and delay tolerance through the implementation of multiple QoS classes.

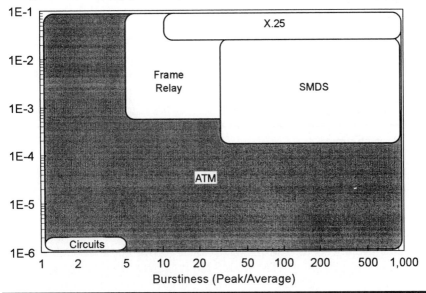

Range of Nodal Delay (seconds)

Figure 22.11 Service Applicability to Delay Tolerance and Burstiness

Table 22.5 Application Traffic Attributes

Application	Burstiness	Delay Tolerance	Response Time (ms)	Throughput (Mbps)
Voice	Medium	Low	Real-Time	.004 to .064
File Transfer	Often High	High	Batch	.01 to 600
CAD/CAM	High	Medium	Near Real-Time	10 to 100
Transaction Processing	High	Low	Near Real-Time	.064 to 1.544
Channel-to-Channel	Low	Low	Real-Time	10 to 600
Imaging	High	Medium	Real-Time	.256 to 25
Business Video	Low	Low	Real-Time	.256 to 16
Entertainment Video	Low	Low	Near Real Time	1.5 to 50
Ubiquitous Video	Low	Low	Real-Time	.128 to 45
Isochronous Traffic	Low	Low	Real -Time	.064 to 2.048
LAN-LAN	High	High	Real- Time	4 to 100
Server Access	Avg.	High	Real- Time	4 to 100
Hi-Fi Audio	Low	Low	Real- Time	.128 to 1

22.7.4 Summary of Application Needs

The above graphs illustrate the power of ATM to serve a broad range of application characteristics. What are examples of applications that require different values of these performance measures? Table 22.5 summarizes the approximate attributes for a number of applications.

22.8 REVIEW

This chapter compared data services from several points of view. Data communications was first categorized based on the method — message, circuit, or packet switching — identifying that most modern methods are based on packet switching. Message switching has largely moved up the protocol stack to the application layer. A detailed comparison of ATM with STM was presented, with the many advantages of ATM over STM (depending on the traffic characteristics). A detailed comparison of the X.25, IP, frame relay, SMDS, and ATM data services was then presented. The comparison covered functional, technical, and business aspects of these services. ATM, FDDI, and the emerging 100-Mbps Ethernet standards for local area applications were compared. The chapter concluded with an analysis of application performance requirements.

22.9 REFERENCES

[1]   S. Saunders, "Choosing High-Speed LANs – Too many technologies, too little time?," *Data Communications*, September 21, 1993.

[2]   L. Gasman, "Don't Count FDDI Out - Yet," *Data Communications*, August 1993.

23

Future Directions
Involving ATM

What will be the future of data communications involving ATM? This chapter does not foretell the future, but it does look at early experiences, missing elements, open issues, and competing technologies to identify some likely scenarios. References are made to other authors who have published their visions of the future. Beginning with a review of the experience of ATM test beds and early adopters, the chapter moves on to explore the range of possible scenarios advocated by various industry experts. The extensive list of challenges ahead in the areas of missing functions, price performance, open traffic management issues, network management, and seamless networking identified earlier are reviewed as factors critical to the success of ATM. ATM's expansion into other transmission media is then covered.

23.1 EARLY DIRECTIONS FOR ATM

One way to predict the future is to extrapolate from early experiences on the leading edge of an emerging technology. The gigabit test beds in North America, the RACE project in Europe, and NTT trials in Japan provide a good source of input in this area. This section also summarizes who the early adopters of ATM are, and why they chose ATM.

23.1.1 ATM Test Bed Experience

The United States began the development of test beds for gigabit networking in 1990 with the National Research and Education Network (NREN) [1]. These test bed projects are sponsored by the United States Government's Advanced Research Projects Agency (ARPA) and the National Science Foundation (NSF). There are six test beds in the United States, as depicted in Figure 23.1: Aurora, Blanca, Casa, Magic, Nectar, and VistaNet. The gigabit rate that has actual been used is the STS-12c rate of 622 Mbps described in Chapters 3 and 8. Some of the links are only DS3 speed. ATM is used in the Aurora, Blanca, Nectar, VistaNet, and Magic test beds. These test beds became operational in 1993 and 1994. Switch manufacturers, application developers, and carriers have all gained experience from these test bed efforts.

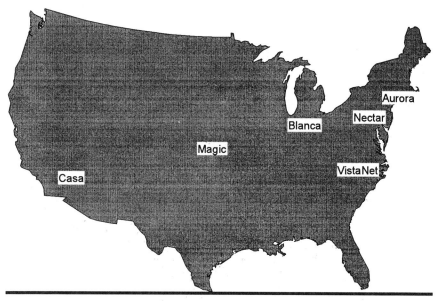

Figure 23.1 United States Gigabit Test Beds

The U.S. government has begun a National Information Infrastructure (NII) initiative targeting deployment of multimedia, interactive, high-performance computing and communications within the educational system and society as a whole beginning in 1994. The funding is largely split across the National Science Foundation (NSF), Defense Advanced Research Projects Agency (DARPA), the

Department Of Energy (DOE), and the National Aeronautics and Space Administration (NASA).

The Research for Advanced Communications in Europe (RACE) project 1022 was begun in 1987 to demonstrate the feasibility of broadband ATM networking [2]. RACE is a consortium with 26 partners, comprised of most European network operators, major companies, and several universities. The R1022 ATM Technology Testbed (RATT) located in Basel, Switzerland, has prototype or pre-production examples of end systems, local switches, multiplexers, and switches. The results achieved to date have exceeded expectation and the RATT is being used to demonstrate basic ATM services.

RACE project 2061, also called EXPLOIT, is a major portion of RACE, with a project goal to achieve Integrated Broadband Communications (IBC) services in the European Community by 1995 [3]. This project, also located in Basel, Switzerland, features 2-Mbps (E1) and 140-Mbps (E4) ATM interfaces, frame relay interworking, and N-ISDN interworking. This activity is focused on interworking with existing frame relay and N-ISDN services and operation over the current digital hierarchy in order to facilitate early introduction of ATM.

The Japanese Nippon Telephone and Telegraph (NTT) Corporation selected several manufacturers to develop an ATM-based network that was to replace the existing trunk infrastructure by 2005 [4]. N-ISDN has seen moderate deployment in Japan, and NTT has been experimenting with extensions to the N-ISDN services. Potential services include video telephony, language lessons, and image retrieval from optical mass storage systems. Increasing competition in Japan and the lack of additional investors to support the tremendous capital outlays required have delayed initial investments [5] and deployment plans. NTT has significant experience in commercializing new services and will be a significant player in the future.

23.1.2 Who Are the Early Adopters?

This section identifies some of the early adopters of ATM, and the sector they come from, and indicates their principal business driver for the use of ATM. The next section gives further details on early uses of ATM.

Major early users of ATM are U.S. government agencies. The Department Of Energy (DOE) and NASA have selected ATM as the basis for a nationwide network. The Defense Advanced Research Project Agency (ARPA) is constructing a high-speed SONET and ATM network in the Washington, D.C., area. The National Science Foundation is planning to deploy a high-speed ATM-based network to interconnect supercomputers as well.

The announcement of the North Carolina Information Highway that will support distance learning and provide access to electronic libraries is based upon ATM.

Some commercial customers have announced plans to adopt ATM, including Hughes, TASC, Bear-Stearns, and SUN.

23.1.3 Why They Chose ATM

Some of the most common reasons for choosing ATM initially are:

- ☞ Construction of a private, high-performance network
- ☞ Enable high-performance, leading edge applications
- ☞ Construct a building or campus backbone
- ☞ Provide the highest speed and performance data communications network
- ☞ Experiment with the latest data communications technology
- ☞ Obtain a national advantage in the competitive arena of technology

The following are reasons why these early adopters believe ATM will meet their future needs:

- ☒ Provide a single, common, simplified, unifying infrastructure for seamless networking
- ☒ Support emerging, productivity-enhancing multimedia workstations and applications
- ☒ Future-proof investments by choice of a scalable, flexible technology
- ☒ Provide a standard upgrade path based upon the rates in the SONET/SDH hierarchy

23.2 POSSIBLE FUTURE DIRECTIONS

What are the possible directions for ATM? What is the best case?
What is the worst case? What is the likely scenario? This section
provides some answers to these questions.

23.2.1 The Universal Network?

The charter of the CCITT for the Broadband Integrated Digital
Services Network (B-ISDN) was to use ATM as the switching and
multiplexing method for carrying all signals in the future [6]. In fact
some industry experts claim that B-ISDN will be the universal
network of the future.

The practicality of a universal B-ISDN vision is questioned by
many however. Experience with technology indicates that someone
will come up with a better idea eventually. Also, since B-ISDN has
been developed in large committees the results are inevitably
weakened by compromise in the standards process. ATM may be a
survivor of the standards process and be the foundation for a
successful data communications infrastructure.

Probably ATM networks will coexist with existing STM based
networks for a long time [6]. The era of the universal adoption of
ATM will likely never occur for various reasons covered later in this
chapter.

23.2.2 A New Star in the Multiprotocol Network?

ATM is the current hype storm of the data communications industry.
Other technologies have been hyped in the past, and others will be in
the future. There is possibly more hype surrounding ATM than has
occurred before in data communications history. It is likely that
ATM will be able to deliver on almost everyone's expectations. We
believe that ATM will be the star in the ever-evolving multiprotocol
network, probably for at least the next five years.

Some bright points of light that ATM promises to deliver on, as
described earlier in the book, are summarized in Figure 23.2. The
remainder of this section provides a description of each application
with references to earlier portions of the book and comments on the
relative potential acceptance of each.

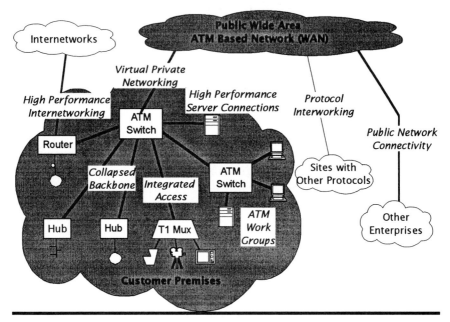

Figure 23.2 Illustration of Likely Applications for ATM

23.2.2.1 Collapsed Backbone

Today's LAN backbone collapsed on the backplane of the router will be collapsed on an ATM switch. Ethernet, Token Ring, and other LAN interfaces will be connected within buildings and campuses by ATM.

23.2.2.2 Integrated Access

Circuit emulation capabilities will allow circuit-based voice, video and data to share expensive access costs. Voice and video integration will flow from the access line, through the LAN, to the desktop, and all the way to the end application, becoming a (virtual) reality.

23.2.2.3 Virtual Private Networking

Wide area connections will be needed between ATM-enabled sites which have applications that require the bandwidth and flexibility of ATM. Carrier-provided virtual networks will be more cost-effective than private networks for many users. Carrier network support for Switched Virtual Connections (SVCs) and support for connectionless traffic over ATM will be developed to meet user needs.

23.2.2.4 High-Performance Internetworking

The advertised death of the router is premature; high-speed connections to internetworks will occur. Support for FDDI, higher layer protocols, and routing functions will be done separately or in hybrid router/switches.

23.2.2.5 Protocol Interworking

ATM-based interworking with other protocols will support enterprise connectivity to the many sites which will not have ATM initially. Virtual public networks will provide this capability to interconnect sites with ATM, frame relay, IP, and SMDS interfaces via a portfolio of interworking services.

23.2.2.6 High-Performance Server Connections

Servers will need the performance of ATM before clients, and will lead the way for ATM to the desktop. FDDI, Switched Ethernet, and Fast Ethernet will be competitors and will be the true market test for acceptance of ATM.

23.2.2.7 ATM Workgroups

ATM adapter cards in high-end workstations and servers will see wider use. Workgroups will be connected by local switches, with access to legacy LANs and other users via a collapsed ATM backbone.

23.2.2.8 Public Connectivity

Standards-based ATM, and public addressing with SVCs, will enable ubiquitous communications, like the Internet and voice networks today. The need for intercarrier business arrangements may delay this goal.

23.2.3 ISDN — Again?

There was a similar vision and a significant amount of hype regarding the Time Division Multiplexing (TDM)-based Integrated Services Digital Network (ISDN) becoming the universal solution for all communications in the late 1970s. The ISDN PBX would handle all voice and data communication in an integrated manner that would work seamlessly in the office, and around the world. The general consensus is that this has not, and will not, come to pass. The fact is that although initial ISDN standards were released in 1984, only a few parties can make an international ISDN call today.

Access to ISDN services is simply not available to most users. The current ITU-T standard has reused this name, and calls it Broadband ISDN (B-ISDN). Is this just an unfortunate choice of names, or a move to avoid admitting defeat? The current ITU-T direction has the same universal vision as summarized previously. Can B-ISDN succeed where ISDN did not?

Let's look at what happened to ISDN to see if the same fate awaits B-ISDN. While the ISDN standards were being developed, Local Area Network (LAN) protocols were developed and being standardized by the IEEE. The most successful LAN data communications protocol today is IEEE 802.3 Ethernet, with IEEE 802.5 Token Ring a distant second. Both of these technologies were invented first, and then formally standardized. The ARPANET was also being developed at this time, and it has blossomed into the most rapidly growing public data communication network in the world today — the Internet. After experiencing such tremendous success, the Internet established its own de facto standards body — the Internet Engineering Task Force (IETF).

The signaling protocol developed for ISDN had as an objective to interwork with any telephone system deployed within the last century, as well as provide new features and functions for applications that were not even envisioned yet. The result was a very complicated signaling protocol specified in CCITT Recommendations Q.921 and Q.931. Achieving interoperable implementations between user devices and network switches has taken years to achieve. These recommendations specified a signaling protocol at a User-Network Interface (UNI). Unfortunately, most of the vendors built the network side of the interface expecting the user side of the interface to be developed. One reason for the failure of ISDN was largely due to never achieving a critical mass of user-side signaling implementations. Another key reason is that Local Exchange Carriers (LECs), IntereXchange Carriers (IXCs) and international Postal Telegraph and Telephone (PTT) operators have not been able to coordinate the offering of a ubiquitous international ISDN service.

However, the situation is different for ATM. As outlined in Chapter 6, the ATM Forum was established with participants from the data communications as well as the telephony carrier sectors — this combination of forces is fundamentally different than what occurred in ISDN, and is the brightest ray of hope for ATM. The choice of the name ATM avoids the four-letter word ISDN in many of their documents, even though the carrier and ITU influence can still be found in many places in the documentation. The ATM Forum has chosen to build the signaling protocol on the ISDN protocol as described in Chapter 9. Herein lies a critical juncture for the success of ATM. If users and networks can build interoperable, high-performance signaling systems, then user needs can be met. The ATM Forum is also blazing new trails in areas traditionally outside the scope of standards bodies in the areas of applications, protocol interworking, routing, and LAN emulation. These are also very different from ISDN.

23.2.4 A Change in Power?

The era of ATM may also usher in a change for those in power. Some industry analysts have published prophesies that ATM will introduce new carriers and service providers into the marketplace. The ones with the most to lose could be the traditional telephone carriers. There is already some evidence of a trend such as this with companies such as Metropolitan Fiber Systems (MFS) using ATM to offer transparent LAN connection service. The introduction of a new technology or paradigm causes chaos and creates the opportunity for a significant change in market share. The explosion of multiprotocol routing in data communication in the early 1990s is evidence of a similar phenomenon. Also, virtual private voice networks have taken a large amount of market from integrated voice/data private network solutions. The adoption of ATM-based technology for video-on-demand and future interactive applications to the home, as described in Chapter 18, also points to a potential shift in power.

23.3 CHALLENGES AHEAD

ATM has several challenges ahead and some key features that are missing. The economics of ATM must prove to be more favorable to spawn widespread use. The critical areas of switching and routing must also be of high performance for ATM to deliver on its promises. The complex area of traffic management must continue to forge ahead to meet user expectations. The critical area of network management must also be addressed. Finally, the goal of seamless networking must be kept in mind by the standards bodies, developers, and users if it is to ever be reached.

23.3.1 Missing Links?

There are still a number of things that users still expect to see from ATM before they are ready to commit their funds. These include the following:

- DS1 and lower speeds for more cost-effective access
- Channel speed connections for supercomputers
- Interoperability and conformance testing
- More economically attractive pricing
- Success stories for ATM
- Technology, standards, and applications training
- Interworking with current LANs and applications
- ATM-enabled application and operating system software
- Application Programming Interfaces (APIs) for ATM

Users cannot afford high-speed access for all of their locations. ATM must provide a range of performance better matched to customers' business needs. This ranges from the very low end to the high end. Better testing for conformance to standards to ensure interoperability is needed; otherwise users will be locked into a proprietary solution. ATM must be economically justified before the majority of corporate users will adopt it, and this must be documented by published success stories. More training on ATM is needed, along with applications that can effectively use ATM. Many users expect that their current LANs will interoperate with ATM.

23.3.2 Proving in on Price

Economics for ATM need to prove favorable in at least the following areas:

- Interface cards for workstations, routers, and hubs
- Switch common equipment and port costs
- Demonstrable reductions in operations and support costs
- Test and analysis tools
- Design tools
- Access costs

The cost of ATM interface cards and switch ports must be competitive with FDDI and Fast Ethernet. The cost savings of integrating all data communications on ATM must be demonstrated through published case studies. More cost-effective test, analysis, and design tools must be made available. The barrier of very high cost of DS3 or SONET/SDH access must be overcome for ATM to be successful in the WAN.

23.3.3 High-Performance Switching and Routing

To date, ATM standards have been defined on the assumption that all bandwidth is reserved — an assumption directly at odds with the shared-bandwidth LAN environment. While this fundamental difference in bandwidth management is natural for some applications, it greatly hampers others.

For example, an engineer easily can determine how much bandwidth is required for a video call and just as easily select an ATM SVC at that bandwidth. Most data applications, which have no concept of making a connection prior to transferring data, also have no clue about the bandwidth required on such a connection.

There is a critical need for high-performance switching in order to support the emulation of connectionless services by what is effectively fast circuit switching using the approaches described in Chapter 11. There is in existence proof that fast circuit switching can be implemented, if the protocol is simple enough. The fast circuit switched X.21 data networks in Scandinavia achieve call setup times on the order of 100 ms or less.

User expectations have been shaped by the types of applications that use whatever bandwidth is available and that provide flow control to adapt for congestion. The astounding increase in computing power and sophistication of applications is fueling the demand for this type of service beyond Ethernets' ability to satisfy it. Thus, a new technology such as ATM will be needed.

23.3.4 Unresolved Traffic Management Issues

A large number of published technical articles describe the complexities, unsolved (or unsolvable) problems, issues, and proposed solutions on the general topic of traffic management (Refs 7, 8, 9, and 10). The fact that this book dedicated five chapters in Part 5 to this subject is evidence of the complexity and importance of this topic. We believe that the basic mechanisms for reserving bandwidth in

ATM are defined sufficiently by the leaky bucket algorithms and requirements defined in Chapters 12 and 13 to achieve interoperability. The problem of achieving LAN-like flow and congestion control over ATM will take longer to solve, and is a critical issue for the success of ATM. Chapter 16 described how the ATM Forum is focusing on this issue as a high priority. If the solution developed by the ATM Forum balances complexity against optimal performance and achieves industry acceptance, then a major step towards the goal of seamless networking using ATM will have been made.

The problem of determining Connection Admission Control (CAC) procedures to implement a network to provide multiple Quality of Service (QoS) classes will be a challenging one. However, this problem is not all that different from providing a Committed Information Rate (CIR) in frame relay. The ability for a network provider to perform this balancing act will be a competitive differentiator.

23.3.5 Managing it All

Network management is a critical feature on any network manager's checklist (see Refs. 13, 14, and 15). That network management usually comes last is due to the fact that the communications technology must first be defined before the means to manage it can be completely determined. Currently, ATM network management standards lag the capabilities required by the end user, creating the need for interim solutions and workarounds.

Part 7 summarized the current ATM standards use of special Operations, Administration, and Maintenance (OAM) ATM cells to provide fault management and performance measurement at the ATM layer. The protocols using these cells can indicate faults, detect faults, and verify continuity via a loopback. They also can be used to estimate the virtual ATM connection's performance in terms of delay, error rate, and loss. Such capabilities are critical in managing an ATM network that supports multiple applications, each of which has different quality requirements. Unfortunately, most users operate at the virtual channel connection (VCC) level, where there still are at least three major limitations:

⊗ There may not be support for the fault-detection capabilities at the VCC level

⊗ Performance measurement probably will be economical only when done across large numbers of VCCs enclosed in an ATM virtual path

⊗ Current standards for delay measurement may be inadequate

Users should consider employing continuity checks and delay measurements in higher layer protocols (such as ICMP Ping) until the ATM standards are completed.

The ATM Forum's Interim Local Management Interface (ILMI), created in 1992, is limited in function, allowing a user to only query statistics on configuration, basic status and physical interface-level usage for an ATM UNI. The ILMI was only supposed to be an interim interface, but many of the ILMI functions have not yet been replaced by a permanent standard. Help may come from a definition by the Internet Engineering Task Force (IETF) of an ATM Management Information Base designated AToMMIB, as Chapter 19 summarized. This includes much of the information users need to manage ATM PVCs, such as detailed status and connection reports, traffic statistics for virtual connections, and ATM Adaptation Layer (AAL) statistics. A great deal more standards work needs to be done before network management meets user and network provider needs in a truly interoperable environment.

The introduction of SVCs will require accounting and such statistics as blocking, call attempt, invalid attempt, and busy counts. To improve cost-effectiveness for both ATM PVCs and SVCs, subscribers need usage statistics and accounting reports on a per-virtual-connection basis.

Two established network management protocol standards serve as the foundation for ongoing efforts to develop ATM network management capabilities; the IETF's highly interoperable Simple Network Management Protocol (SNMP) and the OSI-based Common Management Interface Protocol (CMIP). IETF and ATM Forum activities center on SNMP, while the ITU-T and ANSI are working to adapt the general-purpose Telecommunications Management Network (TMN) architecture, which is based on CMIP. These dual approaches eventually will converge, with SNMP managing private networks and CMIP used within and between public networks. Users should look to SNMP in the interim until this convergence occurs.

The combination of these two protocol spheres with the SVC protocol, along with the development of interim management tools, is critical to ATM. Given the challenges outlined above, it will be at least several years before the overall ATM network management solution arrives.

23.3.6 Seamless Networking

The identification of seamless networking from the LAN across the campus and across the WAN has been consistently identified by users as a key benefit of ATM [11], [12]. Delivering on this vision will be a key challenge for the standards bodies, vendors, network designers, and service providers. Currently this is only a potential result of the successful application of ATM-based technology.

23.4 ATM OVER MULTIPLE MEDIA

The concept of multiple application level media operating over ATM is well known. This section explores the parallel trend of ATM operating over multiple physical layer media. Figure 23.3 illustrates these possibilities graphically.

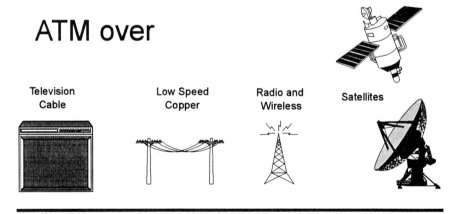

Figure 23.3 ATM over Multiple Media

23.4.1 Satellites

ATM has been successfully demonstrated over satellite links by COMSAT. Standards are being developed that allow operation of

ATM over the over half-second round-trip delays present in geosynchronous satellites. The U.S. Navy is trying ATM over 44.7-Mbps full duplex satellite links.

23.4.2 Lower Speeds

ATM is being defined to operate over lower speeds. ITU-T Recommendation G.804 defines ATM cell mapping over DS1 (1.5 Mbps) and E1 (2 Mbps). There is no reason that the ATM DXI interface cannot operate at even lower speeds.

23.4.3 Wireless

ATM is being considered for use in wireless communications as part of an end-to-end seamless network by the U.S. military. The vision of mobile, radio-based ATM switching systems for field use is being explored by the Air Force.

23.4.4 Cable Systems

ATM is being considered for use as the multiplexing method on cable-based systems. BroadBand Technologies (BBT) has announced a system that uses MPEG over ATM coding that would support over 1500 video signals versus the current 64 in a fiber loop access system [16].

23.5   BETTING ON A WINNER

This book has taken you through the business drivers, technology enablers, a review of data communications, and a detailed look at ATM, spelling out the competing technologies and the challenges ahead. A great deal of ATM is already defined in sufficient detail for interoperability. The needed standards for functions that users need have been identified and are under way. There is a tremendous groundswell of support for ATM across all segments in the data communications industry. These are the ingredients for ATM's success. ATM will be successful; the only questions that remain are how successful, and in which specific areas.

23.6 REVIEW

This final chapter reviewed the state of ATM and expected directions. The research areas of ATM networking were reviewed as an indication of future directions. The early adopters were identified as primarily government agencies and large corporations. The reasons for early adoption of ATM were summarized. Possible scenarios for the fate of ATM were outlined, with the most likely being that ATM will be the star in the multiprotocol network of the 1990s. A number of challenges that ATM must meet in order to meet this goal were identified. The likely applications that ATM will support were reviewed. The other transmission media that ATM is branching out to beyond fiber optics — satellites, low-speed copper, wireless and cable — were reviewed. Finally, reasons for believing that ATM is the winner to bet on were reviewed.

23.7 REFERENCES

[1   R. Karpinski, "Bureaucracy and Politics Aside, Gigabit testbeds Represent Cutting Edge," *Telephony*, July 19, 1993.

[2]   D. De Schoenmacker, P. Verbeeck, "RATT: a glimpse of the broadband future," *Electronics & Communication Engineering Journal*, August 1992.

[3]   M. Potts, "EXPLOITation of an ATM testbed for broadband experiments and applications," *Electronics & Communication Engineering Journal*, December 1992.

[4]   J. Williamson, "Gazing Toward the Broadband Horizon," *Telephony*, 1992.

[5]   N. Gross, P. Coy, "The Ideal Interchange for the Data Superhighway," *Business Week*, October 11, 1993.

[6]   M. Aaron, M. Dècina, "Asynchronous Transfer Mode or Synchronous Transfer Mode or Both?," *IEEE Communications Magazine*, January 1991.

[7]   C. Lea, "What Should Be the Goal for ATM?," *IEEE Network*, September 1992.

[8]   M. Wernik, O. Aboul-Magd, H. Gilbert, "Traffic Management for B-ISDN Services," *IEEE Network*, September 1992.

[9]   M. Decina, "Open Issues Regarding the Universal Application of ATM for Multiplexing and Switching in the B-ISDN," ICC, 1991.

[10]   R. Binder, "Issues in Gigabit Networking," Globecom, 1992.

[11]   J. McQuillan, "Next Generation Attendees Say "Yes — Later"," *Business Communications Review*, January 1994.

[12]   G. Smith, J. P. van Steerteghem, "Seamless ATM Networks," *Telecommunications*, September 1993.

[13]   L. Boxer, "ATM Standards Lag Network Management Needs," *Network World*, March 1994.

[14]   J. McQuillan, "Can ATM Solve the Unsolveable?," *Business Communications Review*, August 1993.

[15]   N. Lippis, "ATM's Biggest Benefit: Better Net Management," *Data Communications*, October 1992.

[16]   R. Karpinski, "BBT to Deliver 1500 Channels," *Telephony*, March 29, 1993.

Acronyms and Abbreviations

AAL	ATM Adaptation Layer
ABM	Asynchronous Balance Mode (HDLC)
ACF	Access Control Field (DQDB)
ACK	Acknowledgment
ADM	Add/Drop Multiplexer
AIS	Alarm Indication Signal
ANS	American National Standard
ANSI	American National Standards Institute
APS	Automatic Protection Switching
ARP	Address Resolution Protocol
ATM	Asynchronous Transfer Mode
AUI	Attachment Unit Interface (Ethernet 802.3)
BCD	Binary Coded Decimal
BECN	Backward Explicit Congestion Notification (FR)
Bellcore	Bell Communications Research
BER	Bit Error Ratio or Rate
BGP	Border Gateway Protocol
BIP	Bit Interleaved Parity
B-ISDN	Broadband Integrated Services Digital Network
B-NT	Broadband Network Terminator
BOC	Bell Operating Company
BOM	Beginning of Message (DQDB)
bps	Bits per second
BRI	Basic Rate Interface (ISDN)

*NOTE : Additional comments in parentheses are a clarification or refer to the standard from which the term is derived. Many acronyms are used by multiple standards and only the most prevalent are mentioned.

BSC	Bisynchronous Communications
B-TA	Broadband Terminal Adapter (ATM)
B-TE	Broadband Terminal Equipment (ATM)
CAD/CAM	Computer Aided Design/ Computer Aided Manufacturing
CBDS	Connectionless Broadband Data Service
CBR	Constant Bit Rate
CCITT	Consultative Committee International Telegraphy and Telephony
CD	CountDown counter (DQDB)
CEPT	Conference on European Post & Telegraph
CIR	Committed Information Rate (FR)
CL	ConnectionLess
CLNS	ConnectionLess Network Service (OSI)
CLLM	Consolidated Link Layer Management (FR)
CLSF	Connectionless Server Function (ITU-T)
CMIP	Common Management Interface Protocol (ISO)
CMIS	Common Management Information Service (ISO)
CMISE	CMIS Element (ISO)
CO	Central Office
COAM	Customer Owned and Maintained
COM	Continuation of Message
CONS	Connection-Oriented Network Service (ISO)
CPE	Customer Premises Equipment
C/R	Command/Response Indicator or bit
CRC	Cyclic Redundancy Check
CS	Convergence Sublayer
CSMA/CD	Carrier Sense Multiple Access with Collision Detection
DA	Destination Address field
DAL	Dedicated Access Line
DCE	Data Communications Terminating Equipment
DCS	Digital Cross-connect System
DE	Discard Eligibility (FR)
DEC	Digital Equipment Corporation
DH	DMPDU Header (DQDB)
DLCI	Data Link Connection Identifier (FR)
DMPDU	Derived MAC PDU (DQDB)
DoD	Department of Defense
DQDB	Distributed Queue Dual Bus (IEEE)
DS0	Digital Signal Level 0
DS1	Digital Signal Level 1

DS3	Digital Signal Level 3
DSG	Default Slot Generator (DQDB)
DT	DMPDU Trailer (DQDB)
DTE	Data Terminal Equipment
DTMF	Dual Tone MultiFrequency
DXC	Digital cross (X)-Connect
EA	Extended Address
ECSA	Exchange Carriers Standards Association
ED	End Delimiter (IEEE 802)
EGP	Exterior Gateway Protocol
EGRP	Exterior Gateway Routing Protocol
EIR	Excess Information Rate (FR)
EOM	End of Message
EOT	End of Transmission
ESF	Extended Super-Frame
ES-IS	End System to Intermediate System protocol (OSI)
ETB	End of Transmission Block
ETX	End of Text
F	Flag
FCS	Frame Check Sequence (FR)
FDDI	Fiber Distributed Data Interface (ANSI)
FDDI-FO	FDDI Follow-On
FDM	Frequency Division Multiplexing
FEBE	Far End Block Error
FEC	Forward Error Correction
FECN	Forward Explicit Congestion Notification (FR)
FERF	Far End Reporting Failure
FM	Frequency Modulation
fps	Frames per second
FR	Frame Relay
FRAD	Frame Relay Assembler/Disassembler, or Access Device
FT1	Fractional T1
Gb	Gigabits (billions of bits)
Gbps	Gigabits per second (10^9bps)
GFC	Generic Flow Control
GFI	General Format Identifier (X.25)
GOSIP	Government Open System Interconnection Profile
HCS	Header Check Sequence (DQDB)
HDTV	High Definition TeleVision

HDLC	High-Level Data Link Control (ISO)
HEC	Header Error Control
HOB	Head of Bus (DQDB) A or B
Hz	Hertz or cycles per second
ICI	InterCarrier Interface
ICMP	Internet Control Message Protocol
IEEE	Institute of Electrical and Electronics Engineers
IETF	Internet Engineering Task Force
IGP	Interior Gateway Protocol
ILMI	Interim Local Management Interface
IMPDU	Initial MAC Protocol Data Unit (DQDB)
IMSSI	Inter-MAN Switching System Interface (DQDB)
IP	Internet Protocol
IPX	Internetwork Packet Exchange protocol (Novell)
IS	Intermediate System (OSI)
ISDN	Integrated Services Digital Network
ISDU	Isochronous Service Data Unit (DQDB)
IS-IS	Intermediate System-to-Intermediate System (OSI)
ISO	International Standards Organization
ISSI	Inter-Switching System Interface
ITU	International Telecommunications Union
ITU-T	ITU - Telecommuncations standardization sector
kb	kilobit (thousands of bits)
kbps	kilobits per second (10^3 bps)
km	kilometers (10^3 meters)
LAN	Local Area Network
LAP-B	Link Access Procedure - Balanced (X.25)
LAP-D	Link Access Procedure - D (ISDN)
LAT	Local Area Transport protocol (DEC)
LATA	Local Access Transport Area
LB	Letter Ballot
LCGN	Logical Channel Group Number
LEC	Local Exchange Carrier
LLC	Logical Link Control (IEEE 802.X)
LME	Layer Management Entity (DQDB)
LMI	Local Management Interface
LSB	Least Significant Bit
LTE	Line Terminating Equipment (SONET)
LT	Line Termination

m	meters
MAC	Media Access Control (IEEE 802.X)
MAN	Metropolitan Area Network (DQDB, FDDI)
Mb	Megabits (millions of bits)
Mbps	Megabits per second (10^6 bps)
MCF	MAC Convergence Function (DQDB)
MCP	MAC Convergence Protocol (DQDB)
MIB	Management Information Base (SNMP)
MID	Multiplexing IDentifier (ATM), Message IDentifier (DQDB)
MIPS	Millions of Instructions Per Second
MPEG	Motion Picture Encoding Group
ms	Millisecond (one-thousandth of a second, 10^{-3} seconds)
MSAP	MAC Service Access Point (SMDS)
MSB	Most Significant Bit
MSDU	MAC Service Data Unit (SMDS)
MSS	MAN Switching System (SMDS)
MTU	Maximum Transmission Unit
MUX	Multiplexer
NE	Network Element
NetBIOS	Network Basic Input/Output System protocol
NFS	Network File Server
nm	Nanometer (10^{-9} meter)
NNI	Network Node Interface, or Network to Network Interface
NP	Network Performance
NOS	Network Operating System
ns	Nanosecond (10^{-9} second)
OAM	Operations Administration and Maintenance
OC-n	Optical Carrier Level n (SONET)
OH	Overhead
OOF	Out Of Frame
OS	Operating Systems
OSI	Open Systems Interconnection
OSIRM	OSI Reference Model
OSPF	Open Shortest Path First
PA	Prearbitrated segment or slot (DQDB)
PABX	Private Automatic Branch Exchange
PAD	Packet Assembler/Disassembler (X.25)

PAF	Prearbitrated Function (DQDB)
PBX	Private Branch Exchange
PDH	Plesiochronous Digital Hierarchy
PDN	Public Data Network
PDU	Protocol Data Unit
Ph-SAP	Physical Layer SAP (DQDB)
PHY	Physical Layer
PL	PAD Length (DQDB)
PLCP	Physical Layer Convergence Protocol (DQDB)
PM	Performance Monitoring or Physcial Medium
PMD	Physical Layer Medium Dependent
POH	Path Overhead (SONET)
POI	Path Overhead Identifier (DQDB)
PON	Passive Optical Network
PoP	Point of Presence
PPP	Point-to-Point Protocol (Internet)
pps	Packets per second
PRI	Primary Rate Interface (ISDN)
PSPDN	Packet Switched Public Data Network
PT	Payload Type
PTE	Path-Terminating Equipment
PTT	Postal, Telegraph & Telephone
PVC	Permanent Virtual Circuit
QA	Queued Arbitrated (DQDB) slot
QAF	Queued Arbitrated Function (DQDB)
QoS	Quality of Service
RBOC	Regional Bell Operating Company
RDI	Remote Defect Indication
REJ	Reject frame
RIP	Routing Information Protocol
RISC	Reduced Instruction Set Computer
RJE	Remote Job Entry
RNR	Receive Not Ready
RQ	Request Counter (DQDB)
s	second
SA	Source Address field
SAP	Service Access Point
SAPI	Service Access Point Identifier (ISO)
SAR	Segmentation And Reassembly
SDH	Synchronous Digital Hierarchy (ITU-T)
SDLC	Synchronous Data Link Control (IBM)

SDU	Service Data Unit
SES	Severely Errored Seconds
SF	SuperFrame
SIP	SMDS Interface Protocol (SMDS)
SIR	Sustained Information Rate (SMDS)
SMDS	Switched MultiMegabit Data Service
SMF	Single-Mode Fiber
SN	Sequence Number
SNA	System Network Architecture (IBM)
SNI	Subscriber Network Interface (SMDS)
SNMP	Simple Network Management Protocol (DOD)
SOH	Section Overhead
SONET	Synchronous Optical Network (ANSI)
SPE	Synchronous Payload Envelope (SONET)
SPF	Shortest Path First protocol
SPM	FDDI-to-SONET Physical Layer Mapping standard (FDDI)
SREJ	Select Reject frame
SRT	Source Route Transparent protocol
SSAP	Source Service Access Point (LLC)
SSCOP	Service-Specific Connection Oriented Protocol
SSCS	Service-Specific Convergence Sublayer
STE	Section Terminating Equipment (SONET)
STM	Synchronous Transfer Mode or Station Management (SDH)
STM-n	Synchronous Transport Module level n (SDH)
STP	Shielded Twisted Pair
STP	Spanning Tree Protocol (IEEE 802.1d)
STS-n	Synchronous Transport Signal Level n (SONET)
STS-Nc	Concatenated Synchronous Transport Signal Level N (SONET)
SVC	Switched Virtual Circuit or Signaling Virtual Channel
SYN	Synchronous Idle
t	Time
TA	Terminal Adapter
TC	Transmission Convergence sublayer of PHY layer
TCP	Transmission Control Protocol Internet
TCP/IP	Transmission Control Protocol/Internet Protocol
TDM	Time Division Multiplexing
TDMA	Time Division Multiple Access
TE	Terminal Equipment
TP4	Transport Protocol Class 4 (ISO)

TR	Technical Report
UDP	User Datagram Protocol (Internet)
UNI	User Network Interface
UTP	Unshielded Twisted Pair
VBR	Variable Bit Rate
VC	Virtual Channel (ATM) or Virtual Call (X.25)
VCI	Virtual Channel Identifier
VC-n	Virtual Container-n (SDH)
VLSI	Very Large Scale Integration
VP	Virtual Path
VPI	Virtual Path Identifier
VT	Virtual Tributary (SONET)
VTx	VT of size "x" (currently x = 1.5, 2, 3, 6)
WAN	Wide Area Network
XNS	Network Services protocol (XEROX)
ZIP	Routing and Management protocol (Apple)
μs	Microsecond (10^{-6} second)

Standards Sources

Alpha Graphics
10215 N. 35th Avenue, Suite A&B, Phoenix, AZ 85051
Ph:602-863-0999 (IEEE P802 draft standards)

American National Standards Institute — ANSI — Sales Department
1430 Broadway, New York, NY 10018
Ph:212-642-4900; FAX:212-302-1286 (ANSI and ISO standards)

Association Francaise de Normalisation, Tour Europe — Cedex 7
92080 Paris La Defense, FR
Ph: 33-1-4-778-13-26; Telex:611-974-AFNOR-F; FAX:33-1-774-84-90

Bell Communications Research — Bellcore Customer Service
60 New England Ave., Room 1B252, Piscataway, NJ 08854-4196
Ph:908-699-5800 / 1-800-521-CORE (1-800-521-2673)
(Bellcore TAs and TRs)

British Standards Institution
2 Park St., London, WIA 2BS England
Ph:44-1-629-9000; Telex:266933 BSI G; FAX:+44-1-629-0506

Canadian Standards Association
178 Rexdale Boulevard, Rexdale, ON M9W 1R9 Canada
Ph:416-747-4363; Telex:06-989344; FAX:1-416-747-4149

Comite Europeen de Normalisation
Rue Brederode 2 Bte 5, 1000 Brussels, Belgium
Ph:32-2-513-79-30; Telex:26257 B

Computer and Business Equipment Manufacturers Association (CBEMA)
311 First Street, N.W., Suite 500, Washington, DC 20001-2178
Ph:202-626-5740; FAX: 202-638-4299/ 202-628-2829 (ANSI X3
secretariat)

Dansk Standardiseringsrad
Aurehojvej 12, Postboks 77, DK-2900 Hellerup, Denmark
Ph:45-1-62-32-00; Telex:15-615 DANSTA DK

DDN Network Information Center — SRI International
333 Ravenswood Avenue, Menlo Park, CA 94025
Ph:415-859-3695 / 1-800-235-3155 / e-mail: NIC@NIC.DDN.MIL
(Requests for Comments [RFC] documents)

Deutsches Institut für Normung
Burggrafenstrasse 4-10, Postfach 1107, D-1000 Berlin 30, Germany
Ph:49-30-26-01-1; Telex:184-273-DIN D; FAX:49-30-260-12-31

Electronics Industries Association (EIA)
Standards Sales, 2001 Eye Street, N.W., Washington, DC 20036
Ph:202-457-4966; Telex:710-822-0148 EIA WSH; FAX:202-457-4985

European Computer Manufacturers Association (ECMA)
Rue du Rhone 114, CH-1204 Geneva, Switzerland
Ph:41-22-735-36-34; Telex:413237 ECMA CH; FAX:41-22-786-52-31

European Conference of Postal and Telecommunications Administrations — CEPT
CEPT Liaison Office, Seilerstrasse 22, CH-3008 Bern, Switzerland
Ph:41-31-62-20-78; Telex:911089 CEPT CH; FAX:41-31-62-20-78

Exchange Carriers Standards Association (ECSA)
5430 Grosvenor Lane, Bethesda, MD 20814-2122
Ph:301-564-4505   (ANSI T1 secretariat)

Global Engineering Documents
15 Inverness Way East, Englewood, CO 80112
Ph:1-800-854-7179; FAX (303)792-2192

Information Handling Services
15 Inverness Way East, Englewood, CO 80112
Ph: (800)447-3352; FAX: (303)397-2599

Institute of Electrical and Electronics Engineers (IEEE) — Standards
Office/Service Center
445 Hoes Lane, Piscataway, NJ 08855-1331
Ph:908-564-3834; FAX:908-562-1571 (IEEE standards)

International Organization for Standardization
1 Rue de Varembe, Case Postale 56, CH-1211 Geneva 20,
Switzerland
Ph:41-22-734-1240; Telex:23-88-1 ISO CH; FAX:41-22-733-3430

International Telecommunications Union — General Secretariat — Sales Service
Place de Nation, CH 1211, Geneva 20, Switzerland
Ph:41-22-730-5860; Telex:421000 UIT CH; FAX:41-22-730-5853
(CCITT and other ITU recommendations)

Japanese Industrial Standards Committee
Standards Department, Agency of Industrial Science & Technology
Ministry of International Trade and Industry
1-3-1, Kasumigaseki, Chiyoda-ku, Tokyo 100 Japan
Ph:81-3-501-9295/6; FAX:81-3-680-1418

National Institute of Standards and Technology
Technology Building 225, Gaithersburg, MD 20899
Ph:301-975-2000; FAX:301-948-1784

National Standards Authority of Ireland
Ballymun Road, Dublin 9, Ireland
Ph:353-1-370101; Telex:32501 IIRS EI; FAX:353-1-379620

Nederlands Normalisatie-Instituut
Kalfjeslaan 2, P.O. Box 5059, 2600 GB Delft, Netherlands
Ph:31-15-61-10-61

Omnicom, Inc.
115 Park St., SE, Vienna, VA 22180-4607
Ph:703-281-1135; Telex:279678 OMNI UR; FAX:703-281-1505

Omnicom International, Ltd.
1st Floor, Forum Chambers, The Forum, Sevenage, Herts, United
Kingdom SG1 1EL
Ph:44-438-742424; Telex:826903 OMNICM G; FAX:44-438-740154

Rapidoc, Technical Indices, Ltd
Willoughby Rd., Bracknelll, Berkshire, RG12 4DW, UK
Ph: (0344) 861666; FAX: (0344) 714440

Saudi Arabia Standards Organization
P.O. Box 3437, Riyadh 11471, Saudi Arabia
Ph:9-661-4793332; Telex:201610 SASO

SRI International
333 Ravenswood Avenue, Room EJ291, Menlo Park, CA 94025
Ph:800-235-3155 (Internet Protocol RFCs)

Standardiseringskommissionen i Sverige
Tegnergatan 11, Box 3 295, S-103 66 Stockholm, Sweden
Ph:468-230400; Telex:17453 SIS S

Standards Association of Australia — Standards House
80-86 Arthur Street, North Sydney N.S.W. 2060 Australia
Ph:61-2-963-41-11; Telex:2-65-14 ASTAN AA

Suomen Standardisoimisliitto
P.O. Box 205, SF-00121 Helsinki 12, Finland
Ph:358-0-645-601; Telex:122303 STAND SF

United Nations Bookstore
United Nations General Assembly Building, Room GA 32B, New
York, NY 10017
Ph:212-963-7680 (CCITT recommendations)

U.S. Department of Commerce — National Technical Information Service
5285 Port Royal Road, Springfield, VA 22161
Ph:703-487-4650 (CCITT recommendations, U.S. Government and
Military standards)

Glossary

address - An identifier of a source or destination in a network. Examples of addresses are IP, E.164, and X.121.

American National Standards Institute (ANSI) - A private, nongovernmental, nonprofit national organization which serves as the primary coordinator of standards within the United States.

Application Layer (OSI) - Layer 7 of the OSIRM. Provides the management of communications between user applications. Examples include e-mail and file transfer.

asynchronous transmission - The transmission of data through start and stop sequences without the use of a common clock.

Asynchronous Transfer Mode (ATM) - A high-speed connection-oriented multiplexing and switching method specified in international standards utilizing fixed-length cells. to support multiple types of traffic. It is asynchronous in the sense that cells carrying user data need not be periodic.

ATM Adaptation Layer (AAL) - A set of internationally standardized protocols and formats that define support for circuit emulation, packet video and audio, and connection-oriented and connectionless data services.

Backward Explicit Congestion Notification (BECN) - Convention in frame relay for a network device to notify the user (source) device that network congestion has occurred.

bandwidth - The amount of transport resource available to pass information (passband), measured in bps for digital systems.

Parts of this glossary were taken from Gary Kessler's book *Metropolitan Area Networks* with the author's permission.

Basic Rate Interface (BRI) - An ISDN access interface type comprised of two B-channels each at 64 kbps and one D-channel at 16 kbps (2B+D).

B-channel - An ISDN bearer service channel which can carry either voice or data at a speed of 64K bps.

Bell Operating Company (BOC) - One of the 22 local telephone companies formed after the divestiture of AT&T (e.g., Illinois Bell, Ohio Bell).

bridge - A LAN/WAN device operating at Layer 1 (physical) and 2 (data link) of the OSIRM.

broadband - A term that refers to channels supporting rates in excess of DS3 (45 Mbps) or E3 (34 Mbps).

Broadband ISDN (B-ISDN) - A set of services, capabilities, and interfaces supporting an integrated network and user interface at speeds greater than that of ISDN. The ITU-T initially decided to develop B-ISDN using ATM in 1988.

broadcast - a transmission to all addresses on the network or subnetwork.

cell - A fixed-length 53-octet packet, or Protocol Data Unit (PDU)used in ATM. The ATM cell has a 5 octet header and a 48 octet payload.

cell header - A 5-octet header that defines control information used in processing, multiplexing, and switching cells.

Central Office (CO) - Telephone company switching office which provides voice, circuit and packet data services.

circuit switching - A connection oriented technique based on either time or space division multiplexing and switching providing minimal delay. Bandwidth is dedicated to the connection.

Committed Information Rate (CIR) - A term defined for frame relay service that defines the average rate a user can send frames and be guaranteed delivery by the network.

Transmissions exceeding the CIR are subject to lower priority treatment or discard.

congestion - The condition where network resources (bandwidth) are exceeded by an accumulation of demand.

Customer Premises Equipment (CPE) - Equipment which resides and is operated at a customer site.

Cyclic Redundancy Check (CRC) - An algorithm which detects bit errors caused in data transmission.

Data Circuit Termination Equipment (DCE) - Data communications equipment defined by the standards as a modem or network communications interface device.

datagram - A packet mode of transmitting data where there is no guaranteed sequential delivery.

Data Link Connection Identifier (DLCI) - A frame relay address designator for each virtual circuit termination point.

data link layer (OSI) - Layer 2 of the OSIRM. Provides for the error-free communications between adjacent network devices over a physical interface. Examples include the LLC and MAC layers which manage LAN and MAN operation.

Data Terminal Equipment (DTE) - Data processing equipment defined by the standards as interfacing to the communications network (DCE).

D-channel - The ISDN out-of-band (16 kbps or 64 kbps, depending on BRI or PRI, respectively) signaling channel which carries the ISDN user signals or can be used to carry packet mode data.

digital - Signals which have discrete values, such as binary bit streams of 0s and 1s.

Digital Signal 0 (DS0) - One 64 kbps digital channel.

Digital Signal 1 (DS1) - The North American standard 1.544 Mbps digital channel.

Digital Signal 3 (DS3) - The North American standard 44.736 Mbps digital channel.

Discard Eligibility (DE) bit - Used in frame relay, this bit signals (when set to 1) that the particular frame is eligible for discard during congestion conditions.

Distributed Queue Dual Bus (DQDB) - The IEEE 802.6 standard for providing both circuit-switched (isochronous) and packet-switched services in a metropolitan area.

E1 - The European standard 2.048-Mbps digital channel.

E.164 - A CCITT Recommendation for defining addresses in a public data international network, varying in size up to 15 digits.

enterprise network - A network which spans an entire organization.

entity - In the OSIRM, a service of management element between peers and within a sublayer or layer.

Ethernet - A LAN which uses the CSMA/CD media access method and operates at 10 Mbps, usually over a coax medium.

fast packet - The generic term used for advanced packet technologies such as frame relay, DQDB, and ATM.

Fiber Distributed Data Interface (FDDI) - Fiber optic LAN operating at 100 Mbps.

fiber optics - Plastic or glass fibers which transmit high data rates through optical signals.

filtering - The selection of frames not to remain at the local LAN but to be forwarded to another network by a network device (i.e., router).

flag - Character which signals a beginning or end of a frame.

Forward Explicit Congestion Notification (FECN) - Convention in frame relay for a network device to notify the user (destination) device that network congestion is occurring.

fractional T1 (FT1) - The transmission of a fraction of a T1 channel, usually based in 64 kbps increments but not less than 64 kbps total.

frame - An OSI data link layer defined unit of transmission whose length is defined by flags at the beginning and end.

Frame Check Sequence (FCS) - A field in an X.25, SDLC, or HDLC frame which contains the result of a CRC error-checking algorithm.

frame relay - An ANSI and CCITT defined LAN/WAN networking standard for switching frames in a packet mode similar to X.25, but at higher speeds and with less nodal processing.

full duplex - The simultaneous bidirectional transmission of information over a common medium.

half duplex - The bidirectional transmission of information over a common medium, but where information may only travel in one direction at any one time.

host - An end-communicating station in a network; also an IP address.

implicit congestion notification - A congestion indication which is performed by upper layer protocols (e.g., TCP) rather than network or data link layer protocol conventions.

Integrated Services Digital Network (ISDN) - CCITT I-series Recommendation defined digital network standard for integrated voice and data network access, services, and user-to-network messages.

Interexchange Carrier (IXC) - The provider of long distance (inter-LATA) service in the United States; also the provider of worldwide switched voice and data services.

interface - In OSI, the boundary between two adjacent protocol layers (e.g., network to transport).

Interim Local Management Interface (ILMI) - An SNMP based management protocol for an ATM UNI defined by the ATM Forum.

interoperability - The ability of multiple, dissimilar vendor devices and protocols to operate and communicate using a standard set of rules and protocols.

layer management - Network management functions which provide information about the operations of a given OSI protocol layer.

Line-Terminating Equipment (LTE) - A device which either originates or terminates an OC-n signal and which may originate, access, modify, and terminate the transport overhead.

Link Access Protocol on the D-channel (LAPD) - CCITT Recommendations Q.920 (I.440) and Q.921 (I.441) defined standards for the data link layer operation of ISDN "D" and frame relay channel.

Local Area Network (LAN) - A MAC level data and computer communications network confined to short geographic distances.

Local Exchange Carrier (LEC) - In the United States, a local phone service provider (cannot provide long distance service).

Local Management Interface (LMI) - A set of user device-to-network communications standards used in ATM DXI and frame relay.

media - The plural form of medium, or multiple mediums (twisted-wire pair, coax cable, fiber, etc.).

medium - The single common access platform, such as a copper wire, fiber, or free space.

Medium Access Control (MAC) - IEEE 802 defined media specific access control protocol.

multicast - A connection type with the capability to broadcast to multiple destinations on the network.

multiplexing - The technique of combining multiple individual channels onto a single aggregate channel for sharing facilities and bandwidth.

network - A system of autonomous devices, links, and subsystems which provide a platform for communications.

Network Layer (OSI) - Layer 3 of the OSIRM. Provides the end-to-end routing and switching of data units (packets), as well as managing congestion control.

network management - The process of managing the operation and status of network resources (e.g., devices, protocols).

node - A device which interfaces with the transmission medium through the physical layer (and often the data link layer) of the OSIRM.

octet - An 8-bit-long transmission unit of measure.

Open Systems Interconnection Reference Model (OSIRM) - A seven-layer model defining the international protocol standards for data communications in a multiple architecture and vendor environment. Both the OSI and CCITT define standards based on the OSIRM.

Optical Carrier level n (OC-n) - The optical carrier level signal in SONET which results from an STS-n signal conversion. In SONET, the basic transmission speed unit is 58.34 Mbps.

Packet Assembler/Disassembler (PAD) - A concentration and network access device which provides protocol conversion into X.25 packet format.

packet switching - A method of switching which segments the data into fixed or variable units of maximum size called packets. These packets then pass the user information (addressing, sequencing, error control, and user-controlled options) in a store-and-forward manner across the network.

path overhead (POH) - Overhead transported with the SONET payload and used for payload transport functions.

payload pointer - Indicates the starting point of a SONET synchronous payload envelope.

Permanent Virtual Circuit (PVC) - A logical dedicated circuit between two user ports in a point-to-point configuration.

Physical Layer (OSI) - Layer 1 of the OSIRM. Provides the electrical and mechanical interface and signaling of bits over the communications medium.

Physical Layer Convergence Protocol (PLCP) - The IEEE 802.6 defined physical layer standard that adapts the actual capabilities of the underlying physical network to provide the services required by the DQDB or ATM layer.

Physical Layer Protocol (PHY) - In FDDI, the medium-independent layer corresponding to the upper sublayer of the OSIRM physical layer.

Presentation Layer (OSI) - Layer 6 of the OSIRM. Identifies the syntax of the user data being transmitted and provides user service functions such as encryption, file transfer protocols, and terminal emulation.

Primary Rate Interface (PRI) - An ISDN T1 access interface type comprised of 23 B-channels each at 64 kbps and one D-channel at 64 kbps (23B+D). The European version operates at 2.048 Mbps (30B+D).

protocol - The rules and guidelines by which information is exchanged and understood between two devices.

Protocol Data Unit (PDU) - The unit of information transferred between communicating peer layer processes.

Regional Bell Operating Company (RBOC) - One of seven U.S. regional holding companies formed by divestiture of AT&T (e.g., Ameritech, Southwestern Bell). The RBOCs also manage the 22 BOCs.

ring - A closed-loop, common bus network topology.

router - A LAN/WAN device operating at Layer 1 (physical), 2 (data link), and 3 (network) of the OSIRM. Distinguished from bridges by its capability to switch and route data based upon network protocols such as IP.

section - A transmission facility between a SONET Network Element and regenerator.

self-healing - The ability for a LAN/MAN to reroute traffic around a failed link or network element to provide uninterrupted service.

service - The relationship between protocol entities in the OSIRM, where the service provider (lower layer protocol) and the service user (higher layer protocol) communicate through a *data service*.

Service Access Point (SAP) - The access point at a network node or station where the service users access the services offered by the service providers.

Service Data Unit (SDU) - Unit of information transferred across the OSI interface between service provider and service user.

Session Layer (OSI) - Layer 5 of the OSIRM. Provides the establishment and control of user dialogues between adjacent network devices.

simplex - One-way transmission of information on a medium.

slot - The basic unit of transmission on a DQDB bus.

SMDS Interface Protocol (SIP) - The three layers of protocol (similar to the first three layers of the OSIRM) which define the SMDS SNI user information frame structuring, addressing, error control, and overall transport.

SNA - IBM's communications networking architecture.

source routing - A routing scheme where the routing of packets is determined by the source address and route to the destination in the packet header.

subnetwork - The smaller units of LANs (called LAN segments) which can be more easily managed than the entire LAN/MAN/WAN.

Subscriber-Network Interface (SNI) - A DQDB user access point into the network or MAN switch.

Switched MultiMegabit Data Service (SMDS) - A MAN service offered at present over the IEEE DQDB bus.

Switched Virtual Circuit (SVC) - Virtual circuits similar to PVCs, but established on a call-by-call basis.

Synchronous Digital Hierarchy (SDH) - The CCITT original version of a synchronous digital hierarchy; based on optical fiber; called SONET in ANSI parlance.

Synchronous Optical Network (SONET) - A U.S. high-speed fiber optic transport standard for a fiber optic digital hierarchy (speeds range from 51.84 Mbps to 2.4 Gbps)

Synchronous Transfer Mode (STM) - The T1 carrier method of assigning time slots as channels within a T1 or E1 circuit.

synchronous transmission - The transmission of frames which are managed through a common clock between transmitter and receiver.

Synchronous Transport Module level N (STM-N) -The SDH line rate of "N" STM-1 signals.

Synchronous Transport Signal level N (STS-N) - SONET transmission signal created with byte interleaving of "N" STS-1 (51.84 Mbps) signals.

Synchronous Transport Signal level Nc (STS-Nc) - Concatenated SONET synchronous payload envelope.

T1 - A 4-wire repeater system. Commonly used to refer to a DS1 signal.

T3 - Commonly used to refer to a DS3 signal.

Time Division Multiplexing (TDM) - The method of aggregating multiple simultaneous transmissions (circuits) over a single high-speed channel by using individual time slots (periods) for each circuit.

time-insensitive - Traffic types whose data is not affected by small delays during transmission. This is also referred to as delay-insensitive.

time-sensitive - Traffic types whose data is affected by small delays during transmission and cannot tolerate this delay (e.g., voice, video, real-time data).

token - A marker which can be held by a station on a token ring or bus indicating that station's right to transmit.

Token Ring - A LAN which uses a token-passing access method for bus access and traffic transport between network elements, where bus speeds operate at either 4 Mbps or 16 Mbps.

Transmission Control Protocol/Internet Protocol (TCP/IP) - The combination of a network and transport protocol developed by ARPANET for internetworking IP-based networks.

Transport Layer (OSI) - Layer 4 of the OSIRM. Provides for error-free end-to-end communications between two "host" users across a network.

transport overhead - In SONET, the line and section overhead elements combined.

twisted pair - The basic transmission media consisting of 22 to 26 American Wire Gauge (AWG) insulated copper wire. TP can be either shielded (STP) or unshielded (UTP).

unshielded twisted pair (UTP) - A twisted-pair wire without the jacket shielding; used for short distances but subject to electrical noise and interference.

user channel - Portion of the SONET channel allocated to the user for maintenance functions.

User-to-Network Interface (UNI) - The point where the user accesses the network.

user-to-user protocols - Protocols which operate between the users and are typically transparent to the network, such as file transfer protocols (e.g., FTP).

Virtual Channel Identifier (VCI) - In ATM, a field within the cell header which is used to switch virtual channels.

virtual circuit - A virtual connection established through the network from origination to destination, where packets, frames, or cells are routed over the same path for the duration of the call. These connections seem like dedicated paths to the users, but are actually network resources shared by all users. Bandwidth on a virtual circuit is not allocated until it is used.

Virtual Path Identifier (VPI) - In ATM, a field within the cell header which is used to switch virtual paths, defined as groups of virtual channels.

Virtual Tributary (VT) - An element that transports and switches sub-STS-1 payloads or VTx (VT1.5, VT2, VT3, or VT6).

Wide Area Network (WAN) - A network which operates over a large region and commonly uses carrier facilities and services.

window - The concept of establishing an optimum number of frames or packets which can be outstanding (unacknowledged) before more are transmitted. Window protocols include X.25, LAP, TCP/IP, and SDLC.

X.25 - CCITT recommendation for the interface between packet switch DTE and DCE equipment.

Index

623

ABOUT THE AUTHORS

DAVID E. MCDYSAN is responsible for the strategic planning of
ATM services at MCI. He has served on the board of The
ATM Forum and is an acknowledged industry expert on
frame relay, SMDS, circuit switching, ATM traffic
engineering, ATM protocols, and internetworking. He is a
widely sought-after conference speaker and leads seminars
on ATM throughout the United States.

DARREN L. SPOHN is Vice President of Operations at
Southwest Network Services (Austin, Texas), a WAN and
facilities joint-sourcing company. He was previously at MCI
where he had experience with every major MCI data
network, from asynchronous terminal access to designing a
100+ node Siemens/Wellfleet cell switch/frame relay
network. He is the author of *Data Network Design*, also
published by McGraw-Hill.